AF793187

Quantum Fields – Algebras, Processes

Edited by Ludwig Streit

Springer-Verlag
Wien New York

Prof. Dr. Ludwig Streit
Fakultät für Physik
Universität Bielefeld
Federal Republic of Germany

Proceedings of the Symposium
"Bielefeld Encounters in Physics and Mathematics II:
Quantum Fields—Algebras, Processes"
with the Workshop "White Noise Approach to Quantum Dynamics"
at the Centre for Interdisciplinary Research,
Bielefeld University, Federal Republic of Germany,
December 1—9, 1978.

This work is subject to copyright.
All rights are reserved,
whether the whole or part of the material is concerned,
specifically those of translation, reprinting, re-use of illustrations,
broadcasting, reproduction by photocopying machine
or similar means, and storage in data banks.
© 1980 by Springer-Verlag/Wien
Softcover reprint of the hardcover 1st edition 1980

With 10 Figures

Library of Congress Cataloging in Publication Data. Main entry under title: Quantum fields. "Proceedings of the symposium, 'Bielefeld Encounters in Physics and Mathematics II: Quantum Fields—Algebras, Processes,' with the workshop, 'White Noise Approach to Quantum Dynamics' at the Centre for Interdisciplinary Research, Bielefeld University, Federal Republic of Germany, December 1—9, 1978." 1. Quantum field theory—Congresses. 2. Algebra—Congresses. 3. Representations of groups—Congresses. I. Streit, Ludwig, 1938— . II. Bielefeld. Universität. Fakultät für Physik. QC174.45.A1Q363. 530.1'43. 80-19953

ISBN-13:978-3-7091-8600-8 e-ISBN-13:978-3-7091-8598-8
DOI: 10.1007/978-3-7091-8598-8

Preface

Are we living in a golden age?

It is now more than half a century that Einstein and Heisenberg have given us the theories of relativity and of quantum mechanics, but the great challenge of 20th century science remains unresolved: to assemble these building blocks into a fundamental theory of matter.

And yet, for anyone watching the interplay of mathematics and theoretical physics to-day, developing symbiotically through the stimulus of a lively, even essential interdisciplinary dialogue, this is a time of fascination and great satisfaction. It is also a time of gratitude to those who had the courage to insist that "a rudimentary knowledge of the Latin and Greek alphabets" was not enough, and tore down the barriers between the disciplines. On the basis of this groundwork there is now so much progress, and, notably, such strengthening of the dialogue with phenomenology that - reaching out for The Great Breakthrough - this may indeed turn out to be the golden age.

The present volume is concerned with recent developments

- in quantum field theory and equilibrium statistical mechanics (their structural similarity provides a fine example of deeper insight gained in the past decade),
- in the theories of normed algebras and of group representations
- and in probability theory, particularly regarding stochastic processes and fields, the concepts of which have recently displayed an enormous unifying power, relating structures as - seemingly - disparate as e.g. the shape of coastlines, the noise bursts in telephone transmission, and gauge field theory.

Bielefeld University, which has hosted the exchange of ideas reproduced in this volume, has done much in fulfilling its pledge to further the exchange of ideas between mathematics and non-mathematical disciplines, and its Centre for Interdisciplinary Research (ZiF) has developed a tradition of sponsoring the dialogue of leading mathematicians and physicists from around the world here in Bielefeld, stimulating at the same time, as it should, local discussions and joint research across the departmental barriers which are often so astonishingly high.

For the success of the "Bielefeld Encounters in Physics and Mathematics II: Quantum Fields - Algebras, Processes", my thanks go to the contributors and participants. They go to T. Hida for organizing the workshop on the "White Noise Approach to Quantum Dynamics" and for his enthusiasm, to S. Albeverio, Ph. Blanchard, E. Brüning, G. Sommer, O. Steinmann, P. Stichel of the Preparatory Committee, to the University of Bielefeld, in particular to the directors and staff of ZiF, and to Mrs. Jegerlehner who did an admirable job of decoding the manuscripts and bringing them into this form.

Ludwig Streit

Bielefeld, July 1980

Contents

The results of the contribution "Stochastic Processes and Commutation Relations" by S.D. Chatterji have been published in Lecture Notes in Mathematics, Vol. 645, pp. 16-26. Berlin-Heidelberg-New York: Springer, 1978.

Causal Analysis in Terms of White Noise

Takeyuki Hida

Department of Mathematics

Nagoya University

Chikusa-ku, Nagoya, 464

Japan

Summary

Functionals of Brownian motion, call them Brownian functionals, are discussed, where development of time is taken into account. In the course of the analysis we are naturally led to introduce classes of generalized Brownian functionals and to discuss differential calculus, where the integral representation of Brownian functionals and Lévy's functional analysis are used.

Contents

§1. Introduction

Let $\{B(t) = B(t,\omega),\ t \in R^1\}$ be a standard Brownian motion. We shall consider functionals such as

$$f(B(t)\ ;\ t \in R^1)$$

and discuss their analysis. Those functionals are simply called Brownian functionals. In order to discuss the analysis we should like to express them in the form

$$\phi(\dot{B}(t)\ ;\ t \in R^1),$$

where the symbol $\dot{B}(t)$ stands for the time derivative of the Brownian motion: $\dot{B}(t) = dB(t)/dt$. We call $\{\dot{B}(t),\ t \in R^1\}$ the white noise. The main reason why the latter expression is more preferable is that white noise is a continuous analogue of a sequence of independent, identically distributed, Gaussian random variables.

Since the white noise is a generalized stochastic process, the probability distribution of $\{\dot{B}(t)\}$ is to be introduced to a space of generalized functions, say to S^* the dual space of the Schwartz space S. This can be done in the following manner. For any $\xi \in S$, $\langle\dot{B},\xi\rangle$ ($\langle\ ,\ \rangle$ is the canonical bilinear form connecting S and S^*) is well defined and it is a Gaussian random variable with mean 0 and variance $\|\xi\|^2$. We now have

$$C(\xi) \equiv E(\exp i\langle\dot{B},\xi\rangle\ ) = \exp[-\|\xi\|^2/2\,]. \tag{1}$$

We then appeal to the Bochner-Minlos theorem to obtain the measure μ on S^* such that

$$C(\xi) = \int_{S^*} e^{i\langle x,\xi\rangle} d\mu(x). \tag{2}$$

This μ is the probability distribution of $\{\dot{B}(t)\}$ and is said to be the measure of white noise. Each member x of S^* with μ is now viewed as a sample function of $\{\dot{B}(t)\}$.

As soon as μ is given, one can form the (complex) Hilbert space

$$(L^2) = L^2(S^*,\mu),$$

which may be thought of as the collection of Brownian functionals with finite variance.

The time evolution can be involved there in the following manner: Let $\{S_t\}$ be the shift which is the one-parameter group of transformations S_t, $t \in R^1$, given by

$$(S_t\xi)(u) = \xi(u - t), \qquad \xi \in S. \tag{3}$$

Let T_t be the adjoint of S_t, which is acting on S^*. Then the T_t is viewed as a realization of the transformation

$$\dot{B}(s) \longrightarrow \dot{B}(s + t). \tag{4}$$

Since $\{T_t\}$ forms a one-parameter group of μ-measure preserving transformations on S^*, a unitary group $\{U_t\}$ is naturally introduced in the usual way, that is

$$(U_t\phi)(x) = \phi(T_t x), \qquad \phi \in (L^2). \tag{5}$$

In §2 we shall observe how the time evolution is involved in the analysis of linear functionals of $\dot{B}(t)$'s. Namely, we shall give a short remark on causal analysis of Gaussian processes. There we shall be able to think of some idea to discuss the analysis of general Brownian functionals in which the development of the time is involved explicitly.

Before doing such analysis we shall have a quick review of the known results on the analysis of Brownian functionals, that is the analysis on (L^2), in §3. There we use some basic tools from functional analysis. First of all the transformation $\mathcal{T}$ should be introduced:

$$(\mathcal{T}\phi)(\xi) = \int_{S^*} \exp[i\langle x,\xi\rangle]\phi(x)d\mu(x), \qquad \phi \in (L^2). \tag{6}$$

It carries the Hilbert space (L^2) onto the reproducing kernel Hilbert space (RKHS) F with kernel $C(\xi-\mu)$, $(\xi,\mu) \in S \times S$ where C is the functional given by (1). With this transformation Brownian functionals in (L^2) become more visualized and hence easier to be dealt with. Then follow the direct sum decomposition of (L^2), called the Wiener-Itô decomposition, and the integral representation of Brownian functionals.

Let the evolution of the time again be in mind, and let $\{\dot{B}(t)\}$ be approximated by $\{\Delta_k B/\Delta_k\}$, where $\Delta_k B$ is the increment of Brownian motion $B(t)$ over the time interval Δ_k, $\{\Delta_k\}$ being a partition of R^1. If we take the system of Hermite polynomials in $(\Delta_k B/\Delta_k)$'s to be a base of nonlinear functionals of $\Delta_k B$'s, and if we let the partition of R^1 finer and finer, then we are naturally led to introduce a class of generalized Brownian functionals (in §4).

The transformation T can be extended to the class of generalized Brownian functionals, and we see in §5 that the image of the class of generalized Brownian functionals under T is a familiar class of functionals on S with regularity. We shall be ready to apply Lévy's functional analysis (see P. Lévy [1]) to those functionals on S and see that the analysis is in agreement with the causal analysis in terms of white noise that we exactly aim at.

As soon as we look at the formulas established by our calculus, we can recognize some interesting connections with quantum dynamics. Having been inspired by the above observation, we shall discuss some developments in this line in the last section.

§2. The Canonical Representation of Gaussian Processes

The simplest class of Brownian functionals is of course that of linear functionals. By definition a linear functional of the $\dot{B}(t)$'s is realized by

$$\langle x,\xi\rangle, \quad x \in S^* \quad \xi \in S, \tag{7}$$

on the probability space (S^*,μ). It extends to

$$\langle x, f\rangle, \qquad f \in L^2(R^1), \tag{8}$$

although it is no more continuous in f. But, it is still Gaussian in distribution and is an (L^2)-functional.

Given a Gaussian Process $\{X(t)\}$ with some reasonable properties, then it is known that $X(t)$ is expressible as

$$X(t) = \int^t F(t,u)\dot{B}(u)du, \tag{9}$$

or in terms of x

$$X(t) = \langle x, F_t\rangle, \tag{10}$$

where $F_t(u) = F(t,u)$ is a Volterra kernel. Needless to say that there are many expressions of the form (9) such that the Gaussian process defined by (9) has the same probability distribution as the given one. Among them there is only one (in fact, essentially unique) representation called canonical or causal and causally invertible.

The canonical property means the following: Let $\mathbb{B}_t(\dot{B})$ be the smallest sigma-field with respect to which all the $\langle \dot{B},\xi\rangle$'s (or $\langle x,\xi\rangle$'s) with $\mathrm{supp}(\xi) \subset (-\infty,t]$ are measurable. Suppose a Gaussian process $\{X(t)\}$ is given by (9). Then we can define $\mathbb{B}_t(X)$ in a similar manner to $\mathbb{B}_t(\dot{B})$. The representation (9) (or (10)) of $X(t)$ in terms of $\dot{B}$ (or of x) is said to be canonical if

$$\mathbb{B}_t(X) = \mathbb{B}_t(\dot{B}) \quad \text{for every } t. \tag{11}$$

The uniqueness as well as the existence of the canonical representations holds for general mean continuous stationary Gaussian processes that are purely non-deterministic (a process $\{X(t)\}$ is purely non-deterministic if $\bigcap_t \mathbb{B}_t(X)$ is trivial).

Example 1 An Ornstein-Uhlenbeck process $\{U_m(t)\}$, $m > 0$, with the characteristic functional

$$C_m(\xi) = \exp\left[-\frac{1}{2}\frac{1}{\sqrt{2\pi}}\int \frac{|\hat{\xi}(\lambda)|^2}{\lambda^2+m^2}d\lambda\right], \quad \hat{\xi}: \text{the Fourier transform of } \xi, \tag{12}$$

has the representation of the form

$$U_m(t) = \int_{-\infty}^{t} \exp[-m(t - u)]\dot{B}(u)du. \tag{13}$$

It is easy to see that the above representation is canonical. Note that the $\{U_m(t)\}$ corresponds to the free particle with mass m in relativistic quantum dynamics.

<u>Example 2</u> Let $\{U_1(t)\}$ and $\{U_2(t)\}$ be Ornstein-Uhlenbeck processes with $m = 1$ and $m = 3$, respectively. We assume that they are represented in the form (13) by the same white noise $\dot{B}(t)$ canonically. Set

$$X_1(t) \equiv 2U_1(t) - U_3(t) \tag{14}$$

$$= \int_{-\infty}^{t} \{2\exp[-(t - u)] - \exp[-3(t - u)]\}\dot{B}(u)du$$

and

$$X_2(t) = 3U_1(t) - 4U_3(t)$$

$$= \int_{-\infty}^{t} \{3\exp[-(t - u)] - 4\exp[-3(t -u)]\}\dot{B}(u)du. \tag{15}$$

Two processes $\{X_1(t)\}$ and $\{X_2(t)\}$ are in fact the same process, since they have the same characteristic functional

$$C(\xi) = \exp[-\frac{1}{2}\frac{1}{\sqrt{2\pi}}\int \frac{|\hat{\xi}(\lambda)|^2(\lambda^2+25)}{(\lambda^2+1)(\lambda^2+9)} d\lambda].$$

It is interesting to point out that the representation (14) for $\{X_1(t)\}$ is canonical, while (15) for $\{X_2(t)\}$ is <u>not</u>. Indeed, we have

$$\mathbb{B}_t(X_2) \subsetneqq \mathbb{B}_t(\dot{B}) \quad \text{for every } t.$$

As for representation (14) the time shift for $\{X_1(t)\}$ implies that for $\{\dot{B}(t)\}$ which is expressed by (4). Hence evolutional phenomena arising from the $\{X_1(t)\}$ can well be expressed in terms of $\{\dot{B}(t)\}$, where the development of the time can also be taken into account. Some other operations such as the time change for $\{X_1(t)\}$ are naturally described by the corresponding operations for $\{\dot{B}(t)\}$, where the latters are much easier than the former. See Hida-Streit [5].

We have realized, by the observation in Example 2, the role of the transformation (4) in describing evolutional phenomena, and at the same time we can see some advantage to express Brownian functionals in terms of white noise. This would be the case for nonlinear Brownian functionals. Before we come to this topic we shall briefly summarize, in the next section, some known results on the analysis of non-linear Brownian functionals.

§3. Nonlinear Brownian Functionals

Let T be the transformation given by (6) which maps the Hilbert space (L^2) to a space of functionals on S. The collection $F = \{T\phi;\ \phi \in (L^2)\}$ obviously forms a vector space. We are able to topologize it in such a way that (L^2) and F are isomorphic under T. This is done as follows: Since $C(\xi-\eta)$, $(\xi,\eta) \in S \times S$ is positive definite, there is a reproducing kernel Hilbert space (RKHS) $F(C)$ with kernel $C(\xi-\eta)$. It is determined by

i) $C(\cdot - \eta) \in F(C)$ for any fixed $\eta \in S$,

ii) $(f(\cdot), C(\cdot - \eta)) = f(\eta)$, $(\ ,\ )$ is the scalar product in $F(C)$.

iii) $\{C(\cdot - \eta);\ \eta \in S\}$ spans the whole space $F(C)$.

With these properties of RKHS and with the evaluations of the $T(\exp[i\langle x,\eta\rangle])$, $\eta \in S$, as well as their linear combinations we can prove the following theorem.

<u>Theorem 1</u> <u>The space</u> F <u>becomes the RKHS</u> $F(C)$ <u>and</u> T <u>gives an isomorphism between</u> (L^2) <u>and</u> F.

We then come to direct sum decompositions of $F(= F(C))$ and (L^2). Start with the expansion

$$C(\xi - \eta) = \sum_{n=0}^{\infty} C_n(\xi,\eta), \tag{16}$$

where

$$C_n(\xi,\eta) = \frac{1}{n!} C(\xi)\langle\xi,\eta\rangle^n C(\eta). \tag{17}$$

It is easy to see that the $C_n(\cdot,\eta)$, $n \geq 0$, are members in F which are mutually orthogonal:

$$(C_m(\cdot,\eta),\ C_n(\cdot,\eta')) = 0, \quad \text{if } m \neq n,$$

and that the series (16) converges in F. In addition, each kernel $C_n(\xi,\eta)$ is positive definite, so that it defines a RKHS, denote it by F_n. Then we have proved

Theorem 2 We have a direct sum decomposition of F:

$$F = \sum_{n=0}^{\infty} \oplus F_n. \tag{18}$$

Introduce a subspace H_n of (L^2) by

$$H_n = T^{-1}(F_n).$$

Then we have the Wiener-Itô decomposition of (L^2):

Corollary We have

$$(L^2) = \sum_{n=0}^{\infty} \oplus H_n. \tag{19}$$

Note The direct sum decomposition (18) as well as the decomposition (19) is said to be the Fock space.

We then proceed to the second step in which the integral representation of H_n-functionals is established. Take a Hermite polynomial in $\langle x,\eta\rangle/\sqrt{2}$ with $\|\eta\| = 1$ and apply the transformation T to obtain

$$\{TH_k(\langle x,\eta\rangle/\sqrt{2})\}(\xi) = C(\xi)(\sqrt{2}i)^k\langle\xi,\eta\rangle^k.$$

Similar but somewhat complicated computations lead us to show that

$$\{T\prod_j H_{k_j}(\langle x,\eta_k\rangle/\sqrt{2})\}(\xi) = C(\xi)(\sqrt{2}i)^n \prod_j \langle\xi,\eta_k\rangle^{k_j}, \quad n = \sum_j k_j, \tag{20}$$

where η_k's are orthonormal in $L^2(R^1)$. The right hand side of (20) is rewritten as

$$C(\xi)(\sqrt{2}i)^n \int \underset{R^n}{\cdots} \int (\eta_1^{k_1 \otimes} \otimes \eta_2^{k_2 \otimes} \otimes \ldots)\xi(u_1)\ldots\xi(u_n)du^n,$$

$\eta^{k\otimes}$ is the k-times tensor product of η.

The tensor product of the integrand may be symmetrized since the same ξ appears as many times as n. Also we note that if $\{\eta_n\}$ is a complete orthonormal system in $L^2(R^1)$, then the system of all possible finite products

$$\prod_j c_{n_j} H_{n_j} (\langle x,\eta_j\rangle/\sqrt{2}), \qquad c_n = 1/\sqrt{n!2^n}, \tag{20'}$$

forms a complete orthonormal system in (L^2). These observations prove

<u>Theorem 3</u> i) <u>For any</u> ϕ <u>in</u> H_n, <u>there is a symmetric</u> $L^2(R^n)$-<u>function</u> F <u>(notation:</u> $F \in \widehat{L^2}(R^n)$) <u>such that</u>

$$(T\phi)(\xi) = i^n C(\xi)U(\xi), \tag{21}$$

<u>where</u>

$$U(\xi) = \int \underset{R^n}{\cdots} \int F(u_1,\ldots, u_n)\xi(u_1)\ldots\xi(u_n)du^n. \tag{22}$$

<u>The mapping</u>

$$T : S \longrightarrow F \in \widehat{L^2}(R^n) = \{F \in L^2(R^n),\ F \ \text{symmetric}\}$$

<u>is surjective</u>.

ii) <u>It holds that</u>

$$\|S\|_{(L^2)} = \sqrt{n!}\, \|F\|_{L^2(R^n)} .$$

The representation of ϕ in H_n in terms of $\widehat{L^2}(R^n)$-function is called the <u>integral representation</u>, and the function F is called the <u>kernel</u> of the representation.

By the integral representation a functional defined on the infinite dimensional space S^* becomes an ordinary function on a finite dimensional Euclidean space, so that the given functional is visualized and is easily analyzed.

Here is an application of the integral representation. Let $\phi(x)$ be a real-valued H_2-functional having $F \in \widehat{L^2}(R^2)$ as the kernel of integral representation. Set

$$f(x) = \exp[i\phi(x)], \tag{23}$$

and expand f(x) in the form

$$f = \sum_{n=0}^{\infty} f_n, \qquad f_n \in H_n.$$

In order to obtain integral representations we introduce the kernel $\tilde{F}$ of the operator

$$iF(-I + 2iF)^{-1}$$

and the modified Fredholm determinant $\delta(z,F)$ of the kernel F. Then we have

$$\begin{aligned} f_0 &= \delta(2i,F)^{-1/2}, \\ f_{2n+1} &= 0, \\ (Jf_{2n})(\xi) &= C(\xi)\delta(2i,F)^{-1/2}\frac{(-1)^n}{n!} \times \\ &\qquad \int\cdots\int_{R^{2n}} \tilde{F}^{n\otimes}(u_1,\ldots,u_{2n})\xi(u_1)\ldots\xi(u_{2n})\,du^{2n}. \end{aligned} \tag{24}$$

This formula gives us a suggestion on how to renormalize the exponential functional f(x) when $\Phi(x)$ in (23) is replaced by a generalized quadratic functional (see §6).

§4. Generalized Brownian Functionals

It is not appropriate to this report to tell the whole story about how and why we have come to introduce generalized Brownian functionals; however one crucial reason should now be mentioned. Our interest is, as was explained before, centered entirely on the causal analysis where the time-dependent random phenomena would be expressed as functionals of $\{\dot{B}(t)\}$ and the time shift is represented in a manner of (4). Therefore, instead of a system of smeared variables $\{\langle\dot{B},\xi_n\rangle\}$ ($\{\xi_n\}$ being a complete orthonormal system in $L^2(R^1)$), the continuum system $\{\dot{B}(t),\ t \in R^1\}$ could be taken to be a variable of Brownian functionals. Such functionals can not always live within the space (L^2), that is, we are requested to extend the class of Brownian functionals.

Unfortunately, $\dot{B}(t)$, for fixed t, has no correct meaning, so that we have to start with an approximation to $\dot{B}(t)$. Let $\Pi = \{\Delta_k\}$ be a partition of R^1, where Δ_k's are arranged from left to right. Then, $\Delta_k B/\Delta_k$, Δ_k containing t, approximates $\dot{B}(t)$, and a discrete version of the transformation (4) is

$$\Delta_k B/\Delta_k \longrightarrow \Delta_{k+1}B/\Delta_{k+1}.$$

We are now able to form the Hilbert space $L^2(\Pi)$ ($\subset(L^2)$) consisting of functionals of the $\Delta_k B$'s. For simplicity assume that $|\Delta_k| = \delta$ for all k. By analogy with (20'), the system of finite products

$$\prod_j d_{n_j} H_{n_j}(\Delta_j B/\Delta_j\ ;\ 1/\Delta_j),\quad d_n = \sqrt{n!}\ \delta^{n/2}, \tag{25}$$

is fitting for a complete orthonormal system in $L^2(\Pi)$, where $H_n(x\ ;\ \sigma^2)$ is the Hermite polynomial with parameter σ and of degree n given by

$$H_n(x\ ;\ \sigma^2) = \frac{(-\sigma^2)^n}{n!} \exp[x^2/(2\sigma^2)]\left(\frac{d}{dx}\right)^n \exp[-x^2/(2\sigma^2)],\ \sigma > 0.$$

We divide the system of Hermite polynomials given in (25) into two classes:

(I) $\quad \prod_j d_1 H_1(\Delta_j B/\Delta_j\ ;\ 1/\Delta_j)$,

(II) $\quad \prod_j d_{n_j} H_{n_j}(\Delta_j B/\Delta_j\ ;\ 1/\Delta_j)$, some $n_j > 1$.

Now let the partition Π be finer and finer in such a way that $\delta \to 0$. Then we are led to formal expressions of the limits of (I) and (II). They are

(i) $\quad \prod_j \dot{B}(t_j), \quad t_j$'s different

(ii) $\quad \prod_j H_{n_j}(\dot{B}(t_j) ; 1/dt_j), \; t_j$'s different, some $n_j > 1$,

where the constants d_n's are ignored.

We can see, still in an intuitive level, that a linear combination of the members in (I) of the same degree, say n, will converge to a member in H_n. While, this would not be the case for a linear combination of members in (II). At least we have to give some interpretation to a formal expression

$$\int f(u)H_n(\dot{B}(u) \; ; \; 1/du)du. \tag{26}$$

Having made these observations, we now come to a rigorous definition of the above formulas and the integral (26). Actually, this can be done by using the integral representation established in §3. Before giving the definition, an illustrative example is presented below.

We now start with our measure space (S^*,μ) of white noise. Let χ_A be the indicator function of a set A. Then, $\langle x,\chi_\Delta\rangle$ is a member of H_1 and is thought of as a realization of ΔB. With this realization $H_n(\Delta B/\Delta; 1/\Delta)$ is realized by $H_n(\langle x,\chi_\Delta\rangle/\Delta \; ; \; 1/\Delta)$. Now apply the transformation T to obtain

$$\begin{aligned} &T(H_n(\langle x,\chi_\Delta\rangle/\Delta \; ; \; 1/\Delta))(\xi) \\ &= \frac{i^n}{\Delta^n n!} C(\xi) \int \cdots \int_{R^n} \chi_{\Delta^n}(u_1,\ldots,u_n)\xi(u_1)\ldots\xi(u_n)du^n. \end{aligned} \tag{27}$$

Take the limit as $\Delta \to \{t\}$, to have

$$i^n C(\xi)\frac{1}{n!}\, \xi(t)^n.$$

It tells us that the kernel of the integral representation of the limit is like the n-dimensional delta-function although the original functional never converges in the ordinary sense. However, $H_n(\langle x,\chi_\Delta\rangle/\Delta;1/\Delta)$ would have a limit, denote it by $H_n(x(t);1/dt)$, if we introduce somewhat weaker topology which is in concordant with the convergence of the corresponding kernel of the integral representation.

Incidentally, we should like to make a short remark. Once a suitable topology is introduced there, one can see that $M_n(dt) = H_n(x(t)\ ;\ 1/dt)dt$ plays a role of random measure. More generally, a product

$$\prod_j M_{n_j}(dt_j), \quad t_j \text{ different},$$

can be made to be a generalized random measure and even the integral with respect to it can well be defined. The integral itself is naturally a generalized functional which is going to be defined in what follows. (For details, see Hida [3].)

Coming back to our main topic to introduce a class of generalized Brownian functionals, we prefer the following diagram rather than a defintion by the use of words.

<u>D I A G R A M</u>

$$\begin{array}{ccccc}
\sqrt{n!}\ \widehat{H^{-(n+1)/2}}(R^n) & \cong & \mathcal{H}_n^{(-n)} & \cong & \mathcal{F}_n^{(-n)} \\
\circlearrowleft & & \circlearrowleft & & \circlearrowleft \\
\sqrt{n!}\ \widehat{L^2}(R^n) & \cong & \mathcal{H}_n & \cong & \mathcal{F}_n \\
\circlearrowleft & & \circlearrowleft & & \circlearrowleft \\
\sqrt{n!}\ \widehat{H^{(n+1)/2}}(R^n) & \cong & \mathcal{H}_n^{(n)} & \cong & \mathcal{F}_n^{(n)}
\end{array}$$

In the diagram the symbol like $\widehat{H^m}(R^n)$ stands for the space $\widehat{L^2}(R^n) \cap H^m(R^n)$, where $H^m(R^n)$ is the Sobolev space of order m over R^n, and $\circlearrowleft$ means a continuous injection from below to above. A member of $\mathcal{H}_n^{(-n)}$ is said to be a <u>generalized (Brownian) functional</u> of degree n, and $\sum_{n=1}^{\infty} \mathcal{H}_n^{(-n)}$ is the class of generalized functionals.

The logical step of the definition, as one can easily guess from the Diagram, is as follows: Let $\mathcal{H}_n^{(n)}$ be the class of $\mathcal{H}_n$-functionals the integral representations of which are given in terms of $\widehat{H}^{(n+1)/2}(R^n)$-kernels, and let it be topologized so as to be isomorphic to the Sobolev space $\sqrt{n!}\,\widehat{H}^{(n+1)/2}(R^n)$. Then find the dual space, denote it by $H_n^{(-n)}$, which is to be isomorphic to $\sqrt{n!}\,\widehat{H}^{-(n+1)/2}(R^n)$.

Our method of introducing the space $H_n^{(-n)}$ enables us to extend the transformation T to $H_n^{(-n)}$. Its image will be denoted by $F_n^{(-n)}$. The space $F_n^{(n)}$ has tacitly been introduced.

§5. Causal Calculus

We are now ready to discuss differential calculus by taking $\dot{B}(t)$'s to be variables of functionals.

Let $\phi(x)$ be a Brownian functional or a generalized Brownian functional. In order to define a partial derivative $\frac{\partial}{\partial \dot{B}(t)}$ or $\frac{\partial}{\partial x(t)}$, t being fixed, we must come back again to the stage of approximation to $\dot{B}(t)$ by $\Delta B/\Delta$, and play a similar game to what was done in the last section. For an illustration of this game we refer to the Lecture Note [2], and here is presented only a simple example which tells our idea.

Example For $\phi(x) = \int f(u)M_2(du)$ $(= \frac{1}{2}\int f(u)(x(u)^2 - 1/du)du)$ we can take $f(t)x(t)$ to be the partial derivative:

$$\frac{\partial}{\partial x(t)}\phi(x) = f(t)x(t).$$

The integral representations of $\phi(x)$ and $\frac{\partial}{\partial x(t)}\phi(x)$ are

$$i^2 C(\xi)\frac{1}{2}\int f(u)\xi(u)^2 du \quad \text{and} \quad iC(\xi)f(t)\xi(t),$$

respectively.

Definition Suppose that $\phi(x)$ be in $H_n^{(-n)}$ and that its integral representation be of the form $i^n C(\xi)U(\xi)$. If the functional derivative $U_\xi'(t)$ (in the sense of Fréchet) exists and if $i^{n-1}C(\xi)U_\xi'(t)$ is in $F_{n-1}^{(-n+1)}$, then we define

$$\frac{\partial}{\partial x(t)}\phi(x) = T^{-1}(i^{n-1}C(\xi)U_\xi'(t)). \tag{28}$$

It is often convenient to introduce smeared differential operators of the form

$$D_\alpha = \int \alpha(t)\frac{\partial}{\partial x(t)}\,dt, \qquad \alpha \in S. \tag{29}$$

Suppose that ϕ is in $H_n^{(n)}$ with the kernel F:

$$U(\xi) = \int\limits_{R^n}\cdots\int F(u_1,\ldots,u_n)\xi(u_1)\ldots\xi(u_n)du^n.$$

Then

$$U_\xi'(t) = n\int\limits_{R^{n-1}}\cdots\int F(t, u_2,\ldots,u_n)\xi(u_2)\ldots\xi(u_n)du^{n-1}.$$

Hence, we see that $\frac{\partial}{\partial x(t)}\phi$ is a stochastic process living in the space $H_{n-1}^{(n-1)}$ and the associated kernel is written as $n\delta_t * F$.

It is straightforward to extend the differential operators to those acting on a subset of $\sum_n H_n^{(-n)}$. Also one can define the adjoint operator $(\frac{\partial}{\partial x(t)})^*$ as well as the multiplication by x(t) (or by $\dot{B}(t)$). A systematic approach has been given by I. Kubo and S. Takenaka [7]. The theorem stated below is due to them, although the choice of the domain is somewhat different. To state the theorem we introduce notations.

Set

$$H = \sum_{n=0}^{\infty} \oplus H_n^{(n)}.$$

For a kernel F we define the symmetrization of $\delta_t \otimes F$ by

$$(\delta_t \hat{\otimes} F)(u_1,\ldots,u_{n+1}) = \frac{1}{n+1}\sum_{k=1}^{n+1} \delta_t(u_k)F(u_1,\ldots,u_{k-1},u_{k+1},\ldots u_{n+1}).$$

Finally, let P_n be the projection operator to the space H_n.

<u>Theorem 4</u> <u>Let ϕ be in H, and let F_n be the kernel of the integral representation of $P_n\phi$. Then,</u>

i) <u>the kernels associated with</u> $P_n(\frac{\partial}{\partial x(t)}\phi)$ <u>and</u> $P_n\{(\frac{\partial}{\partial x(t)})^*\phi\}$

are $(n + 1)\delta_t * F_{n+1}$ and $\delta_t \hat{\otimes} F_{n-1}$,

respectively.

ii) In particular $\frac{\partial}{\partial x(t)} : H_n^{(n)} \longrightarrow H_{n-1}^{(n-1)}$,

$$(\frac{\partial}{\partial x(t)})^* : H_n^{(n)} \longrightarrow H_{n+1}^{-(n+1)}.$$

iii) We have $$[\frac{\partial}{\partial x(t)}, (\frac{\partial}{\partial x(s)})^*] = \delta_{t,s} I. \tag{30}$$

iv) The product $x(t)\phi(x)$ is defined for $\phi \in H_n^{(n)}$

$$x(t)\phi(x) = (\frac{\partial}{\partial x(t)})^*\phi(x) + \frac{\partial}{\partial x(t)} \phi(x). \tag{31}$$

Proof The assertions i) - iii) come from what we have discussed so far. As for the multiplication we again use the idea to observe

$$\lim \frac{\langle x, \chi_\Delta \rangle}{\Delta} \phi(x), \qquad \Delta \to \{t\},$$

for $\phi \in H_n^{(n)}$. The limit is taken in the space of generalized functionals. Actual computation together with the result in i) proves the formula (31).

Further developments in this line are found in [7]. As was mentioned in the Introduction, the formulas in this theorem suggest us to establish interesting connections with quantum dynamics.

§6. Concluding Remarks

This section is devoted to the remarks for further problems arising from our causal analysis in terms of white noise.

i) The first problem is concerned with the infinite dimensional Laplacian operator. (Note that we never aim at the investigation of the infinite dimensional Laplace-Beltrami operator.) Since $\{\dot{B}(t)\}$ is our system of variables of the Brownian functionals, the proposed expression of the Laplacians would be of the form

$$\Delta \equiv \int \frac{\partial^2}{\partial \dot{B}(t)^2}\, dt. \tag{32}$$

Being inspired by the Lévy's approach to the Laplacian on function space (see [1], IIIème Part.), we are able to use $U''_{\xi^2}(t)$ the functional derivative of the second order in F-space, as in (28), to define

$$\int U''_{\xi^2}(t)\, dt,$$

and then transform it to the space (L^2) by T^{-1}. This is a realization of the formal expression (32). For example, for the functional $\phi(x)$ in the Example in §5 we have

$$\Delta\, \phi = \int f(t)dt.$$

It would be fine if we could find good relationships with other concepts in functional analysis such as harmonic functionals, potentials, and so forth like finite dimensional harmonic analysis. Some observation has been made in [7] with the same spirit.

ii) The second problem is how to exponentiate a generalized Brownian functional. This problem is specifically inspired by quantum field theory.

If $\phi(x)$ is an ordinary real-valued Brownian functional, then there is no problem to have $\exp[i\phi(x)]$. However, if $\phi(x)$ is a generalized functional, then we need some renormalization to define the exponential functional. Remind that $\int \dot{B}(t)^2 dt$ is subtle but, after subtract off the expectation, $\int (\dot{B}(t)^2 - 1/dt)dt$ is our best friend, i.e. a typical example of a generalized Brownian functional. We are now in search of a trick with which exponential functionals are well-defined.

A function $\exp[x(t)]$ has no meaning, but

$$\exp[x(t) - \frac{1}{2}\cdot\frac{1}{dt}], \tag{33}$$

which is equal to $\sum_{n=0}^{\infty} H_n(x(t);1/dt)$, is a generalized Brownian functional. This is guaranteed by using the relation (27).

Another example discussed in Streit-Hida [6] is

$$\exp[\int_0^1 M_2(dt)] \equiv \exp[\frac{1}{2}\int_0^1 (x(t)^2 - 1/dt)dt]. \tag{34}$$

There is a way to modify this so that the result is a generalized functional. Namely, in course of approximating $\dot{B}(t)$, the exponential functional should be divided by its expectation. Such a modification is actually in agreement with the method of modifying $\exp[x(t)]$ to have (33). (For details see [6].)

References

[1] P. Lévy, Problèmes concrets d'analyse fonctionnelle. Gauthier-Villars, 1951.

[2] T. Hida, Analysis of Brownian functionals. Carleton Univ. Lecture Notes No. 13, Second Ed. 1978.

[3] T. Hida, Generalized multiple Wiener integrals. Proc. Japan Academy, 54 (1978), 55-58.

[4] T. Hida, Brownian motion. Iwanami 1975, (in Japanese). English edition Springer-Verlag, Applications of Math. 11, 1980

[5] T. Hida and L. Streit, On Quantum Theory in Terms of White Noise. Nagoya Math. J., 68 (1977), 21-34.

[6] L. Streit and T. Hida, Generalized Brownian functionals and Feynman integral, to appear.

[7] I. Kubo and S. Takenaka, Calculus on Gaussian white noise II, to appear.

Introduction to Stochastic Differential Calculus

Paul-André Meyer *) **)

Deptartment of Mathematics

University Louis Pasteur

Strasbourg, France

Abstract

We present an introduction to the recent work on the theory of semimartingales and the stochastic integrals.

*) This lecture is taken from Libero homenajo al Prof. Dr. D. Alfonso Guiraùm Martin, Universidad de Granda, Departamento de Estadistica, February 1979, with the kind permission of the University of Granada.

**) Translated from the French by Ph. Blanchard and D.E. Miller.

This exposition has the goal of explaining the fundamental ideas of stochastic differential calculus by showing how it involves at the foundations a very simple and manageable theory. We have tried to avoid the notion of stopping time, which is somewhat difficult to grasp for non professional audience. This forces us to slightly change the accepted terminology so that the words which have a special meaning in this lecture are written with an asterix (*) the first time they appear, for which a short appendix at the end make clear their relation with the usual expression.

1. An Example: The Langevin Equation

The simplest problem which leads to the consideration of the stochastic integrals is that of Brownian motion from the "physicists" ruled by the Langevin equation. Let us consider a particle with a mass m wich moves in a medium in a state of continued thermal agitation also under the influence of an external force field F. The position of the particle at the time t is described by X_t and its velocity V_t, so that the Langevin equation is written

$$dX_t = V_t dt \qquad m dV_t = F(X_t)dt + dW_t \tag{1.1}$$

to which one must add the initial condition $X_o = x$, $V_o = v$. Here W_t represents the disturbance of the medium which is a random phenomenon, W_t is a function of the time and of a parameter ω, for which the value is unknown but which is drawn by lot in an immense "urn" Ω following a law of probability given by P. In this particular case the law P is that which makes from the process (W_t) a Wiener process (the Brownian motion for the mathematicians). The solution (X_t, V_t) of the equation (1.1) depends themselves on the unknown paramter ω.

The equation (1.1) has no sense, because the function $W_t(\omega)$ are not differentiable with respect to t. One can be reassured of this fact by writing (1.1) in the form

$$X_t = x + \int_0^t V_s ds \qquad mV_t = mv + \int_0^t F(X_s)ds + W_t \tag{1.2}$$

which is not an ordinary differential equation but does have at least an exact meaning. However, one can not do the similar thing for the equation

$$dX_t = V_t dt \qquad mdV_t = F(X_t)dt + K(X_t)dW_t \tag{1.3}$$

which would replace (1.1) in a medium where the temperature varies from one point to another. It is nevertheless necessary to give a meaning to such an equation and to show that it possesses solutions under reasonable conditions. We discuss very briefly the classical theory of Ito while considering from simplicity the case of a single equation in one and not three dimensions.

2. The Classical Theory

We recall the following properties of the Wiener process W,
W _is a process with continuous trajectories_

$$W_0 = 0, \quad E[W_t^2] < \infty \quad \text{for all } t. \tag{2.1(a)}$$

On the other hand, if $s < t$ the increment $W_t - W_s$ is orthogonal not only to all the increments previous to s, but also to the whole σ-algebra generated by these increments. Let us denote it by $\underline{\underline{F}}_s$. Thus $\underline{\underline{F}}_s$ increases in s so that one has

W_s _is_ $\underline{\underline{F}}_s$_-measurable for all_ s _(one says that_ W _is adapted to_ $\underline{\underline{F}}_t$)

$$\text{for } s<t, \quad E[W_t - W_s | \underline{\underline{F}}_s] = 0 \quad (\underline{\text{i.e.}} \int_A (W_t - W_s)P = 0 \ \underline{\text{if}}\ A \in \underline{\underline{F}}_s). \tag{2.1(b)}$$

This property has the name _martingale_ (*). Finally one has

$$\underline{\text{if}}\ s<t, \ E[(W_t - W_s)^2 | \underline{\underline{F}}_s] = t-s \quad (\underline{\text{i.e.}} \int_A (W_t - W_s)^2 P = (t-s)P(A) \ \underline{\text{if}}\ A \in \underline{\underline{F}}_s). \tag{2.1(c)}$$

These three properties permit us to define the stochastic integrals $\int_0^t K_s dW_s$ for a sufficiently rich class of process. First if K

*) See appendix

is a step process on the form

$$K_t = k_0 1_{\{t=0\}} + k_1 1_{]t_0,t_1]} + \dots + k_n 1_{]t_{n-1},t_n]} \tag{2.2}$$

where $k_0, k_1, \dots k_n$ are bounded random variables and $0 = t_0$, $t_1, \dots t_n$ the only possible choice for $\int_0^t K_s dW_s$ is the obvious stochastic integral

$$\int_0^t K_s dW_s = k_1 W_{t_1} + k_2(W_{t_2} - W_{t_1}) + \dots + k_i(W_t - W_{t_{i-1}}) \tag{2.3}$$

$$\text{if} \quad t_{i-1} \leqq t < t_i .$$

Then, if k_0 is $\underline{\underline{F}}_0$-measurable, k_i $\underline{\underline{F}}_{t_{i-1}}$-measurable for $i \geq 1$, we call then K a predictable step process, from we deduce from (2.1(c)) that

$$E[(\int_0^t K_s dW_s)^2] = E[\int_0^t K_s^2 ds] . \tag{2.4}$$

Let us give the name predictable process to the elements of the monotone class spanned by the predictable step processes (for example, all adapted left continuous processes are predictable). The stochastic integral $Y_t = \int_0^t K_s dW_s$ can then be defined for all predictable processes such that $E[\int_0^t K_s^2 \, ds] < \infty$ for all t. Thus we have the following properties

Y is a continuous process, $Y_0 = 0$, $E[Y_t^2] < \infty$ (2.5(a))

Y is adapted, $E[Y_t - Y_s | \underline{\underline{F}}_s] = 0$ if $s < t$ (2.5(b))

If $s < t$, $E[(Y_t - Y_s)^2 | \underline{\underline{F}}_s] = E[B_t - B_s | \underline{\underline{F}}_s]$, with $B_t = \int_0^t K_u^2 du$. (2.5(c))

Thereupon (2.5(a)) and (2.5(b)) are identical to (2.1(a)) and (1.1(b)) in such a way that (2.5(c)) is formally analogous to (2.1(c)) through the replacement of the process $B_t = t$ by an increasing process depending of ω.

This theory of the stochastic integral is sufficient for the study of the stochastic differential equation of the Langevin type

$$dX_t = F(X_t)dt + G(X_t)dW_t \tag{2.6}$$

One can also investigate systems of such equations in order to obtain, in particular, the existence and uniqueness of the solution under very reasonable conditions.

One could, however, put forth various objections to the classical theory. The first one has to do with the fact that there exists a formally analogous equation to (2.6), but where W is replaced by a process including jumps. The other more serious one involves the rôle taken on by the law P. In some problems of the statistics of the processes one can ignore, for example, those which between two laws P and Q is the true law of the observed phenomenon. If W_t is a Wiener process for P, it is not in general for Q. One can in the good cases bring into consideration a process h_t such that $W_t - \int_o^t h_s d_s = W'_t$ be a Wiener process for the law Q. The equation (2.6) is written then for the law Q

$$dX_t = (F(X_t) + h_t G(X_t))dt + G(X_t)dW'_t \tag{2.7}$$

which is not in general of the same type. A basic idea of the stochastic differential calculus consists in constructing a theory which takes the same form whenever one changes one law of probability, or more exactly expressed whenever one replaces the law P by the law Q equivalent to P.

3. Square Integrable Martingales

The preceeding example has shown to us the rôles played

1) by the martingale property of the Wiener process,
2) by the family $\underline{\underline{F}}_t$ considered as an input of the problem. Namely $\underline{\underline{F}}_t$ is at the beginning given by the Wiener process but in (2.6) and (2.7) the family stays the same although we change the process,
3) by the predictable processes.

Now we got out from probability space $(\Omega,\underline{\underline{F}},P)$ on which we give ourselves an increasing family $\underline{\underline{F}}_t \subset \underline{\underline{F}}$ satisfying one or two condition of technical order (*). The predictable step processes and the predictable processes are defined as above, likewise are the adapted processes. Similarly the notion of a square integrable martingale is a process $M = (M_t)$ such that M is a right continuous process with left limits

$M_o = 0$ and $E[M_t^2] < \infty$ for all t (*) (3.1(a))

M is adapted to $(\underline{\underline{F}}_t)$, if s < t $E[M_t - M_s | \underline{\underline{F}}_s] = 0$. (3.1(b))

The concept martingale is obtained by replacing the condition in (3.1(a)) by $E[|M_t|] < \infty$. One may note the provisional absence of a condition corresponding to (2.1(c)) or (2.5(c)).

One calls an increasing process an adapted process A which is increasing. More precisely we require that $A_o = 0$ and that the trajectories of A be increasing functions with right continuity. One says that A is integrable (*) if $E[A_t] < \infty$ for all t. The differences of increasing (integrable) processes are the processes of finite variation.

In this case let M be a square integrable martingale. One can show that there exists at least one increasing integrable process B such that

$$E[(M_t - M_s)^2 | \underline{\underline{F}}_s] = E[B_t - B_s | \underline{\underline{F}}_s] \quad \text{for } s < t. \tag{3.1(c)}$$

There are at least two of these increasing processes which are used, where the limits are taken in L^1. They have the form

$$B_t = \lim_k \sum_{i=0}^{2^k-1} (M_{(i+1)2^{-k}t} - M_{i2^{-k}t})^2 \tag{3.2}$$

and

$$B_t = \lim_k \sum_{i=0}^{2^k-1} E[(M_{(i+1)2^{-k}t} - M_{i2^{-k}t})^2 | \underline{\underline{F}}_{i2^{-k}t}] . \tag{3.3}$$

This result, from the mean difficulty being established, it is only a question of recopying the theory of Ito to construct the stochastic integral $\int_o^t K_s dX_s = Y_t$ for all predictable processes K such that $E[\int_o^t K_s^2 dB_s] < \infty$ for all t-classes which do not depend on the choice of the increasing process B. The process Y is a square integrable martingale so that one can study this stochastic integral in a deeper sense (we shall not do it here). An important point is the following: if X is at the same time a square integrable martingale and a process of finite variation (for example $X_t = P_t - \lambda t$, where P_t is a Poisson process with the parameter λ) and if K is predictible bounded, then Y_t coincides with the integral $\int_0^t K_s dX_s$ understood as an ordinary Stieltjes integral.

Let us go now to the more general case.

4. Semimartingale

Definition: An adapted right continuous process with left limits X is a restricted semimartingale if it can be written as

$$X_t = X_o + M_t + A_t \ , \qquad (4.1)$$

where M is a square integrable martingale, A a process of integrable variation and X_o belongs to L^1.

Definition: An adapted right continuous process with left limits is a semimartingale if there exists a sequence (X^n) of restricted semimartingales and a sequence (S_n) of positive random variables such that it may be written

$$S_n \uparrow \infty \quad \text{a.s.} \ , \qquad X_t = X_t^n \quad \text{for} \quad 0 \le t < S_n . \qquad (4.2)$$

The class of semimartingales is a very interesting one as can be seen from the following properties

- Every process of finite variation is a semimartingale, (4.3) every martingale is a semimartingale (this is only obvious for square integrable martingales); every supermartingale is also a semimartingale (*).
- If $(X^1,\dots,X^k)$ are semimartingales and if $F(x^1\dots x^k)$ (4.4) is a C^2 function of k variables or a convex function, then $F(X^1,\dots,X^k)$ is a semimartingale. In particular, if X and Y are semimartingales, X+Y, XY, $X \wedge Y$ are semimartingales.
- If X is a right continuous adapted process of bounded (4.5) variation (*), that is if one has for all subdivisions $(t_1,\dots, t_n)$

$$X_{t_i} \in L^1 \ , \quad E\Big[\sum_{i=1}^{n-1} |E[X_{t_{i+1}} - X_{t_i} \,|\, \underline{\underline{F}}_{t_i}]| \Big] \leq M$$

independent of the subdivision, then X is a semimartingale.

- Every right continuous process with independent and homogeneous increments is a semimartingale. (4.6)

- If X is an adapted right continuous process with left limits and if there exists some semimartingales X^n and some random variables $S_n \uparrow \infty$ such that $X_t = X_t^n$ for $0 \le t \le S_n$, then X is a semimartingale. (4.7)

However, the most remarkable property of semimartingales is perhaps the following:

- If P and Q are two equivalent probability laws, the semimartingales are the same under the law P as under the law Q. (4.8)

One could say something similar when X is only absolutely continuous with respect to P (see [18], [24]). On the other hand by modifying a theorem of Stricker [23], Dellacherie has shwon that

- If X is a semimartingale for the law P, there exists a law Q equivalent to P such that X be a restricted semimartingale for the law Q. (4.9)

5. Stochastic Integrals with Respect to Semimartingales

We are now about to define the stochastic integral of a bounded predictable process K with respect to a semimartingale X. The hypothesis will be slightly weakened by the following. Although the idea is really the same as that of section 2 one extends the obvious integral by using the convergence in probability and not in L^2. The most agreable consequence of this situation is that if we replace the law P by the equivalent law Q the convergence in probability is not changed, thereby leaving the stochastic integral the same.

In order to facilitate the notation we denote $K \cdot X$ for the process $(\int_o^t K_s dX_s)$.

Theorem 5.1

Let X be a semimartingale. We can define an adapted right continuous process with left limits $K \cdot X$ in a unique way for every bounded predictable process K such that

1) If K stepwise, $K \cdot X$ is the obvious stochastic integral.

2) If some predictable processes K^n uniformly bounded converge simply towards 0, one has for all t $\lim_n K^n \cdot X)_t = 0$ simply in probability measure. Furthermore, $K \cdot X$ is then a semimartingale.

It is necessary to give a little attention to the definition of the obvious stochastic integral, which we have only given in (2.3) for a process such that $W_o = 0$. Here it is necessary to attribute to it the value $K_o X_o$ for all $t = 0$.

Before enumerating the properties of the stochastic integral, let us give a description in passing of a very striking result established quite recently by Dellacherie and Mokobodzki after some partial results of Métivier and Pellaumail. Let us consider a right continuous adapted process X, for which we may define for stepwise predictable K as the obvious stochastic integral, so we may suppose that one has the following property somewhat less strong than 2):

> If from the predictable step process K^n converge uniformly towards 0, one has for all t the $\lim_n (K^n \cdot X)_t = 0$ in probability measure.

Then X is a semimartingale. In other words one could not hope to construct a reasonable theory of the stochastic integral for a class of processes larger that of the semimartingale. This kind of theorems charms the mathematicians but it has clearly no practical applications.

Here are the principal properties of the stochastic integral. For the first one we use the notation ΔY_t for the jump $Y_t - Y_{t-}$ of a right continuous process Y with a left limit in t. One admits that $Y_t = 0$ for $t = 0$ in the way that $\Delta Y_o = Y_o$.

> One has $\Delta(K \cdot X)_t = K_t \Delta X_t$. In particular if X is a continuous semimartingale, so is $K \cdot X$. (5.1)

The second one justifies the differential notation. Since account is taken of the convention of the nullity of the process for

$t < 0$, the relation $Y = K \cdot X$ can be written symbolically as $dY_t = K_t dX_t$. However, this notation is dangerous when one must distinguish $H_t dY_t$ and $H_t K_t dX_t$. Fortunately we know that:

If H and K are bounded predictable, one has $H \cdot (K \cdot X) = (HK) \cdot X$. (5.2)

The following properties make up what one calls the <u>local character</u> of the stochastic integral.

- Let $Y = K \cdot X$. For almost all ω such that $X_{\bullet}(\omega)$ be a function of bounded variation on $[0,\infty[$ then one has the ordinary Stieltjes integral of the form $Y_t(\omega) = \int_0^t K_s(\omega) dX_s(\omega)$. (5.3)

- Let X and $\bar{X}$ be two semimartingales, K and $\bar{K}$ two bounded predictable processes, Y and $\bar{Y}$ the processes $K \cdot X$ and $\bar{K} \cdot \bar{X}$ and S a positive random variable. For almost all ω such that one has (5.4)

$$K_t(\omega) = \bar{K}_t(\omega) \qquad X_t(\omega) = \bar{X}_t(\omega) \qquad \text{for} \quad 0 \leq t \leq S(\omega)$$

  one has also

$$Y_t(\omega) = \bar{Y}_t(\omega) \qquad \text{for} \quad 0 \leq t \leq S(\omega).$$

In particular for almost all ω such that $X_{\bullet}(\omega) = \bar{X}_{\bullet}(\omega)$, $K_{\bullet}(\omega) = \bar{K}_{\bullet}(\omega)$ one has $Y_{\bullet}(\omega) = \bar{Y}_{\bullet}(\omega)$.

An interesting consequence of (5.4) is the following. Let K be a predictable process such that there exist random variables $S_n \uparrow \infty$ as well as bounded predictable processes K^n satisfying $K_t = K_t^n$ for $0 \leq t < S_n$. Then one can define the stochastic integral $K \cdot X$ by sticking together of the $K^n \cdot X$. It is very easy to extend to this situation the properties (5.1), (5.2), (5.3), (5.4), the fact that $K \cdot X$ is a semimartingale, etc. We shall say in this case that K is locally bounded (*) so that we shall use without mention this extension of the stochastic integral.

6. Quadratic Variation and Variable Transformation

It is possible to show that if X is a semimartingale and K an adapted right continuous process with left limits the Riemann sums $\sum_i K_i (X_{t_{i+1}} - X_{t_i})$ over the interval [0,t] converge in probability measure to the stochastic integral $\int_{]0,t]} K_{s-} dX_s$. Now we want to know what will happen if we put $K_{t_{i+1}}$ for K_{t_i}. A partial answer is given by the following theorem:

Theorem 6.1

Let X and Y be two semimartingales. There exist a process with finite variation denoted by [X,Y] so that we have for all t

$$[X,Y]_t = X_0 Y_0 + \lim_k \Sigma_i (X_{t(i+1)2^{-k}} - X_{ti2^{-k}})(Y_{t(i+1)2^{-k}} - Y_{ti2^{-k}}) \tag{6.1}$$

which represents the limit in the probability measure.[1)]

Here again the same limit is obtained if we replace P by an equivalent law. In particular [X,X] is always an increasing process, which we have already encountered for a square integrable martingale. Also we see that the bilinear function [X,Y] can be expressed by polarization of the positive quadratic function [X,X]

$$2[X,Y] = [X+Y,X+Y] - [X,X] - [Y,Y] . \tag{6.2}$$

It is intuitively expected that we have

$$\int_0^t |d[X,Y]_s| \leqq [X,X]_t^{1/1} [Y,Y]_t^{1/2} \tag{6.3}$$

This form is the inequality of Kunita-Watanabe. Like all processes with finite variation [X,Y] are decomposable into a purely discontinuous term and a continuous part. The first term is very easy to calculate. It is simply the sum of the absolutely convergent series of the jumps $\Delta[X,Y]_t$. Thus we have

$$\Delta[X,Y]_t = \Delta X_t \Delta Y_t \quad (\text{ for } t = 0: [X,Y]_0 = X_0 Y_0). \tag{6.4}$$

---

1) [X,Y] is often called the right bracket of X and Y.

The continuous term is denoted by $[X,Y]^c$ (*). For continuous processes X,Y with finite variation we shall always have $[X,Y]^c = 0$. However, the original characteristic feature of the stochastic differential calculus is precisely the fact that $[X,Y]^c$ can be non zero. For the Wiener process W, for example, we find that $[W,W]_t = t$.

If H and K are two predictable locally bounded processes, we have

$$[H\cdot X,\ K\cdot Y]_t = \int_0^t H_s K_s d[X,Y]_s \tag{6.5}$$

where in differential notation we may write

$$d[H\cdot X,\ K\cdot Y] = HKd[X,Y].$$

It would be, however, somewhat more suggestive to write $[dX,dY]$ for $d[X,Y]$.

We are now able to write the formula for the partial integration of the stochastic integrals

Theorem 6.2

If X and Y are two semimartingales, then we have

$$X_t Y_t = \int_0^t (X_{s-}dY_s + Y_{s-}dX_s) + [X,Y]_t \tag{6.6}$$

with the convention that $X_{0-} = Y_{0-} = 0$. Furthermore, in differential notation we write

$$d(XY) = X_-dY + Y_-dX + d[X,Y]. \tag{6.7}$$

For example, if W is the Wiener process, we obtain

$$\int_0^t W_s dW_s = W_t^2 - [W,W]_t = W_t^2 - t. \tag{6.8}$$

For the calculations on the continuous semimartingales it is in many cases useful to introduce the Stratonovič integral

$$\int_0^t X_s * dY_s{}^{(*)} = \int_0^t X_{s-}dY_s + \frac{1}{2}[X,Y]_t,$$

which is the limit of Riemann sums of the type

$$\Sigma_i X_{\frac{t_i+t_{i+1}}{2}} (Y_{t_{i+1}} - Y_{t_i}).$$

This integral has the advantage of verifying the usual formula

$$X_t Y_t - X_0 Y_0 + \int_0^t (X_s * dY_s + Y_s * dX_s) \tag{6.9}$$

as well as even more generally all the rules of the ordinary differential calculus. However, it presents nothing interesting when the processes have jumps.

The formula for the partial integration can be seen as a special case of the formula for the variable transformation. Let us consider a two times continuously differentiable function $F(X_1, X_2)$ with X_1 and X_2 being two semimartingales. Furthermore, let us propose that $X_t = (X_t^1, X_t^2)$. Then we have

$$F(X_t) = F(X_0) + \int_{]0,t]} D_1 F(X_{s-}) dX_s^1 + \int_{]0,t]} D_2 F(X_{s-}) dX_s^2 \tag{6.10}$$

$$+ \sum_{0<s\leqq t} \{F(X_s) - F(X_{s-}) - D_1 F(X_{s-}) \Delta X_s^1 - D_2 F(X_{s-}) \Delta X_s^2\}$$

$$+ \tfrac{1}{2} \Sigma_{ij} \int_{]0,t]} D_{ij} F(X_{s-}) d[X^i, X^j]_s^c \quad (i,j = 1.2).$$

In this formula the second line is missing if X^1 and X^2 are continuous processes. Similarly the third line is not there if X^1 and X^2 are processes with finite variation. When X^1 and X^2 are simultaneously both continuous and of finite variation, it can be written simply as $dF(X) = F_1'(X)dX^1 + F_2'(X)dX^2$, which comes as a surprise to nobody.

7. Conservation of the Martingale Property

We have mentioned in §3 that if X is a square integrable martingale and K is a predictable process such that $E[\int_0^t K_s^2 d[X,X]_s < \infty$ for all t, then $K \cdot X$ is also a square integrable martingale. In particular, since the property $X_0 = 0$ is a part of our definition of martingales, we have $K_0 X_0 = 0$ so that for all t

$$E[\int_0^t K_s dX_s] = 0 . \tag{7.1}$$

We want to know what is left over from all this in the general case. As a first step we are able to define with very little effort a class of semimartingales, which contains the martingales and which is stable with respect to the stochastic integrals.

Definition: A semimartingale Y is a transform of a martingale if there exists a martingale X and a predictable locally bounded process K, such that $Y = K\cdot X$. Then $Y_0 = K_0 X_0 = 0$.

Theorem 7.1

Let $Y = K\cdot X$ be a transform of a martingale as discussed above. In order that Y be a square integrable martingale, it is necessary and sufficient that for all t

$$E[[Y,Y]_t] = E[\int_0^t K_s^2 d[X,X]_s] < \infty \tag{7.2}$$

In order that Y be a martingale, it is sufficient that

$$E[[Y,Y]_t^{1/2}] = E[(\int_0^t K_s^2 d[X,X]_s)^{1/2}] < \infty . \tag{7.3}$$

Moreover, if Y is a transform of a martingale and if Y is majorized in absolute value by a positive martingale, then Y is also a martingale.

8. Stochastic Differential Equations

Catherine Doléans-Dade has studied the differential equations of the type

$$Y_t = H_t + \int_0^t f(.,s,Y_{s-}) dX_s . \tag{8.1}$$

In this equation the unknown quantity Y is an adapted right continuous process with left limits with the given quantities H,f,X, so that it admits a unique solution under the following very reasonable conditions: H is an adapted right continuous process with left limits, X is a semimartingale; $f(\omega,s,x)$ is a function of the three variables $\omega \in \Omega$, $s \in \mathbb{R}_+$, and $x \in \mathbb{R}$. We list the properties of $f(\omega,s,)$ under the following stated conditions:

a. for given s and x, $f(\cdot,s,x)$ is $\underline{\underline{F}}_s$-measurable,

b. for given x and ω, $f(\omega,\cdot,x)$ is left continuous with right limits

c. for fixed ω and s, $f(\omega,s,\cdot)$ is Lipschitz in x, whereby the Lipschitz constant can be dependent on ω,s, as Lenglart has later shown.

Moreover, if H is a semimartingale, then is Y also one. Independently Protter has studied a little less flexible equation where no process H replaced by an initial value and f dependent only on s and x. Also Emery has studied a little later a somewhat more convenient form for the theory. There is also on this subject some recent work by Métivier-Pellaumail. One knows a number of things about these equations. For example, the method of the successive approximations as well as the method of linear interpolation converge always in probability measure. The stability of the solutions have been also well established by Emery and Protter. All that we have said for a single equation can be extended to a system of differential equations.

Of course the principle Lipschitz differential equation is that which defines the exponential function

$$Y_t = 1 + \int_0^t Y_{s-}dX_s \qquad (8.2)$$

which has been solved by C. Doléans-Dade before the whole general theory. The solution is explicitly written as

$$Y_t = e^{X_t-\frac{1}{2}[X,X]_t^c} \prod_{0\le s\le t} (1+\Delta X_s)e^{-\Delta X_s} . \qquad (8.3)$$

In return we have no explicit solution for the analogous equation ([21]) in which X is a semimartingale with values in the set of square matrices, whereby $\mathbb{1}$ represents the unit matrix and the unknown quantity Y is a process with values in the same set. We can formally write Y as a multiplicative integral (non commutative)

$$Y_t = \prod_0^t (1+dX_s). \qquad (8.4)$$

Many important works have been devoted to estimate the solution (integrability of Y_t, etc.).

Finally if we replace the law P by an equivalent one, likewise the form of the differential equation (8.1) as the solutions stay unaltered.

Appendix

It should be noted that in this lecture all martingales are supposed to be zero in O, which is not a usual convention. In the literature the square integrable martingales and the increasing integrable prosesses are supposed to be integrable up to infinity. Not only $E[M_t^2]<\infty$ for all t or even $E[A_t]<\infty$ for all t are required but also sup $E[M_t^2]<\infty$ or sup $E[A_t]<\infty$.

The assumptions imposed on the family $(\underline{\underline{F}}_t)$ are the "usual conditions", namely, the family must be right continuous and each σ-algebra $\underline{\underline{F}}_t$ must contain all the sets of measure zero.

The class of restricted semimartingales has no particular importance though it is quite simple.

Our definition of the semimartingales is not "classical" although it is equivalent to the standard one. The use of the arbitrary random variables in place of the stopping time is an illusion. Thus if two adapted right continuous processes X and Y with left limits are equal up to the time S the stopping time $T = \inf\{t: X_t \neq Y_t\}$ is $\geq S$ and X and Y are equal for $0 \leq t \leq T$. Our definition of locally bounded predictable processes is a little bit weaker than the usual definition, which assumes the existence of stopping times $T_n \uparrow \infty$ and constants M_n such that $|K_t| \leq M_n$ for $t \leq T_n$ with $T_n > 0$. However, the two definitions coincide for the predictable right or left continuous processes.

Since we do not use stopping times, we were not able to speak about <u>local martingales</u>. We replace them by the transforms of martingales, which form undoubtedly a bit larger class, but otherwise void of interest other than pedagogical. If the word locally bounded hat its usual sense, the two classes would be identical. Since we do not speak about local martingales, we cannot define the partial locally continuous martingale X^c of a semimartingale X. This is why the continuous part of the right bracket [X,Y] is denoted $[X,Y]^c$ and not $[X^c,Y^c]$ as usual. We have entirely omitted the mention of the pointed brackets <X,Y>, which the reader will find in the literature.

The notation $\int_0^t X_s * dY_s$ for the Stratonovič integration is never usual and never especially recommendable.

The word <u>supermartingale</u> is used one time in (4.3). A supermartingale is an adapted right continuous process X such that $X_t \in L^1$ for all and $E[X_t - X_s \mid \underline{\underline{F}}_s] \leq 0$ for $s < t$.

The "process of bounded stochastic variation" of (4.5) are most often called <u>quasimartingales</u>.

References

1. General references

Here are some very good surveys accessible to the non professionals

[1] M. Métivier, Reelle und vektorwertige Quasimartingale und die Theorie der stochastischen Integration. Lecture Notes in M. 607, Springer 1977.

[2] A.U. Kussmaul, Stochastic integration and generalized martingales. Research Notes in M. no. 11, Pitman, London, 1977.

The following reference is not commercialy accessible: write to INSA BP 14 A, 35031 Rennes Cedex, France. It does not require any preliminar knowledge.

[3] M. Métivier et J. Pellaumail, A basic course on stochastic integration. Séminaire de Probabilité 1, Université de Rennes 1 (1977).

For the readers woh are familiar with the theory of processes see

[4] P.A. Meyer, Un cours sur les intégrables stochastique. Sém. Prob. X, Université de Strasbourg. Lecture Notes in M. 511, Springer 1976.

2. The History of the Subject[1)]

[5] K. Ito, Stochastic integral. Proc. Imperial Acad. Tokyo, 20, 1944, p. 519-524.

[6] K. Ito, On stochastic differential equations. Memoirs Amer. M. Soc. no. 4, 1951.

[7] J.L. Doob, Stochastic processus. Wiley, NY, 1953.

[8] P.A. Meyer, A decomposition theorem for supermartingales. III. J.M. 6, 1962, p. 193-205.

[9] Ph. Courrège, Intégrales stochastiques et martingales de carré intégrable. Séminaire Brelot-Choquet-Deny, 1962-63.

1) We quote here only the important steps in the history of the ideas exposed above without pretending to write the history of the stochastic integral in general. Otherwise,we would be very unfair toward the school of McShane or the Russian school, whose publications are concerned with other aspects of the theory.

[10] H. Kunita et S. Watanabe, On square integrable martingales. Nagoya M.J. 30, 1967, p. 209-245.

[11] P.W. Millar, Martingale integrals. Trans. Amer. M. Soc. 133-1, 1968, p. 145-166.

[12] K. Ito et S. Watanabe, Transformation of Markov processes by multiplicative functionals. Ann. Inst. Fourier 15, 1965, p. 15-30.

[13] P.A. Meyer, Intégrales stochastiques I, II, III, IV. Sém. Prob. I, Université de Strasbourg. Lect. Notes in M. 39, Springer 1967.

[14] C. Doléans-Dade et P.A. Meyer, Intégrales stochastiques par rapport aux martingales locales. Sem. Prob. IV, Université de Strasbourg, Lect. Notes in M. 124, Springer 1970.

The references [5], [6], [7] present the "classical period", whose definitive form appears in

[15] H.P. McKean, Stochastic integrals, Academic Press, 1969.

The articles [8] [9] [10] extend the theory to the square integrable martingales; [11] is a first attempt to go beyond these constraints. The local martingales appear in [12]. The synthesis of the ideas of [12] and [10] is made in [13] and the semimartingales appearing in [14] all the essential ideas are brought in order.

The invariance of the property of martingale by a change of law has been elaborated in three essential papers (see also [24])

[16] I.V. Girsanov, On transforming a certain class of stochastic processes by absolutely continuous substitutions of measures. Teor. Prob. Appl. 5, 1960, p. 285-301.

[17] J.H. van Schuppen et E. Wong, Transformations of local martingales under a change of law. Ann. Prob. 2, 1974, p. 879-888.

[18] J. Jacod et J. Mémin, Caractéristiques locales et condition de continuité absolue pour les semimartingales. ZfW 35, 1976. p. 1-37.

About the differential stochastic equations after partial results by Kazamaki (Tohoku M.J. 1974; Sém. Prob. VI, 1976) the modern theory is developed in

[19] C. Doléans-Dade, Existence and unicity of solutions of stochastic differential equations. ZfW 36, 1976, p. 93-102.

[20] Ph. Protter, Right continuous solutions of stochastic integral equations. J. of Multivariate analysis, 7, 1977, 204-214.

[21] M. Emery, Stabilité des solutions des équations différentielles stochastiques. Application aux intégrales multiplicatives stochastiques. ZfW 41, 1978, p. 241-262. Voir aussi [26].

A Few Recent Articles (quoted in the text)

[22] C. Dellacherie, Quelques applications du lemme de Borel-Cantelli à la théorie des semimartingales. Sém. Prob. XII, Lect. Notes 649, Springer 1978, p. 742-745.

[23] C. Stricker, Quasimartingales, martingales locales, semimartingales et filtrationnaturelle. ZfW. 39, 1977, p. 55-64.

[24] E. Lenglart, Transformation des martingales locales par changement absolument continu de probabilités. ZfW 39, 1977, p. 65-70.

[25] Ph. Protter, H^p-stability of stochastic differential equations. ZfW. 44, 1978, p. 337-352.

[26] M. Métivier et J. Pellaumail, On a stopped Doob's inequality and general stochastic equations. Rapport interne no. 28, Ecole Polytechnique. To appear.

Finally a quite modern and very readable account will appear in the Proceedings of the Seminar on measure theory Oberwolfach, July 2-6, 1979, led by Professor D. Kölzow (Lecture Notes in Mathematics). This is

[27] C. Dellacherie, Un survol de la théorie de l'intégrale stochastique.

We recommend it very highly.

A Generalized Stochastic Calculus in Homogenization

Masatoshi Fukushima
Osaka University, Toyonaka, Osaka
Japan

Abstract

Take a formally self-adjoint second order partial differential operator with the coefficients $a_{ij}(x)$ being periodic in x. When a_{ij} are smooth, Bensoussan-Lions-Papanicolaou showed that the diffusion process corresponding to the coefficients $a_{ij}(\frac{x}{\varepsilon})$ converges as $\varepsilon \downarrow 0$ to a diffusion with constant coefficients. We show that their method still works without any smoothness assumption by using a stochastic calculus relevant to the Dirichlet forms.

§1. Introduction

We consider a family $\{A^{(\varepsilon)}, \varepsilon > 0\}$ of differential operators of the divergence form:

$$A^{(\varepsilon)}u = \sum_{i,j=1}^{n} \frac{\partial}{\partial x_j} \left(a_{ij}\left(\frac{x}{\varepsilon}\right) \frac{\partial u}{\partial x_j}\right) . \tag{1}$$

Here we assume

$$a_{ij} = a_{ji} \in L^{\infty}(R^n), \quad 1 \leqq i, \quad j \leqq n, \tag{2}$$

uniform ellipticity:

$$\lambda_0 |\xi|^2 \leqq \sum_{i,j=1}^{n} a_{ij}(x)\xi_i\xi_j \leqq \Lambda_0 |\xi|^2 , \tag{3}$$

$\xi \in R^n$, $x \in R^n$, for some positive constants λ_0 and Λ_0,

periodicity:

$$a_{ij}(x_1, \ldots, x_k, \ldots, x_n) = a_{ij}(x_1 \ldots, x_k+1, \ldots, x_n), \tag{4}$$

$$1 \leqq i, \; j, k \leqq n.$$

In 1973, E. De Giorgi and S. Spagnolo [5] proved that $A^{(\varepsilon)}$ is G-convergent as $\varepsilon \downarrow 0$ to an operator $A^{(0)}$ in the sense that

$$(\lambda - A^{(\varepsilon)})^{-1} f \xrightarrow[\varepsilon \downarrow 0]{} (\lambda - A^{(0)})^{-1} f \quad \text{on } L^2(R^n) \tag{5}$$

for any $\lambda > 0$ and $f \in L^2(R^n)$. Here the "homogenized" operator

$$A^{(0)}u = \sum_{i,j=1}^{n} q_{ij} \frac{\partial^2 u}{\partial x_i \partial x_j} \tag{6}$$

possesses constant coefficients q_{ij} that will be specified later. The general theory for the Dirichlet spaces [4], [11] enables us to associate with each operator $A^{(\varepsilon)}$ a conservative diffusion process on R^n. A natural question then arises: are these diffusions convergent as $\varepsilon \downarrow 0$ to the diffusion associated with $A^{(0)}$? The first affirmative answer to this question was given in 1975 by A. Bensoussan, J.L. Lions and G. Papanicolaou [2]. However some smoothness requirement for the coefficients a_{ij} including their continuous differentiability was imposed mainly because it was necessary to represent the diffusion process associated with $A^{(\varepsilon)}$ as a solution of a stochastic differential equation before Ito's calculus was called for.

Our present aim is to show that such smoothness assumptions are entirely dispensed with once one uses the general decomposition theorem developed in [3] in connection with the Dirichlet form and its associated diffusion. Let $P_x^{(\varepsilon)}$ and $P_x^{(0)}$ be the probability measures on the space $W = C([0,\infty) \longrightarrow R^n)$ induced by the diffusions starting at $x \in R^n$ associated with $A^{(\varepsilon)}$ and $A^{(0)}$ respectively. We use the notation $P_h^{(\varepsilon)}(\Lambda) = \int_{R^n} P_x^{(\varepsilon)}(\Lambda)h(x)dx$, $\Lambda \in \mathcal{B}(W)$, $\varepsilon \geqq 0$, for a function $h \geqq 0$ on R^n. Only by assuming (2), (3) and (4), we assert as follows:

<u>Theorem</u> For any non-negative bounded Borel h with compact support

$$P_h^{(\varepsilon)} \longrightarrow P_h^{(0)} \quad , \ \varepsilon \to 0, \text{ on } W. \tag{7}$$

By making use of the probabilistic method as mentioned above, we prove the relative compactness of the family $\{P_h^{(\varepsilon)}, \varepsilon > 0\}$ but we only mention the idea to get the convergence of the finite dimensional distributions. The detail of the latter will be published elsewhere.

<u>Remark</u> If we invoke the powerful apriori estimates of J. Nash [6] on the fundamental solution ((12), (13) below), we can give a direct (but non-probabilistic) proof of the stronger result.

$$P_x^{(\varepsilon)} \longrightarrow P_x^{(0)} \quad , \ \varepsilon \to 0, \qquad \text{on } W \quad \text{for each } x \in R^n. \tag{8}$$

In fact, the compactness of $\{P_x^{(\varepsilon)}, \varepsilon > 0\}$ is immediate from (13), while (5) and (12) combined with Trotter's result [10] imply the uniform convergence of $p_t^{(\varepsilon)} f$ as $\varepsilon \to 0$. In applications the discontinuity of coefficients a_{ij} actually occurs: see [8].

§2. Harmonic Functions and Local Martingales

Given $a_{ij}(x)$, $1 \leqq i, j \leqq n$, $x \in R^n$, satisfying the conditions (2) and (3), we let

$$\varepsilon(u,v) = \sum_{i,j=1}^{n} \int_{R^n} \frac{\partial u}{\partial x_i} \frac{\partial v}{\partial x_j} a_{ij}(x)\,dx, \quad u,v \in H^1(R^n). \tag{9}$$

ε is then a Dirichlet form on $L^2(R^n)$ and consequently decides uniquely a strong continuous semigroup $\{T_t, t > 0\}$ of symmetric operators on $L^2(R^n)$ such that $0 \leqq T_t f \leqq 1$ whenever $0 \leqq f \leqq 1$, $f \in L^2(R^n)$, and that the generator A of $\{T_t, t > 0\}$ is characterized by

$$\varepsilon(u,v) = -(Au,v), \quad u \in \mathcal{D}(A), \quad v \in H^1(R^n), \tag{10}$$

$(\ ,\ )$ being the inner product of $L^2(R^n)$.

The theory of J. Nash [6] asserts much more than this: $T_t f$ for $f \in L^2(R^n) \cap L^\infty(R^n)$ admits a smooth version $p_t f(x)$, $x \in R^n$, expressible as $p_t f(x) = \int_{R^n} p_t(x,y) f(y)\,dy$ with the kernel $p_t(x,y)$ possessing the following properties:

$$p_t(x,y) > 0,\ p_t(x,y) = p_t(y,x),\ \int_{R^n} p_t(x,y) p_s(y,z)\,dy = \\ = p_{t+s}(x,z)\ \int_{R^n} p_t(x,y)\,dy = 1 \tag{11}$$

$$|p_t f(x) - p_t f(x')| \leqq \|f\|_\infty A\left(\frac{|x - x'|}{\sqrt{t}}\right)^\alpha \tag{12}$$

$$p_t(x,y) \leqq kt^{-\frac{n}{2}} \exp\left[-k \frac{|y - x|}{\sqrt{t}} \log k \frac{|y - x|}{\sqrt{t}}\right] \tag{13}$$

where $0 < \alpha < 1$, $A > 0$ and $k > 0$ are apriori constants, namely, constants depending only on λ_0, Λ_0 and n.

In view of (11), (12) and (13), there exists a conservative diffusion process $\underset{\sim}{M} = (\Omega, M, X_t, P_x)$ on R^n with $p_t(x,y)$ being its transition probability density. We now apply the general theory of [3] to the symmetric diffusion $\underset{\sim}{M}$ on R^n and its Dirichlet form (9) on $L^2(R^n)$.

For two AF's (additive functionals) $A^{(1)}$ and $A^{(2)}$ of the process $\underset{\sim}{M}$, we write $A_t^{(1)} = A_t^{(2)}$ whenever the equality holds P_x-a.s. for q.e. $x \in X$, namely, for every $x \in X$ except on a set of capacity zero. The energy e(A) of an AF A is defined by $e(A) = \lim_{t \downarrow 0} \frac{1}{2t} E_m(A_t)$. An AF M is said to be a MAF (martingale AF) if for each $t > 0$ $E_x(M_t^2) < \infty$, $E_x(M_t) = 0$ for q.e. $x \in X$. The class of MAF's of finite energy is denoted by $\mathring{\mathcal{M}}$. The class of continuous AF's of zero energy is denoted by $\mathcal{N}_c$. According to [3], any function $u \in H^1(R^n)$ admits a unique decomposition

$$\tilde{u}(X_t) - \tilde{u}(X_0) = M_t^{[u]} + N_t^{[u]}, \quad M^{[u]} \in \mathring{\mathcal{M}}, \ N^{[u]} \in \mathcal{N}_c. \qquad (14)$$

<u>Lemma 1</u>

(i) The quadratic variation $\langle M^{[u]} \rangle$ of $M^{[u]}$ has the following expression

$$\langle M^{[u]} \rangle_t = 2 \int_0^t \sum_{i,j=1}^n u_{x_i}(X_s) u_{x_j}(X_s) a_{ij}(X_s) ds. \qquad (15)$$

(ii) Let G be an open set. $N^{[u]}$ does not increase on G, that is, $N_t^{[u]} = 0$ $\forall t < \tau_G$, τ_G being the first leaving time from G, if and only if $\varepsilon(u,v) = 0$ for any C^∞-function v with compact support in G.

The first assertion follows from [3; Lemma 4]. The second assertion is clear when u is a potential of a signed Radon measure of finite energy integral in view of [3; Lemma 2]. For general $u \in H^1(R^n)$, it suffices to use the approximation [3; Corollary 1 to Theorem 2] and the spectral synthesis due to Beurling and Deny.

We say that an AF M is locally in $\mathring{\mathcal{M}}$ ($M \in \mathcal{M}_{loc}$) if there exist a suequence of relatively compact open sets G_k such that $\bar{G}_k \subset G_{k+1}$,

$G_k \uparrow R^n$, $k \to \infty$, and a sequence of $M^{(k)} \in \mathcal{M}$ such that $M_t = M_t^{(k)}$, $\forall t < \tau_{G_k}$.

A function $u \in H^1_{loc}(R^n)$ is said to be harmonic (with respect to the form (9)) if

$$\varepsilon(u,v) = 0 \quad \text{for any } v \in C_0^\infty(R^n), \tag{16}$$

namely, $A^{(1)}u = 0$ in the sense of distribution.

As an immediate consequence of Lemma 1, we get

Lemma 2 Each $u \in H^1_{loc}(R^n)$ admits a unique functional $M^{[u]} \in \mathcal{M}_{loc}$ such that the identity (15) holds. The equality

$$\tilde{u}(X_t) - \tilde{u}(X_0) = M_t^{[u]} \tag{17}$$

takes place if and only if u is harmonic.

Lemma 3 Any harmonic function $u \in H^1_{loc}(R^n)$ is Holder continuous. Furthermore there exists a universal constant $Q > 2$ ($Q \to \infty$ as $\frac{\lambda_0}{\Lambda_0} \to 1$ and $Q \to 2$ as $\frac{\lambda_0}{\Lambda_0} \to 0$) such that

$$\frac{\partial u}{\partial x_i} \in L^p_{loc}(R^n) \quad \forall p < Q \quad (1 \leq i \leq n). \tag{18}$$

See G. Stampacchia [9] for the first assertion. The second assertion is due to G. Meyers [7]. (See also [11; Example 5.4.2] for Lemma 2).

§3 Relative Compactness of $\{P_h^{(\varepsilon)}\}_{\varepsilon > 0}$

Given $a_{ij}(x)$, $1 \leq i, j \leq n$, satisfying (2), (3) and the periodicity (4) , we still consider the Dirichlet form (9) on $L^2(R^n)$ and the associated diffusion process $\underset{\sim}{M} = (\Omega, M, X_t, P_x)_{x \in R^n}$ of the preceding section. For $\varepsilon > 0$, we perform the scaling

$$Y_t^{(\varepsilon)} = \varepsilon X_{t/\varepsilon^2}, \quad \underset{\sim}{M}^{(\varepsilon)} = (\Omega, M, Y_t^{(\varepsilon)}, P_{x/\varepsilon})_{x \in R^n}, \tag{19}$$

then the Dirichlet form on $L^2(R^n)$ of the process $\underset{\sim}{M}^{(\varepsilon)}$ is

$$\varepsilon^{(\varepsilon)}(u,v) = \sum_{i,j=1}^{n} \int_{R^n} \frac{\partial u}{\partial x_i} \frac{\partial v}{\partial x_j} a_{ij}\left(\frac{x}{\varepsilon}\right) dx, \quad u,v \in H^1(R^n). \tag{20}$$

In fact, the transition semigroup of $\underset{\sim}{M}^{(\varepsilon)}$ being expressed as $p_t^{(\varepsilon)}u(x)\ E_{x/\varepsilon}(uY_t^{(\varepsilon)})) = (p_{t/\varepsilon^2}u^{(\varepsilon)})(\frac{x}{\varepsilon})$ with $u^{(\varepsilon)}(x) = u(\varepsilon x)$, we have $\varepsilon^{(\varepsilon)}(u,v) = \lim_{t\downarrow 0}\frac{1}{t}(u-p_t^{(\varepsilon)}u,v) = \lim_{t\downarrow 0}\frac{\varepsilon^n}{t}(u^{(\varepsilon)}-p_{t/\varepsilon^2}u^{(\varepsilon)},\ v^{(\varepsilon)})$ $= \varepsilon^{n-2}\varepsilon(u^{(\varepsilon)},\ v^{(\varepsilon)})$ which equals to the right hand side of (20). Let $P_x^{(\varepsilon)}$ be the image measure on the space of W of $P_{x/\varepsilon}$ by the map $Y_{\cdot}^{(\varepsilon)}(\omega)$ from Ω to W. Thus we are concerned with the tightness of the family $\{P_x^{(\varepsilon)}\}_{\varepsilon > 0}$ on W.

Consider the unit cube $T = (0,1)^n$ and the space $W(T) = \{u \in H^1(T):$ $\int_T u(x)dx = 0,\ u(x_1,..,x_{k-1},\ 0+,x_{k+1},..,x_n) = u(x_1,..,x_{k-1},1-,x_{k+1},..,x_n)$ for a.e. $(x_1,..,x_{k-1},\ x_{k+1},..,x_n),\ 1 \leqq k \leqq n\}$. We can see that the space W(T) is a real Hilbert space with inner product

$$\varepsilon^T(u,v) = \sum_{i,j=1}^{n}\int_T \frac{\partial u}{\partial x_i}\frac{\partial v}{\partial x_j}a_{ij}(x)dx. \tag{21}$$

Owing to the Riesz representation theorem, there exists a unique function $\chi_k \in W(T)$ such that

$$\varepsilon^T(\chi_k,\ v) = \varepsilon^T(x_k,\ v) \quad \forall\ v \in W(T),\quad 1 \leqq k \leqq n, \tag{22}$$

x_k being the k-th coordinate function.

Lemma 4 Extend χ_k to R^n periodically and denote the extended function by the same notation χ_k. We then set

$$\psi_k(x) = x_k - \chi_k(x),\quad x \in R^n,\quad 1 \leqq k \leqq n. \tag{23}$$

Then $\psi_k \in H^1_{loc}(R_n)$ and ψ_k is harmonic with respect to the form ε.

This is clear when $a_{ij} \in C^\infty(R^n)$. In general, we approximate a_{ij} by the convolution $a_{ij}^{(\delta)} = \rho_\delta * a_{ij}$ with a mollifier and utilize the apriori estimates of the local boundedness and Hölder continuity of harmonic functions due to G. Stampacchia [9] to get Lemma 4.

Introducing the vectors $\chi(x) = (\chi_1(x),\ldots,\chi_n(x))$,

$\psi(x) = (\psi_1(x),\ldots,\psi_n(x))$ and $M_t^{[\psi]} = (M_t^{[\psi_1]},\ldots,M_t^{[\psi_n]})$, we get from Lemma 2, Lemma 3 and Lemma 4,

$$\psi(X_t) - \psi(X_s) = M_t^{[\psi]} - M_s^{[\psi]}, \quad t > s > 0. \tag{24}$$

Since the process $Y_t^{(\varepsilon)}$ of (20) is expressed as

$$Y_t^{(\varepsilon)} - Y_s^{(\varepsilon)} + \varepsilon[\psi(X_{t/\varepsilon^2}) - \chi(X_{s/\varepsilon^2})] + \varepsilon[\psi(x_{t/\varepsilon^2}) - \psi(X_{s/\varepsilon^2})] \tag{25}$$

and χ is uniformly bounded, the compactness question of $(P_x^{(\varepsilon)})_{\varepsilon>0}$ reduces to a uniform estimate of a moment of the second term of the right hand side of (25) with respect to $P_{x/\varepsilon}$.

Note that the equality (24) holds P_x-a.s. for every $x \in R^n$ because this is true up to a polar set and then the Markovity and the additivity apply. According to a well known inequality for continuous local martingale, there exists a universal constant C such that

$$E_{x/\varepsilon}(|M_t^{[\psi]} - M_s^{[\psi]}|^{2p}) \leqq C \sum_{i=1}^{n} E_{x/\varepsilon}((<M^{[\psi_k]}>_t - <M^{[\psi_k]}>_s)^p).$$

Hence we have from (15) and (24)

$$E_{x/\varepsilon}(|\psi(X_{z/\varepsilon^2}) - \psi(X_{s/\varepsilon^2})|^{2p}) \leqq C\, E_{x/\varepsilon}[(\int_{s/\varepsilon^2}^{t/\varepsilon^2} f(X_s)ds)^p]$$

with

$$f(x) = tr(I - \frac{\partial\chi}{\partial x}\cdot a\cdot(I - \frac{\partial\chi}{\partial x})^*(x), \quad x \in R^n, \tag{26}$$

where I is the identity matrix, $\frac{\partial\chi}{\partial x} = (\frac{\partial\chi_i}{\partial x_j})_{1\leqq i,j\leqq n}$ and $a = (a_{ij})_{1\leqq i,j\leqq n}$. The Hölder inequality then leads us to for $p > 1$

$$E_{x/\varepsilon}[\varepsilon^{2p}|\psi(X_{t/\varepsilon^2})-\psi(X_{s/\varepsilon^2})|^{2p})\leqq C\, E_{x/\varepsilon}(\int_s^t f(X_{u/\varepsilon^2})^p du)(t-s)^{p-1}. \tag{27}$$

<u>Lemma 5</u> For any non-negative bounded function h with compact support, $\{P_h^{(\varepsilon)}\}_{\varepsilon > 0}$ is relatively compact on W.

In fact it is easy to get from (27)

$$\int_{R^n} h(x) E_{x/\varepsilon}[\varepsilon^{2p} |\psi(X_{t/\varepsilon^2}) - \psi(X_{s/\varepsilon^2})|^{2p}] dx \leq C(h) \, \|f\|^p_{L^p(T)} (t-s)^p, \tag{28}$$

and it suffices to choose $p > 1$ such that $2p < Q$.

§4. Convergence of Finite Dimensional Distributions

Let us define the $n \times n$ matrix $q = (q_{ij})$ by

$$q = \int_T (I - \frac{\partial \chi}{\partial x}) \cdot a \cdot (I - \frac{\partial \chi}{\partial x})^*(x) dx. \tag{29}$$

We set

$$\varepsilon^{(0)}(u,v) = \int_{R^n} \frac{\partial u}{\partial x_i} \frac{\partial v}{\partial x_j} q_{ij} dx \tag{30}$$

and consider the measure $P_x^{(0)}$ on W of the corresponding diffusion starting at $x \in R^n$.

<u>Lemma 6</u> Under the condition of Lemma 5, each finite dimensional distribution relative to $P_h^{(\varepsilon)}$ converges as $\varepsilon \downarrow 0$ to that relative to $P_h^{(0)}$ for each $x \in R^n$.

This lemma together with Lemma 5 completes the proof of our theorem stated in the introduction. Lemma 6 amounts to the assertion

$$E_{x/\varepsilon}[\exp i\{\lambda_1 \cdot Y_{t_1}^{(\varepsilon)} + \dots + \lambda_\ell \cdot Y_{t_\ell}^{(\varepsilon)}\}]$$

$$\xrightarrow[\varepsilon \downarrow 0]{} \exp\{i(\lambda_1 + \dots + \lambda_\ell) \cdot x\} \exp\{- \sum_{1 \leqq i, j \leqq \ell} (\lambda_i \cdot q \cdot \lambda_j^*) \, t_i \wedge t_j$$

for any $\lambda_1, \dots, \lambda_\ell \in R^n$. In view of the expression (25), this is equivalent to

$$E_{x/\varepsilon}[\exp i\varepsilon\{\lambda_1(\psi(X_{t_1/\varepsilon^2}) - \psi(X_0)) + \dots + \lambda_\ell \cdot (\psi(X_{t_\ell/\varepsilon^2}) - \psi(X_0)]$$

$$\xrightarrow[\varepsilon \downarrow 0]{} \exp\{- \sum_{1 \leqq i, j \leqq \ell} (\lambda_i \cdot q \cdot \lambda_j) \, t_i \wedge t_j\}, \tag{31}$$

which can be proved by adopting a method in [2] together with an approximation of ψ by those functions of bounded derivatives.

In order to illustrate how the coefficients q_{ij} come out, we indicated a proof of the identity

$$\lim_{\varepsilon\downarrow 0} E_0[\exp i\varepsilon(\lambda\cdot\psi(X_{t/\varepsilon^2}) - \lambda\cdot\psi(X_{s/\varepsilon^2}))] = \exp(-\lambda q\lambda^* (t-s)). \tag{32}$$

Applying a well known representation theorem to the continuous local martingale $\lambda\cdot\psi(X_t) - \lambda\cdot\psi(X_s)$, $t > s$, we can see that there exists a Brownian motion $B(t)$ and

$$\varepsilon(\lambda\cdot\psi(X_{t/\varepsilon^2}) - \lambda\cdot\psi(X_{s/\varepsilon^2})) = B(2\varepsilon^2\int_{s/\varepsilon^2}^{t/\varepsilon^2} g(X_u)du) \tag{33}$$

with $g(x) = \lambda\cdot\frac{\partial\psi}{\partial x}\cdot a\cdot(\frac{\partial\psi}{\partial x})^*\cdot\lambda^*$.

Since $g(x)$ is T-periodic, the ergodic theorem related to the strongly recurrent diffusion $\dot{X}_t$ on the torus $[0,1)^n$ induced by $\underset{\sim}{M}$ implies

$$\lim_{\varepsilon\downarrow 0} 2\varepsilon^2\int_{s/\varepsilon^2}^{t/\varepsilon^2} g(X_u)du$$

$$= \lim_{\varepsilon\downarrow 0} 2\varepsilon^2\int_{s/\varepsilon^2}^{t/\varepsilon^2} g(\dot{X}_u(du = 2(t-s)\int_T g(x)dx = 2(t-s)\cdot\lambda\cdot q\cdot\lambda^*,$$

P_0-a.s.

(32) follows this way.

References

[1] A. Bensoussan, J.L. Lions and G. Papanicolaou, Asymptotic analysis for periodic structures, North-Holland, 1978, Amsterdam-New York-Oxford.

[2] A. Bensoussan, J.L. Lions and G. Papanicolaou, Sur quelques phénomènes asymptotiques stationnaire, C.R. Acad. Sc. Paris, 281 (1975) Ser A 89-94.

[3] M. Fukushima, A decomposition of additive functionals of finite energy, Nagoya Math. J. 74 (1979), 137-168.

[4] M. Fukushima, Dirichlet spaces and additive functionals of finite energy, to appear in Proc. International Congress of Mathematicians, Helsinki 1978.

[5] E. De Giorgi and S. Spagnolo, Sulla convergenza degli integrali dell'energia per operatori ellittici del secondo ordine, Boll. Un. Mat. Ital. 8 (1973), 391-411.

[6] J. Nash, Continuity of solutions of parabolic and elliptic equations, Amer. J. Math. 80 (1958), 931-953.

[7] M. Meyers, An L^p-estimate for the gradient of solutions of second order elliptic divergence equations, Ann. Sc. Norm. Sup. Pisa 17 (1963), 189-206.

[8] E. Sanchez-Palencia, Comportements local et macroscopique d'un type de milieux physiques hétérogènes, Int. J. Engng. Sci. 12 (1974), 331-351.

[9] G. Stampacchia, Equations elliptiques du second ordre à coefficients discontinus, Séminaire sur les équations aux derivées partielles, Collège de France (Mai 1964).

[10] H.F. Trotter, Approximation of semi-groups of operators, Pacific J. Math. 1958, 887-919.

[11] M. Fukushima, Dirichlet forms and Markov processes, Kodansha Ltd. and North Holland Publ. Co. 1980.

Interaction Picture for Stochastic Differential Equations*)

John R. Klauder
Bell Laboratories
Murray Hill, NJ 07974, USA

Abstract

Stochastic differential equations are discussed in which part of the interaction is accounted for by a non-Wiener process stochastic driving force while the remainder appears in the drift term.

---

*) Talk presented at T. Hida's Workshop on White Noise Approach to Quantum Dynamics.

Introduction

Quantum Theory

In the general context of quantum theory, one of the more useful notions is the concept of the interaction picture (I) which stands partway between the Schrödinger picture (S) and the Heisenberg picture (H). In such a picture part of the dynamics is already taken into account while the remainder may be dealt with explicitly. If $H = H_0 + H_1$ denotes a quantum Hamiltonian, then in the three pictures, respectively, the time dependence of states $|\psi\rangle$ and operators O is given by [1]

$$|\psi_S(t)\rangle = e^{-iHt}|\psi\rangle; \quad O_S(t) = O, \tag{S}$$

$$|\psi_I(t)\rangle = U_I(t)|\psi\rangle; \quad O_I(t) = e^{iH_0t}Oe^{-iH_0t}, \tag{I}$$

$$|\psi_H(t)\rangle = |\psi\rangle; \quad O_H(t) = e^{iHt}Oe^{-iHt}, \tag{H}$$

where, in terms of the time-ordered product,

$$U_I(t) \equiv Te^{-i\int_0^t H_1(t')dt'},$$

$$H_1(t') \equiv e^{iH_0t'}H_1e^{-iH_0t'} .$$

These relations are determined of course so that various matrix elements $\langle\psi_1|O|\psi_2\rangle$ are the same in all pictures. It is important to emphasize that the division of H into H_0 and H_1 is absolutely arbitrary, and if $H_0 = 0$, then (I) = (S) while if $H_1 = 0$, then (I) = (H).

Stochastic Processes

A corresponding set of pictures arises for the stochastic processes characterized by imaginary-time quantum mechanics (generalized diffusion equations) in which $t \to -it$ in the preceding expressions [2]. Corresponding to the previous relations we now have

$$|\psi_S(t)\rangle = e^{-\mathcal{H}t}|\psi\rangle; \quad \mathcal{O}_S(t) = \mathcal{O}, \tag{S}$$

$$|\psi_I(t)\rangle = V_I(t)|\psi\rangle; \quad \mathcal{O}_I(t) = e^{\mathcal{H}_0 t}\,\mathcal{O}e^{-\mathcal{H}_0 t}, \tag{I}$$

$$|\psi_H(t)\rangle = |\psi\rangle; \quad \mathcal{O}_H(t) = e^{\mathcal{H}t}\mathcal{O}e^{-\mathcal{H}t}, \tag{H}$$

where

$$V_I(t) = Te^{-\int_0^t \mathcal{H}_1(t')dt'},$$

$$\mathcal{H}_1(t') = e^{\mathcal{H}_0 t'}\,\mathcal{H}_1 e^{-\mathcal{H}_0 t'}.$$

In the present case these relations are determined so that various matrix elements $\langle\psi_1|\mathcal{O}|\psi_2\rangle$, now in a real Hilbert space, are the same in all pictures.

Functional Integral Formulation

An even clearer characterization of the different pictures may be given in a functional integral formulation. To be specific let $\mathcal{H}_0 = \frac{1}{2}P^2 = -\frac{1}{2}\partial^2/\partial x^2$ and $\mathcal{H}_1 = V(Q) = V(x)$ appropriate to a single degree of freedom. Then the three expressions for the same characteristic functional given by

$$\mathcal{N}\int e^{i\int s(t)x(t)dt - \int[\frac{1}{2}\dot{x}^2(t) + V(x(t))]dt}\mathcal{D}x, \tag{S}$$

$$N\int e^{i\int s(t)x)t)dt - \int V(x(t))dt}d\mu_W(x), \tag{I}$$

$$\int e^{i\int s(t)x(t)dt}d\mu_y(x), \tag{H}$$

represent the analogue of the three pictures [2]. Here μ_W denotes the measure for a standard Wiener process (Gaussian process on continuous paths $x(t)$ for which $x(0) = 0$, $\langle x(t)\rangle = 0$ and $\langle x(t)x(s)\rangle = \min(t,s)$). Thus the paths in the formal path integral are also constrained to satisfy $x(0) = 0$. Finally, the last expression involves a generally non-Gaussian measure μ_y

that incorporates the effects of the potential, and for which $x(0) = 0$ as well.

In the preceding example we have illustrated one particular form of interaction picture, but there are many others generically given by

$$\tilde{N} \int e^{i\int s(t)x(t)dt - \int v(x(t))dt} d\bar{\mu}(x), \qquad (\bar{I})$$

where

$$v(x) \equiv V(x) - \bar{V}(x),$$

$$d\bar{\mu}(x) \equiv \bar{N} e^{-\int \bar{V}(x(t))dt} d\mu_W(x)$$

for any reasonable $\bar{V}(x)$.

Stochastic Differential Equations

An alternative fashion to characterize stochastic processes is by means of stochastic differential equations [2], for example, the Itô stochastic differential equation

$$dY(t) = a(Y(t))dt + dW(t),$$

where $W(t)$ denotes a Wiener path and $a(Y)$ represents the drift term. Solution of this equation implicitly provides a one-to-one map $R\colon W \mapsto Y$ with the property that

$$d\mu_Y(Rx) = d\mu_W(x).$$

The drift term $a(x)$ is related to the potential $V(x)$ in the Hamiltonian according to

$$a(x) \equiv (\partial/\partial x)\ln\phi_0(x) = \phi_0'(x)/\phi_0(x),$$

where $\phi_0(x)$ is the "eigenstate" of lowest energy, which is sufficiently general to cover cases such as $\phi_0(x) = 1$, $a(x) = 0$, when $V(x) = 0$. But as it stands the stochastic differential equation is always driven by a Wiener Process and the drift term accounts for the entire potential V. In order to obtain an analogue of the interaction picture one would like to have an equation such as

$$dY(t) = b(Y(t))dt + dX(t)$$

where X is a process intermediate to W and Y that takes part of the potential (say $\bar{V}$) into account according to

$$dX(t) = \bar{a}(X(t))dt + dW(t),$$

and the drift b takes into account the rest of the potential $(V-\bar{V})$. While such a stochastic differential equation for Y undoubtedly holds, it is not clear how to give an explicit expression for b [3].

Vector Formulation

One approach to bring such a relation to the desired form is to use a two-component stochastic variable $Z \equiv (Y,X)$ and the two-component stochastic differential equation

$$dZ(t) = \alpha(Z(t))dt + dV(t),$$

where $V \equiv (X,X)$ and

$$\alpha(Z) \equiv (a(Y)-\bar{a}(X),0),$$

which follows from combining the equations for Y and X as driven by W [4]. This relation has the virtue that an explicit prescription for $\alpha(Z)$ can be given, but it will not be the method we shall pursue.

Interaction Picture

To achieve a single relevant equation we give up the canonical form for a stochastic differential equation. To motivate our approach consider the alternative to the usual equation based on

$$Y(t) = W(t) + \hat{\Delta}(t),$$

$$\hat{\Delta}(0) = 0,$$

in which case it follows that

$$d\hat{\Delta}(t) = a(W(t) + \hat{\Delta}(t))dt,$$

which is a differential relation that implicitly determines $\hat{\Delta}$ for each Wiener path W.

In like fashion we let

$$Y(t) = X(t) + \Delta(t),$$

$$\Delta(0) = 0,$$

in which case the relevant equations lead to the stochastic differential equation

$$d\Delta(t) = [a(X(t) + \Delta(t)) - \bar{a}(X(t))]dt$$

which implicitly determines Δ for each path X. Here the drift a is determined by the potential V while the drift $\bar{a}$ is determined by the potential $\bar{V}$, which also determines the X process; hence the equation for Δ implicitly incorporates $\bar{V}$ into X and and it makes explicit the contribution of the remainder of the potential, $v = V - \bar{V}$. This stochastic differential equation constitutes the analogue of the interaction picture that we shall adopt.

Remarks

Several general remarks may be stated initially: (1) if $\bar{a} = a$, then it follows that $\Delta(t) = 0$ for all $t \geq 0$ due to the initial condition $\Delta(0) = 0$ and the Itô definition for stochastic equations; (2) if $\bar{a} \neq a$ then it generally follows that $\Delta(t) \neq 0$ for almost all $t > 0$; and (3) if $a - \bar{a}$ is small in some sense, then it can be expected that Δ also remains small.

Harmonic Oscillator

A particularly simple example is that old standby, the harmonic oscillator. In this case we choose

$$H = -\frac{1}{2}\frac{\partial^2}{\partial x^2} + \frac{1}{2}\omega^2 x^2 - \frac{1}{2}$$

for which $\phi_0(x) \propto \exp(-\frac{1}{2}\omega x^2)$, and thus $a(x) = -\omega x$. The solution of the stochastic differential equation

$$dY(t) = -\omega Y(t)dt + dW(t)$$

is straightforward [2] and is given by

$$Y(t) = \int_0^t e^{-\omega(t-t')}dW(t')$$

$$= W(t)-\omega\int_0^t e^{-\omega(t-t')}W(t')dt'.$$

For our example we choose $\bar{V} = \frac{1}{2}x^2$ (i.e., $\omega = 1$) so that

$$dX(t) = -X(t)dt + dW(t)$$

with solution

$$X(t) = \int_0^t e^{-(t-t')}dW(t')$$

$$= W(t)-\int_0^t e^{-(t-t')}W(t')dt'.$$

It follows that the difference variable $\Delta(t) = Y(t)-W(t)$, $\Delta(0) = 0$, satisfies the stochastic differential equation

$$d\Delta(t) = -\omega[X(t) + \Delta(t)]dt + X(t)dt$$

$$= -\omega\Delta(t)dt - (\omega-1)X(t)dt.$$

The solution to this equation is given by

$$\Delta(t) = -(\omega-1)\int_0^t e^{-\omega(t-t')}X(t')dt'$$

which expresses the difference variable Δ in terms of the stochastic variable X [5].

This expression vanishes when $\omega = 1$ as it must, and it is non-zero but small when $0 < |\omega-1| \ll 1$.

Discontinuous Perturbations

Of greater interest are singular potentials that lead to discontinuous perturbations [6]. Suppose as an example we choose $V(x) = \lambda(x-c)^{-2}$, where we assume $c < 0$ without loss of generality. In this case

$$dY(t) = \frac{\theta dt}{Y(t)-c} + dW(t),$$

where

$$2\lambda \equiv \theta(\theta-1).$$

We are interested here in the case that $\lambda \to 0^+$ when $\theta \to 1^+$, and thus

$$\theta = \frac{1}{2}(1 + \sqrt{1+8\lambda}).$$

Moreover, we shall take the nonvanishing limit of the drift term to define our interaction stochastic variable by

$$dX(t) = \frac{dt}{X(t)-c} + dW(t).$$

Even though the potential has "vanished" it has changed the system (from W to X), characteristic of a discontinuous perturbation [6]. It is typical of such processes that $X(t) > c$ for all time with probability one, and in fact the X process can be expressed in terms of the radius of a vector composed of three distinct Wiener processes [7]. In particular, if W_j, $j = 1,2,3$, denote three, independent, standard Wiener processes, then

$$X(t) = c + \{W_1^2(t) + W_2^2(t) + [W_3(t) - c]^2\}^{\frac{1}{2}}$$

provides a realization of the interaction stochastic variable.

The appropriate interaction picture stochastic differential equation for this example reads

$$d\Delta(t) = \left\{\frac{\theta}{X(t)+\Delta(t)-c} - \frac{1}{X(t)-c}\right\} dt$$

$$= \frac{(\theta-1)dt}{X(t)+\Delta(t)-c} - \frac{\Delta(t)dt}{[X(t)+\Delta(t)-c][X(t)-c]} .$$

In general, it is not known how to solve this equation exactly.

One approach to proceed further is based on an iteration scheme in which $(\theta-1)$--hence Δ-- is regarded as a small quantity. In this approach one studies the set of equations

$$d\Delta_{k+1}(t) = \frac{(\theta-1)dt}{X(t)+\Delta_k(t)-c} - \frac{\Delta_{k+1}(t)dt}{[X(t)+\Delta_k(t)-c][X(t)-c]}$$

for $k = 0,1,2,\ldots$, where $\Delta_0(t) \equiv 0$, and postulates the convergence

$$\Delta(t) = \lim_{k\to\infty} \Delta_k(t).$$

The solution to the iteration equation may easily be found by analogy to the harmonic oscillator case to be

$$\Delta_{k+1}(t) = (\theta-1)\int_0^t \exp\{-\int_{t'}^t [X(t'')+\Delta_k(t'')-c]^{-1}[X(t'')-c]^{-1}dt''\}$$

$$\times [X(t')+\Delta_k(t')-c]^{-1}dt' .$$

In particular, since $\Delta_0(t) \equiv 0$ the first iterative approximation reads

$$\Delta_1(t) = (\theta-1)\int_0^t \exp\{-\int_{t'}^t [X(t'')-c]^{-2}dt''\}[X(t')-c]^{-1}dt' ,$$

where to this order $\theta-1 = 2\lambda$. Moreover, provided the iteration scheme converges, then the solution Δ may also be expressed as the solution of the integral equation

$$\Delta(t) = (\theta-1)\int_0^t \exp\{-\int_{t'}^t [X(t'')+\Delta(t'')-c]^{-1}[X(t'')-c]^{-1}dt''\}$$

$$\times [X(t')+\Delta(t')-c]^{-1}dt' .$$

In each case one has the realization of the X paths as the radius of three-dimensional Wiener process as already outlined.

General Case

The iteration scheme indicated above to treat the special case $V(x) = \lambda(x-c)^{-2}$ can be generalized to a wide class of potentials. The general form of the interaction picture stochastic differential equation, as noted before, is given by

$$d\Delta(t) = [a(X(t) + \Delta(t)) - \bar{a}(X(t))]dt$$

$$= [a(X(t) + \Delta(t)) - \bar{a}(X(t) + \Delta(t))]dt$$

$$+ \Delta(t)\{[\bar{a}(X(t) + \Delta(t)) - \bar{a}(X(t))]/\Delta(t)\}dt.$$

The iteration scheme that naturally suggests itself is given by

$$d\Delta_{k+1}(t) = [a(X(t) + \Delta_k(t)) - \bar{a}(X(t) + \Delta_k(t))]dt$$

$$+ \Delta_{k+1}(t)\{[\bar{a}(X(t) + \Delta_k(t)) - \bar{a}(X(t))]/\Delta_k(t)\}dt$$

$$\equiv A(t,\Delta_k(t))dt + \Delta_{k+1}(t)B(t,\Delta_k(t))dt,$$

which is solved by

$$\Delta_{k+1}(t) = \int_0^t \exp[\int_{t'}^t B(t'',\Delta_k(t''))dt'']A(t',\Delta_k(t'))dt' ,$$

and one postulates the convergence

$$\Delta(t) = \lim_{k\to\infty} \Delta_k(t).$$

In particular since $\Delta_0(t) \equiv 0$ it follows that the first iteration is given by

$$\Delta_1(t) = \int_0^t \exp[\int_{t'}^t \bar{a}'(X(t''))dt''][a(X(t'))-\bar{a}(X(t'))]dt' ,$$

where $\bar{a}'(x) = \partial\bar{a}(x)/\partial x$. Once again, provided such an iteration scheme converges, then the solution $\Delta(t)$ fulfills the integral equation

$$\Delta(t) = \int_0^t \exp[\int_{t'}^t B(t'',\Delta(t''))dt'']A(t',\Delta(t'))dt' .$$

Conclusion

In this work we have endeavored to extent the valuable interaction picture of quantum theory to the context of stochastic differential equations. One natural formulation suggests itself based on the difference stochastic variable. The utility of such a picture stems from a desire to make perturbations about Hamiltonians other than the free-particle Hamiltonian, and especially in the context of discontinuous perturbations such a scheme would prove extremely significant. In certain cases the iteration scheme we have proposed may yield the answer. Further work on the validity of the proposed iteration scheme needs to be done.

Acknowledgements

It is a pleasure to thank Vincenzo Mirelli for several discussions relating to the subject of this work.

References

[1] See, e.g., A.S. Davydov, Quantum Mechanics, translated by D. ter Haar (Pergamon, London, 1965), Chap. IV, Sec. 31.

[2] See, e.g., H. Ezawa, J.R. Klauder and L.A. Shepp, Annals of Physics **88**, 588 (1974); S. Albeverio, R. Høegh-Krohn, and L. Streit, J. Math. Phys. **18**, 907 (1977).

[3] Vincenzo Mirelli (private communication).

[4] Thanks are expressed to Vincenzo Mirelli for suggesting this formulation. See also I.I. Gihman and A.V. Skorohod, Stochastic Differential Equations (Springer-Verlag, Berlin-Heidelberg-New York 1972).

[5] A related example is treated by T. Hida and L. Streit, Nagoya Math. J. **68**, 21 (1977), especially Eqs. (65)-(69).

[6] See, e.g., H. Ezawa, J.R. Klauder, and L.A. Shepp, J. Math. Phys. **16**, 783 (1975); J.R. Klauder, Science **199**, 735 (1978).

[7] H.P. McKean, Stochastic Integrals (Academic Press, New York 1969), p. 47, problems 6 and 7; p. 80, section 3.10c.

Path Integrals, Stationary Phase Approximations and Complex Histories

John R. Klauder

Bell Laboratories

Murray Hill, NJ 07974,USA

Abstract

Path integrals for canonical and spin variables are formulated in terms of generalized coherent states in a fashion suitable for stationary phase approximations. Some remarks about complex histories and their role for scattering matrix approximations are also made.

Introduction

Path integrals provide a natural formalism with which to study stationary phase approximations, and this is particularly true for coherent state representations and their generalizations [1,2]. If the unit vectors $|\ell\rangle$ for all $\ell = (\ell',\dots,\ell^L)$ in an L-dimensional space denote generalized coherent states for which

$$\mathbb{1} = \int |\ell\rangle\langle\ell|\,\delta\ell\ ,$$

for a suitable measure $\delta\ell$, then the appropriate (lattice-space) path integral is given by

$$\langle\ell'',t''|\ell',t'\rangle \equiv \langle\ell''|e^{-i(t''-t')\mathcal{H}}|\ell'\rangle$$

$$= \lim_{\varepsilon\to 0}\int\!\cdots\!\int \prod_{k=0}^{N}[\langle\ell_{k+1}|\ell_k\rangle e^{-i\varepsilon H(\ell_{k+1};\ell_k)}]\prod_{k=1}^{N}\delta\ell_k,$$

where $\varepsilon \equiv (t''-t')/(N+1)$, $\ell_0 \equiv \ell'$, $\ell_{N+1} \equiv \ell''$, and

$H(\ell_{k+1};\ell_k) \equiv \langle\ell_{k+1}|\mathcal{H}|\ell_k\rangle/\langle\ell_{k+1}|\ell_k\rangle$. A formal functional integral emerges as

$$\langle\ell'',t''|\ell',t'\rangle = \mathcal{N}\int e^{i\int_{t'}^{t''}[i\langle\ell|\dot{\ell}\rangle - H(\ell)]dt}\,\mathcal{D}\ell,$$

where $\ell(t') \equiv \ell'$, $\ell(t'') = \ell''$, $|\dot{\ell}\rangle = d|\ell\rangle/dt$, $\mathcal{D}\ell \equiv \Pi_t\delta\ell(t)$, $H(\ell) = \langle\ell|\mathcal{H}|\ell\rangle$, and $\mathcal{N}$ is a suitable normalization constant. Here the expression

$$I = \int[i\langle\ell|\dot{\ell}\rangle - H(\ell)]\,dt \equiv \int[y_a(\ell)\dot{\ell}^a - H(\ell)]dt,$$

plays the role of the classical action [3], and the Euler-Lagrange equations that arise are the first-order equations

$$A_{ab}(\ell)\dot{\ell}^b - \partial H(\ell)/\partial \ell^a = 0,$$

where $A_{ab} = -A_{ba}$ and is given by

$$A_{ab}(\ell) = \frac{\partial y_b(\ell)}{\partial \ell^a} - \frac{\partial y_a(\ell)}{\partial \ell^b} \quad .$$

When A_{ab} has an inverse (thus L is even dimensional) the extremal equations have a unique solution $\ell(t)$ evolving from the initial condition $\ell(t') \equiv \ell'$. But for use in the stationary phase approximation to the path integral the solution must also satisfy $\ell(t") = \ell"$, and this is generally impossible.

This paradox is resolved by use of the proper, lattice-space expression and by making a stationary-phase approximation to each of the finite-dimensional integrals involved [2]. Essentially what emerges is the need to develop the kinematic term in the action to one higher order. In certain cases a formal functional integral can be written in the form

$$\langle \ell",t"|\ell',t'\rangle = \lim_{\varepsilon\to 0} \; \mathcal{N} \int e^{iI_\varepsilon}\, \mathcal{D}\ell,$$

where

$$I_\varepsilon = \int [i\langle\ell|\dot{\ell}\rangle + \tfrac{1}{2}i\varepsilon\langle\dot{\ell}|(1-|\ell\rangle\langle\ell|)|\dot{\ell}\rangle - H(\ell)]dt$$

$$\equiv \int [y_a(\ell)\dot{\ell} + \tfrac{1}{2}i\varepsilon g_{ab}(\ell)\dot{\ell}^a\dot{\ell}^a - H(\ell)]dt.$$

In this case the extremal equation are second order and in general they are now compatible with the requisite initial and final boundary conditions. Such a procedure yields the correct extremal solution, and provided the ε term in the action makes no explicit contribution in the limit the dominant stationary phase approximation is also correctly given. The examples we shall consider fulfill this criterion. Further details may be found in Ref. [2].

Canonical Variables

Single Degree of Freedom

The familiar coherent states appropriate to a single degree of freedom are given by

$$|p,q\rangle = e^{i(pQ-qP)}|0\rangle,$$

for all real p and q, where $[Q,P] = i$ and $|0\rangle$ satisfies $(\omega Q+iP)|0\rangle = 0$ for some real ω. In addition the resolution of unity

$$\mathbb{1} = \int |p,q\rangle\langle p,q|\,(dpdq/2\pi)$$

holds. The formal expression for the coherent-state path integral then becomes [2]

$$\langle p'',q'',t''|p',q',t'\rangle \equiv \langle p'',q''|e^{-i(t''-t')H}|p',q'\rangle$$

$$= \lim_{\varepsilon\to 0} \mathcal{N} \int \exp\{i \int_{t'}^{t''} [\tfrac{1}{2}(p\dot{q}-q\dot{p}) + \tfrac{1}{4}i\varepsilon(\omega^{-1}\dot{p}^2 + \omega\dot{q}^2)$$
$$- H(p,q)]dt\}\mathcal{D}p\mathcal{D}q.$$

A stationary phase approximation to this expression leads to the extremal equations

$$\dot{q} - \partial H(p,q)/\partial p = \tfrac{1}{2}i\varepsilon\omega^{-1}\ddot{p}.$$

$$\dot{p} + \partial H(p,q)/\partial q = -\tfrac{1}{2}i\varepsilon\omega\ddot{q}.$$

For small ε the solution reads

$$q(t) = \bar{q}(t) + (q'-\bar{q}')e^{-2(t-t')/\varepsilon} + (q''-\bar{q}'')e^{-2(t''-t)/\varepsilon}$$

$$p(t) = \bar{p}(t) + (p'-\bar{p}')e^{-2(t-t')/\varepsilon} + (p''-\bar{p}'')e^{-2(t''-t)/\varepsilon} ,$$

where $\bar{q}' \equiv \bar{q}(t')$, $\bar{p}' \equiv \bar{p}(t')$, $\bar{q}'' \equiv \bar{q}(t'')$, $\bar{p}'' \equiv \bar{p}(t'')$, and $\bar{p}(t)$, $\bar{q}(t)$ satisfy the usual equations of motion

$$\dot{q}(t) - \partial H(p,q)/\partial p = 0,$$

$$\dot{p}(t) + \partial H(p,q)/\partial q = 0,$$

subject to the boundary conditions

$$\omega\bar{q}' + i\bar{p}' = \omega q' + ip',$$

$$\omega\bar{q}'' - i\bar{p}'' = \omega q'' - ip''.$$

One may choose $\bar{q}' = q' + w$ and $\bar{p}' = p' + i\omega w$ initially where w is a complex parameter adjusted to fit the single complex final boundary condition.

The solution above characteristically divides into three temporal regions: a rapid change near t', a normal evolution for intermediate times, and another rapid change near t". Generally the solution is complex, because the integration variables become complexified in the search for saddle points of the integrand. When the action I_ε is evaluated for this solution the ε term in the action makes no explicit contribution due to the boundary conditions. The limit $\varepsilon\to 0$ may readily be taken to yield the dominant approximation

$$\langle p'',q'',t''|p',q',t'\rangle \approx \exp\{\tfrac{1}{2}i(q''\bar{p}''-p''\bar{q}''+\bar{q}'p'-\bar{p}'q') + i\int_{t'}^{t''}[\tfrac{1}{2}(\bar{p}\dot{\bar{q}} - \bar{q}\dot{\bar{p}}) - H(\bar{p},\bar{q})]dt\}.$$

For any quadratic H this expression is actually exact, and in particular for $H = \frac{1}{2}(p^2 + \omega^2 q^2)$ it follows that

$$\langle p'',q'',t''|p',q',t'\rangle = \exp\{\tfrac{1}{2}i(q''_{-t''}\, p'_{-t'} - p''_{-t''}\, q'_{-t'}) - \tfrac{1}{4}[\omega^{-1}(p''_{-t''}-p'_{-t''})^2 + \omega(q''_{-t''}-q'_{-t'})^2]\},$$

where

$$q'_{t'} = q'\cos\omega t' - \omega^{-1}p'\sin\omega t' \quad ,$$

$$p'_{t'} = p'\cos\omega t' + \omega q'\sin\omega t' \quad ,$$

and similarly for $q''_{t''}$ and $p''_{t''}$.

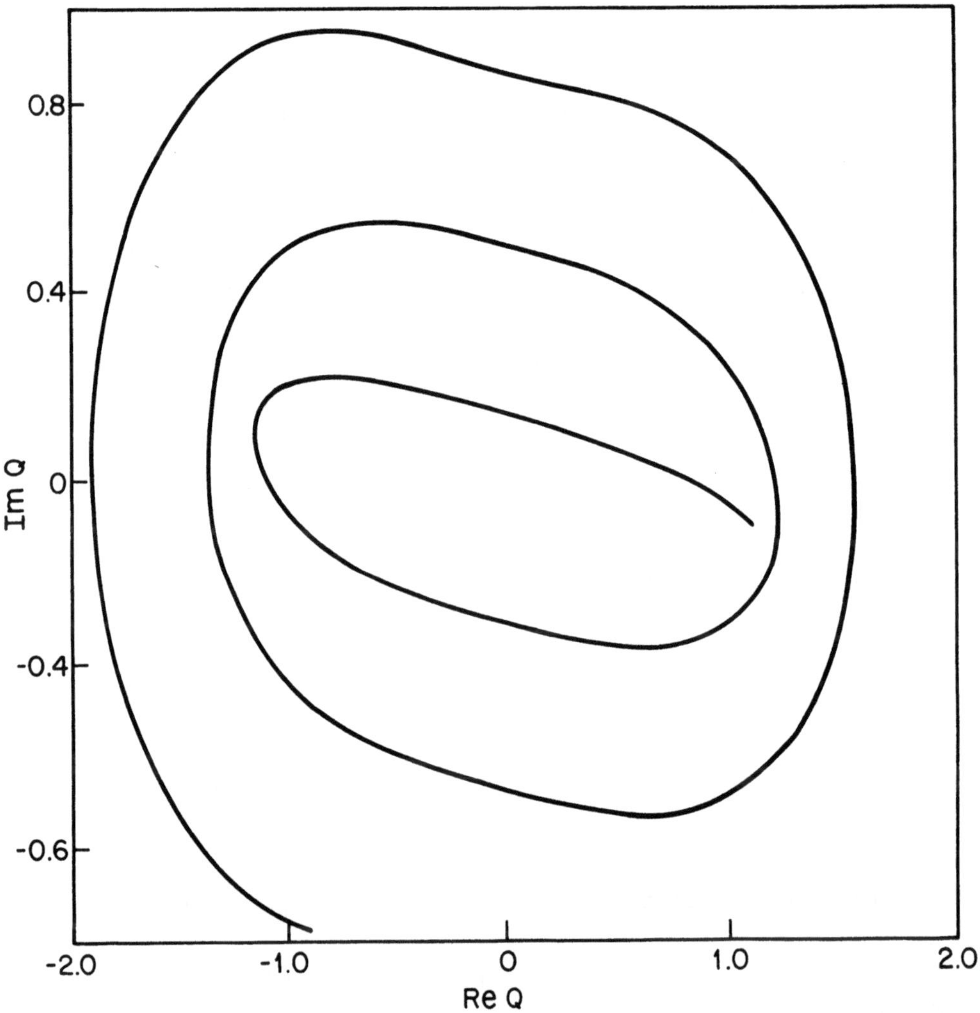

Fig. 1 Real and imaginary parts of $\bar{q}(t)$ in the interval $0 \leq t \leq 10$ as determined by solution of the anharmonic oscillator with initial values $\bar{q}' = 1.1-0.1\,i$ and $\bar{p}' = 0.1+0.1\,i$.

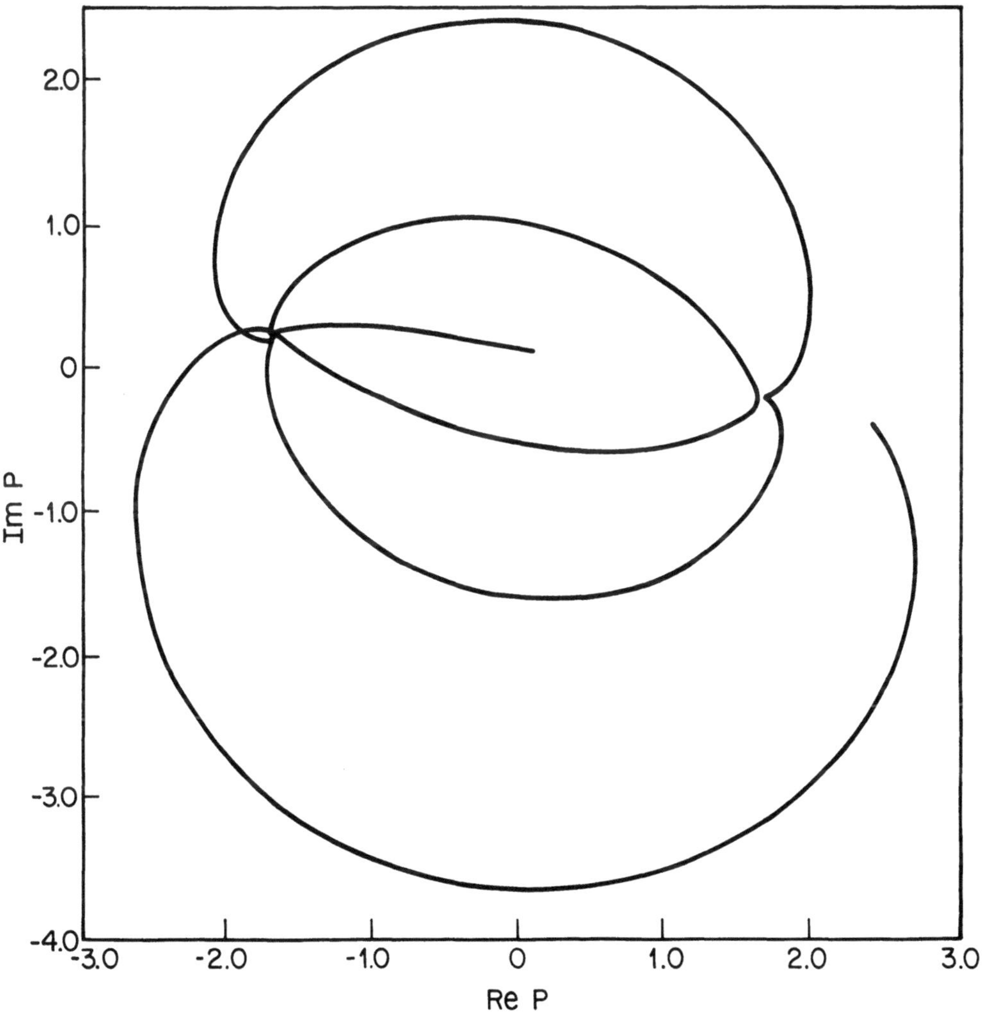

Fig. 2 Real and imaginary parts of $\bar{p}(t)$ in the interval $0 \leq t \leq 10$ as determined by solution of the anharmonic oscillator with initial values $\bar{q}' = 1.1-0.1\ i$ and $\bar{p}' = 0.1+0.1\ i$.

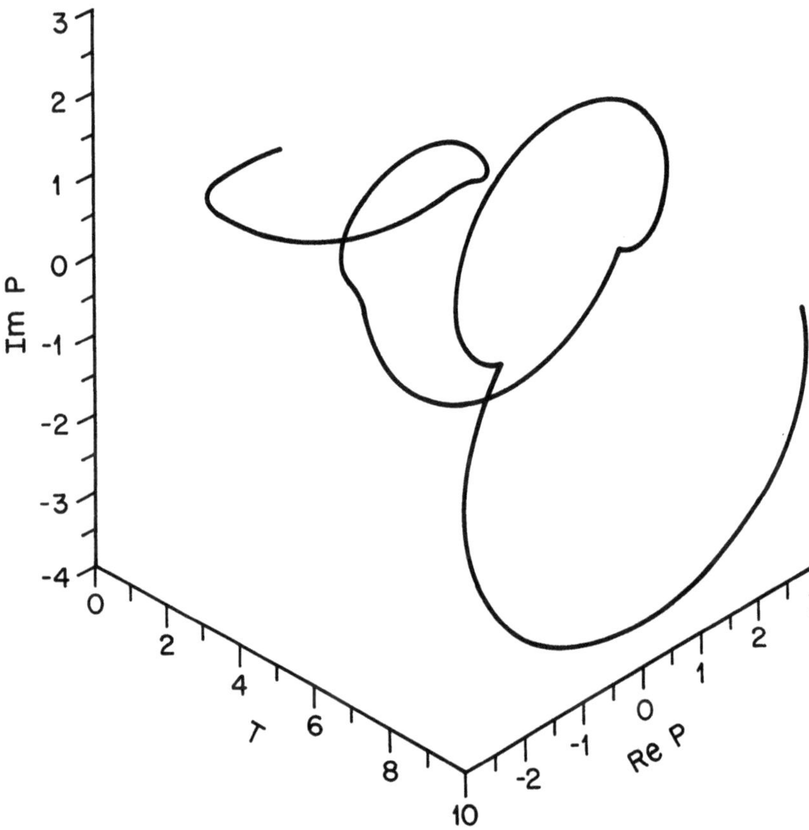

Fig. 3 Projected three-dimensional curve showing the real and imaginary parts of $\bar{p}(t)$ as a function of t, $0 \le t \le 10$, as determined by solution of the anharmonic oscillator with initial values $\bar{q}' = 1.1-0.1\ i$ and $\bar{p}' = 0.1+0.1\ i$.

One feature of complex extremal histories should be emphasized. Consider the classical Hamiltonian

$$H = \frac{1}{2}(p^2 + q^2) + \lambda q^4$$

for $\lambda \geq 0$. The usual trajectories for this example are confined to a finite region of phase space fixed by the energy, but this situation may change for complex initial values. Figures 1 and 2 illustrate the time evolutions of the complex solution for the case $\lambda = \frac{1}{2}$ in the interval $0 \leq t \leq 10$ when $\bar{q}' = q' + w = 1 + (1-i)/10$ and $\bar{p}' = p' + iw = 0 + (i+1)/10$. Figure 3 shows a projected three-dimensional view that clarifies the behavior of $\bar{p}(t)$.

It is likely that the solution eventually diverges in a finite time for almost all complex initial conditions whenever $\lambda > 0$. Of course, there is no inconsistency in this situation whatsoever. On the other hand it is noteworthy that for the harmonic oscillator ($\lambda=0$) the solution is periodic hence bounded for arbitrary complex initial conditions.

Scalar Field Theory

The lessons of the preceding example may be extended to a non-linear self-interacting scalar field. Let

$$|f,g\rangle = e^{i\int[f(\underset{\sim}{x})\phi(\underset{\sim}{x}) - g(\underset{\sim}{x})\pi(\underset{\sim}{x})]d\underset{\sim}{x}}|0\rangle \quad ,$$

denote a coherent state for a scalar field, where

$$[\phi(\underset{\sim}{x}), \quad \pi(\underset{\sim}{y})] = i\delta(\underset{\sim}{x}-\underset{\sim}{y}),$$

and $|0\rangle$ is annihilated by each of the destruction operators. The coherent state matrix elements of the S matrix are given by

$$\begin{aligned}\langle f'',g''|S|f',g'\rangle &= \lim_{\substack{t''\to\infty \\ t'\to-\infty}} \langle f'',g''|e^{it''H_0} e^{-i(t''-t')H} e^{-it'H_0}|f',g'\rangle \\ &= \lim_{\substack{t''\to\infty \\ t'\to-\infty}} \langle f''_{t''},g''_{t''}|e^{-i(t''-t')H}|f'_{t'},g'_{t'}\rangle \ ,\end{aligned}$$

where $f'_{t'}$, $g'_{t'}$ represent time-evolved solutions of the classical free field equations of motion based on the Hamiltonian

$$H_0(f,g) \equiv \langle f,g|\mathcal{H}_0|\, f,g\rangle = \frac{1}{2}\int [f^2+(\nabla g)^2+m^2g^2]d\underset{\sim}{x} \;,$$

for some m (analogous to the single degree of freedom case).

The S-matrix expression can be recast into the form of a path integral and a stationary phase approximation may be carried out. The result is given by a straightforward generalization of that for a single degree of freedom. In particular, the dominant stationary phase approximation leads to the relation

$$\langle f'',g''|S|f',g'\rangle$$

$$\approx \lim_{\substack{t''\to\infty \\ t'\to-\infty}} \exp\{\tfrac{1}{2}i\int_{(t'')} (g''_{t''}\bar{f}'' - f''_{t''}\bar{g}'')d\underset{\sim}{x} - \tfrac{1}{2}i\int_{(t')} (g'_{t'}\bar{f}'$$

$$- f'_{t'}\bar{g}')d\underset{\sim}{x} + i\int_{t'}^{t''} [\tfrac{1}{2}(\bar{f}\,\dot{\bar{g}} - \bar{g}\,\dot{\bar{f}}) - \mathcal{H}(\bar{f},\bar{g})]d\underset{\sim}{x}dt\}.$$

Here $\bar{f}'(\underset{\sim}{x}) \equiv \bar{f}(\underset{\sim}{x},t')$, $\bar{f}''(\underset{\sim}{x}) \equiv \bar{f}(\underset{\sim}{x},t'')$, etc., and $\bar{f},\bar{g}$ denotes a complex solution of the interacting classical equations of motion (with Hamiltonian density $\mathcal{H}$) subject to the boundary conditions

$$\sqrt{-\Delta+m^2}\; g'_{t'}(\underset{\sim}{x}) + if'_{t'}(\underset{\sim}{x}) = \sqrt{-\Delta+m^2}\;\bar{g}'(\underset{\sim}{x}) + i\bar{f}'(\underset{\sim}{x}) \;,$$

$$\sqrt{-\Delta+m^2}\; g''_{t''}(\underset{\sim}{x}) - if''_{t''}(\underset{\sim}{x}) = \sqrt{-\Delta+m^2}\;\bar{g}''(\underset{\sim}{x}) - i\bar{f}''(\underset{\sim}{x}) .$$

Since $f = \partial g/\partial t$ the first and second expressions in the exponent have the form

$$\int_{(t)} (g_t \frac{\partial\bar{g}}{\partial t} - \frac{\partial g_t}{\partial t}\bar{g})d\underset{\sim}{x}.$$

For asymptotically large positive times $\bar{g}$ tends to a solution of the free equation, and for large negative times $\bar{g}$ tends to another solution of the free equation. This ensures that even a complex solution remains bounded for asymptotically large positive and negative times. Moreover since g_t is already a solution of the free equation, the integrals become asymptotically independent of time and thus they converge. A similar argument about the asymptotic fall off of the solution also ensures the convergence of the action integral for the extremal solution.

Spin Variables

For spin degrees of freedom suitable generalized coherent states are given by

$$|\theta,\phi\rangle = e^{-i\phi\hat{S}_3} e^{-i\theta\hat{S}_2}|0\rangle \quad ,$$

where $\hat{S}_j$, $j = 1,2,3$, determine an irreducible representation of the spin algebra,

$$[\hat{S}_j,\hat{S}_k] = i\varepsilon_{ik\ell}\hat{S}_\ell \ ,$$

$$\Sigma\, \hat{S}_j^2 = s(s+1),$$

and $\hat{S}_3|0\rangle = s|0\rangle$, for $s = \frac{1}{2},1,\frac{3}{2}\ldots$. The resolution of unity for such states reads

$$\mathbb{1} = \int |\theta,\phi\rangle\langle\theta,\phi|\,[(2s+1)/4\pi]\sin\theta d\theta d\phi.$$

The path integral for spin variables now becomes [4]

$$\langle\theta'',\phi'',t''|\theta',\phi',t'\rangle \equiv \langle\theta'',\phi''|e^{-i(t''-t')H}\,|\theta',\phi'\rangle$$

$$= \lim_{\varepsilon\to 0} \mathcal{N}\int \exp\{i\int_{t'}^{t''}[s\cos\theta\dot{\phi}+\tfrac{1}{4}is\varepsilon(\dot{\theta}^2+\sin^2\theta\dot{\phi}^2) - H(\theta,\phi)]dt\}\mathcal{D}\Omega,$$

where $\mathcal{D}\Omega \equiv \Pi_t\sin\theta(t)d\theta(t)d\phi(t)$. The stationary phase approximation leads to the extremal equations

$$s\sin\theta\,\dot{\phi} + \partial H(\theta,\sigma)/\partial\theta = -\tfrac{1}{2}\,i\,s\,\varepsilon\,(\ddot{\theta}-\sin\theta\cos\theta\,\dot{\phi}^2),$$

$$s\sin\theta\,\dot{\theta} - \partial H(\theta,\phi)/\partial\phi = \tfrac{1}{2}\,i\,s\,\varepsilon\,(\sin^2\theta\,\ddot{\phi} + 2\sin\theta\cos\theta\,\dot{\theta}\,\dot{\phi}).$$

For small ε the solution to these equations reads

$$\cos\theta\,(t) = \cos\bar{\theta}\,(t) + (\cos\theta'-\cos\bar{\theta}')e^{-2(t-t')/\varepsilon}$$

$$+ (\cos\theta''-\cos\bar{\theta}'')e^{-2(t''-t)/\varepsilon}$$

while

$$\phi(t) = \bar{\phi}(t) - \bar{\phi}' + \phi' + \tfrac{1}{2}i\ln\left(\frac{1+\cos\theta'}{1-\cos\theta'}\right)$$

$$- \tfrac{1}{2}i\ \ln\left(\frac{1+\cos\bar{\theta}'+(\cos\theta'-\cos\bar{\theta}')e^{-2(t-t')/\varepsilon}}{1-\cos\bar{\theta}'-(\cos\theta'-\cos\bar{\theta}')e^{-2(t-t')/\varepsilon}}\right)$$

for t near t' or intermediate times, and

$$\phi(t) = \bar{\phi}(t) - \bar{\phi}'' + \phi'' - \tfrac{1}{2}i\ \ln\left(\frac{1+\cos\theta''}{1-\cos\theta''}\right)$$

$$+ \tfrac{1}{2}i\ \ln\left(\frac{1+\cos\bar{\theta}''+(\cos\theta''-\cos\bar{\theta}'')e^{-2(t-t'')/\varepsilon}}{1-\cos\bar{\theta}''-(\cos\theta''-\cos\bar{\theta}'')e^{-2(t''-t)/\varepsilon}}\right)$$

for t near t" or intermediate times. Here $\bar{\theta}' = \bar{\theta}(t')$, $\bar{\phi}' = \bar{\phi}(t')$ etc., and $\bar{\theta}(t)$, $\bar{\phi}(t)$ are complex solutions of the equations

$$s\sin\theta\ \dot{\phi} + \partial H(\theta,\phi)/\partial\theta = 0,$$

$$s\sin\theta\ \dot{\theta} - \partial H(\theta,\phi)/\partial\phi = 0,$$

subject to the boundary conditions

$$\tan\tfrac{1}{2}\theta' e^{i\phi'} = \tan\tfrac{1}{2}\bar{\theta}' e^{i\bar{\phi}'}$$

$$\tan\tfrac{1}{2}\theta'' e^{-i\phi''} = \tan\tfrac{1}{2}\bar{\theta}'' e^{-i\bar{\phi}''}\ .$$

One may choose $\tan\frac{1}{2}\bar{\theta}' = e^{w}\tan\frac{1}{2}\theta'$ and $\bar{\phi}' = \phi' + iw$ initially where w is a complex parameter adjusted to fit the single complex final boundary condition.

When the action I_ε is evaluated for this solution the ε term in the action makes no explicit contribution due to the boundary conditions. The limit $\varepsilon\to 0$ yields the dominant approximation

$$\langle\theta'',\phi'',t''|\theta',\phi',t'\rangle = \left(\frac{\sin\theta''\sin\theta'}{\sin\bar{\theta}''\sin\bar{\theta}'}\right)^{s} \exp\{i\int_{t'}^{t''}[s\cos\bar{\theta}\dot{\bar{\phi}} - H(\bar{\theta},\bar{\phi})]dt\}.$$

If $\mathcal{H} = h\hat{S}_3$, then $H = hs\cos\theta$ and the dominant approximation is actually exact for any s, and reads

$$\langle\theta'',\phi'',t''|\theta',\phi',t'\rangle$$

$$= [\cos\tfrac{1}{2}\theta''\cos\tfrac{1}{2}\theta' e^{i(\phi''_{t''}-\phi'_{t'})/2} + \sin\tfrac{1}{2}\theta''\sin\tfrac{1}{2}\theta' e^{-i(\phi''_{t''}-\phi'_{t'})/2}]^{2s} ,$$

where

$$\phi'_{t'} = \phi' + ht' ,$$

$$\phi''_{t''} = \phi'' + ht'' .$$

It is rather remarkable that the dominant approximation yields the exact result here since none of the integrals involved are Gaussian. Further discussion of the spin case may be found in Ref. [4].

It may be expected, as before, that the general solution of the classical equations with complex initial conditions may eventually diverge in a finite time. An exception arises for $H = h\, s \cos\theta$ which leads to periodic hence bounded solutions for the propagator for arbitrary complex initial conditions.

For the case of an extension to a field of spin variables (see, for example, Ref. [5]) the existence of an asymptotic Hamiltonian for which the propagator solution remains bounded is important for the existence of an S matrix.

Acknowledgment

It is a pleasure to thank Adrian Simons for computation of the complex solutions of the anharmonic oscillator.

References

[1] J.R. Klauder, Annals of Physics 11, 123 (1960).

[2] J.R. Klauder, in Path Integrals, Proceedings of the NATO Advanced Summer Institute, edited by G.J. Papadopoulos and J.T. Devreese (Plenum, 1978), p. 5.

[3] J.R. Klauder, J. Math. Phys. 4, 1058 (1963).

[4] J.R. Klauder, Phys. Ref. D 19, 2349 (1979).

[5] A. Jevicki and N. Papanicolaou, Annals of Physics 120, 107 (1979).

Stochastic Dynamics and the Semiclassical Limit of Quantum Mechanics*)

G. Jona-Lasinio

Istituto di Fisica dell'Università - Roma

Gruppo G.N.S.M. - Roma, Italy

Abstract

We outline a new approach to tunnelling problems in quantum mechanics based on the so called stochastic quantization. The starting point of the method is the observation that in the limit $\frac{\hbar}{m} \to 0$ one can approximate the stochastic process underlying the quantum motion with a Markov chain making transitions at random times from one classical equilibrium position to another. This picture is suggested by the theory of small random perturbations of dynamical systems developed by Ventzel and Freidlin which provides the natural mathematical ideas for our purpose.

As an example of application we discuss the instanton problem for the double well anharmonic oscillator. The flexibility and wide applicability of the present approach is pointed out.

---

*) Talk presented at T. Hida's Workshop on the White Noise Approach to Quantum Dynamics.

1. Motivation

The content of this paper is the outcome of an attempt to construct a rational theory of instantons in quantum mechanics. A posteriori, however, it appears that one can be more ambitious and use the point of view developed here as the basis for a general theory of the semiclassical limit of quantum mechanics. We shall comment on this aspect later.

Instantons have been introduced by Polyakov and t'Hooft [1] as solutions of the classical equations of motion in the Euclidean region and quantum mechanically are associated with tunnelling between different classical equilibria.

Effects which are presumably dominated by instantons are calculated by expanding the Euclidean functional integral representation of physically interesting quantities, e.g. correlation functions, around instantons. These calculations are usually painful, there are complications like zero modes or zero eigenvalues appearing in the evaluation of functional determinants, and generally speaking one has very little control over the errors of the approximation [2].

The point of view presented here is essentially different and apparently avoids some of the drawbacks indicated above at least in the case of systems with a finite number of degrees of freedom. The starting point is the "stochastic quantization" in the form developed by Nelson some years ago [3]. If for illustrative purposes we restrict ourselves to one dimensional systems the scheme of "stochastic quantization" for stationary states goes as follows. The position of a particle in a potential V(x) obeys the stochastic differential equation

$$dx = b(x, \frac{\hbar}{m})dt + \sqrt{\frac{\hbar}{m}}dW \qquad (1.1)$$

where dW is the Wiener process defined by $E(dW)^2 = dt$ and $b(x)$ a drift term determined by the equation

$$\frac{1}{2}b^2 + \frac{\hbar}{2m}\frac{db}{dx} - \frac{V}{m} = -\frac{E}{m} . \tag{1.2}$$

The connection with the Schroedinger equation is straight forward if we note that (1.2) is a Riccati equation which can be linearized in the usual way by putting

$$b = \frac{\hbar}{2m}\frac{d}{dx}\ln\psi^2 \tag{1.3}$$

from which the equation for ψ follows

$$\frac{1}{2}\frac{\hbar^2}{m}\frac{d^2\psi}{dx^2} - V\psi = -E\psi .$$

The description of time dependent states is more complicated but shall not be needed here. The relevant point for us is that through equations (1.3), (1.2), (1.1) to each stationary state, in particular to the ground state, there is associated a stochastic process. The bridge to imaginary time quantum mechanics is provided by the important remark of Guerra and Ruggiero [4] that the process corresponding to the ground state according to (1.3), (1.2), (1.1) is the same as the stochastic process generated by the imaginary time functional integral.

Let us now examine the structure of equation (1.1). When $\frac{\hbar}{m} \to 0$ i.e. in the semiclassical limit, it describes a dynamical system subject to a small random perturbation. This is a situation for which a powerful theory has been developed by Khasminski [5], Ventzel and Freidlin [6]. The conclusion is that the structure of the stochastic process generated by (1.1) is practically determined by a few qualitative properties of the deterministic system

$$\frac{dx}{dt} = b(x,\frac{\hbar}{m}) \simeq b(x,0) .$$

Clearly $b(x,0)$ is a quantity which can be easily calculated in many cases.

In the next Section we briefly review some of the ideas developed by the above authors.

In Section 3 we discuss the instanton problem for the double well anharmonic oscillator.

In Section 4 we review the more general implications of our point of view.

A detailed presentation with additional applications will be given in a series of forthcoming papers in collaboration with F. Martinelli and E. Scoppola.

2. Small Random Perturbations of Dynamical Systems

The typical situation we want to consider is an n-dimensional system of the form

$$d\underline{x} = \underline{b}(\underline{x},\epsilon)dt + \epsilon d\underline{W} \tag{2.1}$$

where ϵ is a small positive number(+). For the time being we also assume that $\underline{b}(\underline{x},\epsilon)$ tends uniformly (in x) to $\underline{b}(\underline{x}, 0) = \underline{b}_0(\underline{x})$ as $\epsilon \to 0$. This will not be generally true in the quantum mechanical applications we have in mind. However the non uniformities encountered in concrete cases are of such a nature that the picture we are going to describe is not altered. Obviously in the quantum mechanical case $\epsilon = \sqrt{\frac{\hbar}{m}}$.

The second assumption is that the deterministic dynamical system

$$\frac{d\underline{x}}{dt} = \underline{b}_0(\underline{x}) \tag{2.2}$$

has only a finite number of equilibrium points or limit cycles. Again this may not be true in some applications and one has to check in every case the applicability of Ventzel and Freidlin theory.

The intuitive description which emerges from the theory is as follows. Any process generated by (2.1) will reach very quickly

(+) In order to include also motions on a Riemannian manifold the authors of Ref. [6] start from the more general equation $d\underline{x} = \underline{b}(\underline{x},\epsilon)dt + \epsilon\, \sigma(\underline{x})d\underline{W}$ where $\sigma(x)$ is an nxn matrix. Equation (2.1) will be sufficient for the purposes of the present paper.

the vicinity of a stable equilibrium orbit of (2.2). The stationary motion will consist mainly of long periods around such orbits (points or limit cycles) and rare passages between them. But a passage from one stable equilibrium to another must necessarily go "against the flow" of the velocity field $\underline{b}_o$ at least for a part of the trajectory. This is the analog of tunnelling as already emphasized in [6].

In the one dimensional case and also in the many dimensional case if $\underline{b}$ is a gradient (and this is the case for the $\underline{b}(\underline{x},\frac{\hbar}{m})$ of stochastic quantization) one can show that the invariant measure of the process concentrates as $\epsilon \to 0$ on the absolute minima of the corresponding potential.

Quantitatively the above picture is substantiated through a series of estimates obtained by direct probabilistic arguments. A fundamental role in the sequel is played by the "action integral"

$$I_{T_1T_2}(\phi) = \int_{T_1}^{T_2} dt|\dot{\underline{\phi}} - \underline{b}_o(\underline{\phi})|^2$$

$$= \int_{T_1}^{T_2} dt[\dot{\underline{\phi}}^2 + \underline{b}_o^2(\phi)] + 2[\mathcal{U}(\phi(T_2)) - \mathcal{U}(\phi(T_1))] \tag{2.3}$$

where

$$\underline{b}_o = - \underline{\nabla}\,\mathcal{U}. \tag{2.4}$$

Referring to the ground state quantum mechanical process the reader will easily recognize that the first term in the second line of (2.3) is $\frac{2}{m}$ the action appearing in the usual Euclidean functional integral. In the one dimensional case this follows from (2.1) and in the multidimensional case from the more general equations that one can find in [3]. Strictly speaking there is also the constant $E(\hbar = 0)$ corresponding to the classical ground state energy, but we can normalize it to zero.

We are now in a position to formulate the basic estimates of Ventzel and Freidlin. The first one is concerned with the probability that a trajectory $\underline{X}^\epsilon(t)$ of the process does not deviate in the interval $0 \leq t \leq T$ by more than δ from a function $\underline{\phi}(t)$, where $\underline{\phi}(0) = \underline{X}$ is the initial point of the trajectory. The precise result is as follows:

For any $h > 0$ and sufficiently small δ and $\epsilon > 0$ the probability in question is bounded above and below by

$$\exp\left[\frac{-I_{OT}(\phi) \pm h}{2\,\epsilon^2}\right] . \tag{2.5}$$

This estimate makes it likely that tunnelling from one equilibrium position $\underline{X}_1$ to another $\underline{X}_2$ takes place along a trajectory $\bar{\phi}$ that minimizes I_{OT} (ϕ), $\underline{\phi}(0) = \underline{X}_1$, $\underline{\phi}(T) = \underline{X}_2$, T arbitrary. However, in general there will be infinitely many trajectories different from $\bar{\phi}$ connecting $\underline{X}_1$, with $\underline{X}_2$. One has to be sure that the sum of the probabilities of all these unlikely trajectories does not exceed the probability of $\bar{\phi}$. The answer to this question is provided by a second estimate. Let $\Phi_o(X)$ be the set of curves $\underline{\phi}(t)$ such that $\underline{\phi}(0) = \underline{X}$ and $I_{OT}(\phi) \le I_o$. Then for any $h > 0$ $\delta > 0$ and sufficiently small ϵ the probability that a trajectory $\underline{X}(t)$ with initial point $\underline{X}$ has a maximum distance from $\Phi_o(X)$ greater than δ is bounded by

$$\exp\left[\frac{-I_o + h}{2\epsilon^2}\right] . \tag{2.6}$$

While (2.5) is not entirely unfamiliar to physicists (for example I_{OT} is known in physics as the Onsager-Machlup function. For recent work see [7]), the second estimate (2.6) seems to go beyond what people have done by functional integral methods. The proof of (2.6) uses in an essential way the strong Markov property of the process.

With the help of (2.5) and (2.6) Ventzel and Freidlin show that the behaviour of the diffusing particle for small ϵ is well approximated by a Markov chain with a number of states equal to the number of stable equilibrium orbits of the system (2.2). The transition probabilities p_{ij} of this chain are bounded above and below by the expression (as usual for any $h > 0$ and sufficiently small ϵ)

$$\exp\left[\frac{-V_{ij} \pm h}{2\epsilon^2}\right] \tag{2.7}$$

where

$$V_{ij} = \inf_{\substack{\underline{\phi}(0)=\underline{X}_i \\ \underline{\phi}(T)=\underline{X}_j \\ T}} I_{OT}(\phi) \tag{2.8}$$

$\underline{X}_i$ and $\underline{X}_j$ are points belonging to the stable orbits i and j.

It is a continuous time Markov chain governed by a random clock.

3. The Anharmonic Oscillator with two Symmetric Equilibrium Positions

We consider the quantum mechanical system described by the potential

$$V(X) = \lambda(X^2-1)^2 \qquad \lambda > 0. \tag{3.1}$$

From equation (1.2) setting $\hbar = 0$, $m = 1$ we obtain

$$b_o^2(X) = 2V(X) \quad ;$$

our drift term is therefore

$$b_o(X) = \sqrt{2\lambda}(1-X^2) \qquad X > 0$$

$$b_o(X) = -\sqrt{2\lambda}(1-X^2) \qquad X < 0 . \tag{3.2}$$

At X = 0 we shall by symmetry put $b_o(0) = 0$. The situation is illustrated in the picture

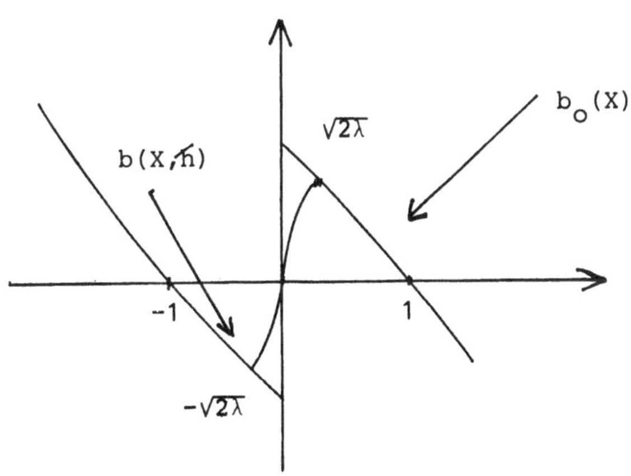

Corresponding to $b_o(X)$ we have a potential $\mathcal{U}(X)$

$$\mathcal{U}(X) = \sqrt{2\lambda}\left(\frac{X^3}{3} - X\right) \qquad X > 0$$
$$\mathcal{U}(X) = -\sqrt{2\lambda}\left(\frac{X^3}{3} - X\right) \qquad X < 0. \tag{3.3}$$

The discontinuity of the drift term at zero is not a serious difficulty. The mathematical problems connected with this point as well as other aspects related to the behaviour at infinity will be dealt with in a subsequent paper. Here let us simply observe that the application of the previous theory requires that we follow only the motion of the particle against the flow, therefore up to the point zero.

Let us now calculate a quantity which has relevant physical information, i.e. the autocorrelation function of the stationary process corresponding to the drift term (3.2). By definition we have

$$E(X(0)X(t)) = \int \mu(dx)P_x(t,dy)xy \tag{3.4}$$

where μ is the invariant measure and $P_x(t,dy)$ is the transition function. We are mainly interested in the large t behaviour of (3.4) as this is connected with the energy difference between the lowest levels of our quantum system. How large t has to be will be apparent in a moment. Intuitively it must be large enough to allow jumps between the classical equilibrium positions at ± 1. Deferring again the mathematical details, a direct estimate of (3.4) can be obtained as follows. According to the discussion of the previous Section for $\hbar$ small enough all the relevant contributions to the integrals come from small regions (exponentially small) around ± 1. We have therefore

$$E(X(0)X(t)) \simeq P_{11}(t) - P_{12}(t) \tag{3.5}$$

where P_{11} and P_{12} are the probabilities of remaining in the same minimum or passing to the other one respectively after time t. We decompose now P_{11} or P_{12}

$$P_{11}(t) = \sum_{n \text{ even}} P_n(t)$$
$$P_{12}(t) = \sum_{n \text{ odd}} P_n(t) \qquad (3.6)$$

where $P_n(t)$ is the probability of n jumps at time t. Due to the symmetry of the problem the transition times from -1 to 1 and from 1 to -1 will have the same distribution. $P_n(t)$ is therefore a Poisson process, i.e. $P_n(t) = \frac{e^{-\Lambda t}}{n!} (\Lambda t)^n$ where Λ^{-1} is the mean transition time. From (3.5) we obtain

$$E(X(0)X(t)) \simeq e^{-2\Lambda t} \qquad (3.7)$$

The final step consists in calculating Λ. This is an easy matter. Λ^{-1} in fact can be identified with the mean exit time of a process starting at +1 from the interval (0, L) where L is a large positive number.

Using the standard formula [8]

$$E\left[\tau_1(0,L)\right] = -2 \int_0^1 e^{\frac{2}{\hbar}\mathcal{U}(y)} \int_0^y \frac{dz}{\hbar} e^{-\frac{2}{\hbar}\mathcal{U}(z)} dy$$
$$+ 2 \int_0^L e^{\frac{2}{\hbar}\mathcal{U}(y)} \int_0^y \frac{dz}{\hbar} e^{-\frac{2}{\hbar}\mathcal{U}(z)} dy \; \frac{\int_0^1 e^{\frac{2}{\hbar}\mathcal{U}(z)} dz}{\int_0^L e^{\frac{2}{\hbar}\mathcal{U}(z)} dz} . \qquad (3.8)$$

in the small $\hbar$ limit we obtain

$$\Lambda^{-1} = E[\tau_1(0,L)] \simeq C \sqrt{\hbar}\, e^{\frac{4}{3}\frac{\sqrt{2\lambda}}{\hbar}} \qquad (3.9)$$

C is a constant which can be estimated. Our final result is then

$$E(X(0)X(t)) \simeq \exp\left[- \frac{2}{C\sqrt{\hbar}}\, e^{-\frac{4}{3}\frac{\sqrt{2\lambda}}{\hbar}}\, t\right]. \qquad (3.10)$$

This expression is well known to physicists and is the same as that obtained by approximate functional integral methods. The factor $\sqrt{\hbar}$ which in the physicists language is connected with the so called zero modes, in our approach came out automatically and it is easily seen to originate from the mean square fluctuation around the classical equilibrium positions. In this connection, however, we have to make a remark.

An expression like (3.9) in some sense goes beyond the approximation of Ventzel and Freidlin whose point of view consists in extracting in every situation the main exponential contribution. The $\sqrt{\hbar}$ means that the $\hbar$ dependence of the error term in $E(\tau_1(O.L))$ has been made explicit.

4. Outlook

In this Section we analyse more general aspects of the approach proposed here. We begin with some remarks on the application of Section 3. We see that to a certain extent the logic of the usual functional integral approach to the instanton problem has been reversed. To arrive at the correlation formula (3.10) we never had to calculate explicitly the solution which minimizes the action and we used only the gross structure of the process as revealed by the Ventzel and Freidlin theory. This means that also in cases much more complicated than the one we have considered we may expect to go a long way even if we are not able to solve explicitly the minimum problem. This flexibility as it will be shown elsewhere becomes crucial for instance in discussing asymmetrical double well potentials, situations with more than two minima or in multidimensional problems.

Generally speaking the Markow chain with a finite number of states corresponding to stable equilibria proves to be a very powerful intermediate concept between quantum mechanics and classical mechanics. The approach works very well also in cases without tunnellings and gives the proper information about rare fluctuations. All this makes it reasonable to pursue a systematic study of the semiclassical limit starting from stochastic dynamics and using the Ventzel and Freidlin techniques as a substitute for functional integration.

The next natural question is the possibility of extending this point of view to field theory. It is easy to see that many of the steps we have described admit a formal generalization to the infinite dimensional case. In particular one could reinterpret several heuristic results in gauge theories along the lines described here. So this is an interesting direction to explore which presumably would fit quite well within the Albeverio-Høegh-Krohn [9] description of field theory.

Another point, a technical one, that we would like to mention is the following. It is clear from the arguments in Section 2 that the strong Markov property plays an important role in the whole Ventzel-Freidlin construction. This property is not a usual tool of constructive field theory. My suggestion is that the analog of the strong Markov property could have an interesting role also in Euclidean field theory where it would be connected with conditioning on regions defined by random boundaries. On the other hand this notion is quite natural in the approach of Ref. [9].

In closing I would like to mention some papers [10] of which I became aware only recently, related to some of the problems touched upon in this work.

Acknowledgements

The investigations presented in this talk were first stimulated by the preparation of a report for the meeting on the Interdisciplinary Aspects of Modern Physics held in Parma, May 1976. In this connection I would like to thank Prof. M. Fontana for having given me such an opportunity. After resuming more recently this line of thought I benefited from a conversation with Profs. Ventzel and Freidlin on certain aspects of their theory. Finally I would like to thank the organizers and the participants of the Bielefeld Encounter and of the Workshop on the White Noise Approach to Quantum Dynamics for their interest and for stimulating remarks.

References

[1] A.M. Polyakov, Nucl.Phys. B121, 429 (1977); G. t'Hooft, Phys.Rev.Lett. 37, 8 (1976), Phys.Rev. D14, 3432 (1976).

[2] For a clear exposition of the functional integral approach see S. Colemann, "The Uses of Instantons", Lectures at the 1977 International School "Ettore Majorana".

[3] E. Nelson, Phys.Rev. 150, 1079 (1966) and "Dynamical Theories of Brownian Motion", Princeton 1967.

[4] F. Guerra, P. Ruggiero, Phys.Rev.Lett. 31, 1022 (1973).

[5] R.Z. Khasminski, "Stability of Systems of Differential Equations under Random Perturbations of their Parameters", Moscow 1969 (in Russian).

[6] A.D. Ventzel, M.I. Freidlin, Uspehi Math.Nauk 25, 3 (1970) (English translation Russian Mathematical Surveys 25, 1 (1970)).

[7] D. Dürr, A. Bach, Comm.Math.Phys. 60, 153 (1978).

[8] I.I. Gihman, A.V. Skorohod, "Stochastic Differential Equations", Springer Verlag, Berlin 1972, pag. 109.

[9] S. Albeverio, R. Høegh-Krohn, in "Les Méthodes Mathématiques de la Théorie Quantique des Champs" CNRS, Marseille 1975, pag. 11-59.

[10] K. Yasué, Phys.Rev.Lett. 40, 665 (1978), Phys.Rev. D18, 532 (1978); D.L. Weaver, Phys.Rev.Lett. 40, 1473 (1978).

Asymptotic Expansion of Fresnel Integrals Relative to a Non-Singular Quadratic Form

Ph. Blanchard

Fakultät für Physik

Universität Bielefeld

Federal Republic of Germany

Abstract

We give an exposition of some of the methods of the mathematical theory of oscillatory integrals in infinitely many dimensions. In particular for a class of phase functions we extend the method of stationary phase and the corresponding asymptotic expansions to the infinite dimensional case.

I. Introduction

I would like to report here on some recent work done by S. Albeverio, R. Høegh-Krohn and myself on the study of asymptotic expansions of Fresnel integrals.

Let me remind you shortly to the formal definition of Feynman path integrals. Consider for simplicity a non relativistic particle of mass m in d-dimensional Euclidean space $\mathbb{R}^d$. Suppose the particle is under the influence of a bounded and continuous potential $V(x)$. The dynamics is given by the unitary groups $e^{-\frac{i}{\hbar} tH}$ where $H = H_o + V$, H_o being the free kinetic energy operator $-\frac{\hbar^2}{2m}\Delta$, where Δ is the Laplace operator in $\mathbb{R}^d$, $\hbar$ is Planck's constant divided by 2π. By Lie-Trotter's product formula we have

$$e^{-\frac{i}{\hbar} t H} = \underset{n \to +\infty}{s - \lim} \left(e^{-\frac{i}{\hbar}\frac{t}{n} V} e^{-\frac{i}{\hbar}\frac{t}{n} H_o} \right)^n . \tag{1.1}$$

If ϕ is for example in $S(\mathbb{R}^d)$ we have, on the other hand

$$(e^{-\frac{i}{\hbar}\frac{t}{n} H_o} \phi)(x) = (2\pi i \frac{\hbar}{m}\frac{t}{n})^{-\frac{d}{2}} \int_{\mathbb{R}^d} e^{i \frac{m(x-y)^2}{2\hbar \frac{t}{n}}} \phi(y) dy . \tag{1.2}$$

Thus from (1.1) and (1.2) we obtain

$$(e^{-\frac{i}{\hbar} t H} \phi)(x) = \underset{n\to+\infty}{s\text{-}\lim} \, (2\pi i \frac{\hbar}{m}\frac{t}{n})^{-\frac{nd}{2}}$$

$$\int_{\mathbb{R}^{nd}} e^{-\frac{i}{\hbar} S_t(x_n \ldots x_o)} \phi(x_o) \, dx_{n-1} \ldots dx_o \tag{1.3}$$

where $x_n \equiv x$ and

$$S_t(x_n \ldots x_o) = \sum_{j=1}^{n} \left[\frac{m}{2} \frac{(x_j - x_{j-1})^2}{(t/n)^2} - V(x_j)\right] \frac{t}{n} . \tag{1.4}$$

Consider on the other hand the classical action integral along the path $\gamma(\tau)$ $0 \le \tau \le t$, supposed to be absolutely continuous, starting at time $\tau = 0$ from the point x_o in $\mathbb{R}^d$ and ending at time $\tau = t$ in the point $x_n = x \in \mathbb{R}^d$. This classical action is equal to

$$S_t(\gamma) = \int_o^t \frac{m}{2} \dot{\gamma}^2(\tau)\, d\tau - \int_o^t V(\gamma(\tau))\, d\tau \qquad (1.5)$$

where $\dot{\gamma}(\tau) = \frac{d}{dt}\gamma(\tau)$. For any piecewise linear path $\gamma_o(\tau)$ such that $\gamma_o(\tau_j) = x_j$ for $\tau_j = \frac{jt}{n}$, $j = 1\ldots n$ with x_j any given points in $\mathbb{R}^d$ and with $x_n = x, \gamma(o) = x_o$ we have from (1.5) and (1.4)

$$S_t(\gamma_o) = S_t(x_n \ldots x_o).$$

On the other hand the piecewise linear path $\gamma_o(\tau)$ can be looked upon, for $n \to +\infty$, as an "approximation" of an absolutely continuous path $\gamma(\tau)$ and $S_t(\gamma_o)$ can be considered as a formal Riemann sum approximation for $S_t(\gamma)$. In the same limit formally $dx_{n-1}\ldots dx_o$ becomes

$$d\gamma = \prod_{0 \le \tau \le t} d\gamma(\tau)$$

hence formally $(2\pi i \frac{\hbar}{m} \frac{t}{n})^{-\frac{nd}{2}} \int_{\mathbb{R}^{nd}} e^{\frac{i}{\hbar} S_t(x_n \ldots x_o)} \phi(x_o) dx_{n-1}\ldots dx_o$

becomes, as $n \to +\infty$

$$N \int_{\gamma(t) = x} e^{\frac{i}{\hbar} S_t(\gamma)} \phi(\gamma(o))\, d\gamma \qquad (1.6)$$

where now the integrals should be over all paths which start anywhere at time $\tau = 0$ and end at time $\tau = t$ in x. N should be a suitable normalization. This is Feynman's formal expression for the solution of Schrödinger's equation $(e^{-\frac{it}{\hbar} H} \phi)(x)$ and the basis for Feynman's more general formulation of quantum mechanics. The problem of giving a mathematical meaning to this heuristic expression consists in giving an interpretation of the integral as a linear functional on a space of paths. For more details we refer to [1]. The definition should be a useful one in the sense of e.g. allowing a mathematical control on the approach to the classical limit $\hbar \to 0$, which, according to the original Dirac's and Feynman's ideas should be determined by the path which makes $S_t(\gamma)$ stationary i.e. according to Hamilton's principle, the trajectory of the classical motion.

We can write (1.6)

$$I(\hbar) = \widetilde{\int_{\mathcal{H}}} e^{\frac{i}{2\hbar}|x|^2} e^{-\frac{i}{\hbar}w(x)} dx$$

where $\mathcal{H}$ is a real separable Hilbert space with inner product (x,y) and norm $|x|^2 = (x,x)$. In [2] the asymptotic behaviour as $\hbar \to 0$ of $I(\hbar)$ was studied. This study in the finite dimensional case $\dim \mathcal{H} < +\infty$ is well known from the work of Maslov, Hörmander, Arnold, Duistermaat and others see e.g. the references in [2].

2. Fresnel Integrals Relative to a Non Singular Quadratic Form

What is needed to treat many physical systems is a general theory of integrals of the form $\int_{\mathcal{H}} e^{i(x,Bx)} f(x)\, dx$, where B is a linear operator which is not necessarily positive definite. The definition of the Fresnel integral relative to a non singular, not necessarily positive quadratic form is given in [1]. This definition involves the concept of what we call a Fresnel fourtuple defined as follows:

A Fresnel fourtuple is a fourtuple $(\mathcal{H}, D, B, \Delta)$ where

a) $\mathcal{H}$ is a real separable Hilbert space with norm $|\cdot|$
b) D is a densely defined symmetric operator on $\mathcal{H}$
c) D is a dense linear subset of $\mathcal{H}$ containing the range of B

(i.e. Ran B $\subset$ D) and such that D is a separable Banach space with respect to a norm $\|\cdot\|_D$ which is stronger than the norm $|\cdot|$ of $\mathcal{H}$, in the sense that

$$|x| \leq a\|x\|_D \tag{2.1}$$

where a is a constant. This implies $D \subset \mathcal{H} \subset D'$, D' being the topological dual of D, with continuous injections.

d) Δ is a complex-valued bilinear, symmetric continuous form on D x D with $\mathrm{Im}\Delta(x,x) \leq 0$. Moreover Δ should be an inverse of B in the following sense that

$$\Delta(x,By) = (x,y) \tag{2.2}$$

$\forall\, x \in D \quad \forall\, y \in D(B)$. From (2.2) for $x \in D$ and any $y \in D(B)$ we get that $By = 0$ implies that $y = 0$ since D is dense in $\mathcal{H}$. This implies that B is one to one so that $\Delta = B^{-1}$ on $D \times B\,D(B)$. We define now F(D') - the space of "Fresnel integrable functions" - as the set of all functions on D' which are Fourier transforms of bounded measures in D

$$F(D') = \{f : D' \to \mathbb{C} \quad | \quad f(x) = \int_D e^{i<x,\alpha>} d\mu(\alpha) \tag{2.3}$$

$$\text{for some complex bounded measure } \mu \text{ on } D\}$$

where <,> is the canonical pairing of D with its topological dual D'. Note that for $x \in \mathcal{H}$ $f(x) = \int_D e^{i(x,\alpha)} d\mu(\alpha)$ where (,) is again the scalar product of $\mathcal{H}$.

Defining as norm in F(D') the norm $\|\cdot\|_0$ given by $\|f\|_0 = \|\mu\| = \int d|\mu|$ we have that F(D') is a Banach function algebra with unit, which is contained in $F(\mathcal{H})$.

We are ready now to state the definition of <u>Fresnel integrals relative</u> to $(\mathcal{H},B,D,\Delta)$:

For $f \in F(D')$ we define

$$F_\Delta(f) = \int_{\mathcal{H}}^{\Delta} e^{\frac{i}{2}(x,Bx)} f(x)\, dx \equiv \int_D e^{-\frac{i}{2}\Delta(\alpha,\alpha)} d\mu(\alpha). \tag{2.4}$$

The integral on the right hand side is well defined since $\mathrm{Im}\Delta(\alpha,\alpha) \leq 0$ and $\Delta(\alpha,\alpha)$ is continuous on D. This normalized integral $(F_\Delta(1) = 1)$ is a bounded continuous linear functional on F(D') such that

$$|F_\Delta(f)| \leq \|f\|_0 = \|\mu_f\| = \int d|\mu_f|.$$

<u>Remark I</u>

In the general case Δ is not uniquely determined by B. However, if B is non degenerate in the stronger sense that it has a bounded continuous inverse B^{-1} on $\mathcal{H}$, then it follows easily that $D = \mathcal{H}$ since D contains the range of B and that $\Delta(x,y) = (x,B^{-1}y)$. Hence Δ is in this situation unique and real.

Remark 2

If $\mathcal{H}$ is finite dimensional then B^{-1} is bounded and Δ is uniquely given by $\Delta(x,y) = (x,B^{-1}y)$ for all x and y, and with $\mathcal{H} = \mathbb{R}^n$

$$\int_{\mathcal{H}}^{\Delta} e^{\frac{i}{2}(x,B\ )} f(x)\ dx = \left|\frac{1}{2\pi i}\ B\right|^{1/2} \int_{\mathbb{R}} e^{\frac{i}{2}(x,Bx)} f(x)dx \quad (2.5)$$

for any $f \in S(\mathbb{R}^n)$, where $\left|\frac{1}{2\pi i}\ B\right|^{1/2}$ is given by the formula

$$\left|\frac{1}{2\pi i}\right| B^{\ 1/2} = \left(\frac{1}{2\pi}\right)^{\frac{n}{2}} \|B\|^{1/2}\ e^{-i\frac{\pi}{4}\sigma(B)} \quad (2.6)$$

where $\|B\|^{1/2}$ is the positive root of the absolute value of the determinant of B and $\sigma(B)$ is the signature of the form B. This is according to the formula

$$\left|\frac{1}{2\pi i}\ B\right|^{1/2} \int_{\mathbb{R}^n} e^{\frac{i}{2}(x,Bx)} dx = 1.$$

The integral on the right hand side of (2.5) is the Lebesgue integral. Let now $f \in S(\mathbb{R}^n)$ then

$$\int_{\mathcal{H}}^{\Delta} e^{\frac{i}{2}(x,Bx)} f(x)\ dx = \int_{\mathbb{R}^n} e^{-\frac{i}{2}(x,B^{-1}x)} \tilde{f}(x)\ dx$$

with $f(x) = \int e^{i(x,y)}\ \tilde{f}(y)\ dy$.

On the other hand one verifies easily that, for $f \in S(\mathbb{R}^n)$,

$$\int_{\mathbb{R}^n} e^{-\frac{i}{2}(x,B^{-1}x)} \tilde{f}(x)\ dx = \left|\frac{1}{2\pi i}\ B\right|^{1/2} \int_{\mathbb{R}^n} e^{\frac{i}{2}(x,Bx)} f(x)\ dx$$

and this proves the relation (2.5). Thus the Fresnel integral reduces to $\left|\frac{1}{2\pi i}\ B\right|^{1/2}$-times the ordinary integral in the finite dimensional case; this is the reason for the notation (2.4).

Example 1 (Systems of anharmonic oscillators)

We consider the system of finitely many quantum mechanical oscillators described by the Schrödinger operator

$$H_N = \sum_{j=1}^{N} \frac{-\hbar^2}{2m} \Delta_{x_j} + \frac{1}{2} \Sigma x_j A^2{}_{jk} x_k + V(x_1 \ldots x_n)$$

where A^2_{jk} is a strictly positive definite symmetric N×N matrix. Let t be so that the matrices cos At and sin At are both non singular.

Let ($\mathcal{H}$,B,D,Δ) be the following Fresnel fourtuple.

$$\mathcal{H} = \{\gamma(\cdot) : [0,t] \to \mathbb{R}^d, \gamma(\cdot) \text{ absolutely continuous and}$$

$$\text{such that } \int_0^t (\dot{\gamma}^2 + \gamma^2) \, d\tau < +\infty\}.$$

Let D = $\mathcal{H}$ with the same norm. Let B be the operator given by

$$(\gamma_1, B\gamma_2) = \frac{1}{\hbar} \int_0^t (m \dot{\gamma}_1 \dot{\gamma}_2 - \gamma_1 A^2 \gamma_2) \, d\tau$$

i.e. by the harmonic part $S_t^o(\gamma)$ of the classical action

$$S_t(\gamma) = \frac{m}{2} \int_0^t \dot{\gamma}^2 d\tau - \frac{1}{2} \int_0^t (\gamma A^2 \gamma) \, d\tau - \int_0^t V(\gamma(\tau)) \, d\tau$$

$$= S_t^o(\gamma) - \int_0^t V(\gamma(\tau)) \, d\tau \quad .$$

This implies that B^{-1} is bounded [1], symmetric, defined on $\mathcal{H}$, and we take $\Delta = B^{-1}$.

Relative to this Fresnel fourtuple the integral

$$\int_{\mathcal{H}}^{\Delta} e^{\frac{i}{\hbar} S_t(\gamma)} \bar{\phi}_1(\gamma(t)) \, \phi_2(\gamma(0)) \, d\gamma$$

is well defined for any $\phi_1, \phi_2 \in F(\mathbb{R}^d)$ and we have

$$(\phi_1, e^{-\frac{i}{\hbar} t H_N} \phi_2) = \left| \frac{1}{2\pi} A \sin A t \right|^{-1/2} \int_{\mathcal{H}}^{\Delta} e^{\frac{i}{\hbar} S_t(\gamma)} \bar{\phi}_1(\gamma(t)) \phi_2(\gamma(0)) d\gamma$$

provided $\phi_1, \phi_2 \in F(\mathbb{R}^d) \cap L^2(\mathbb{R}^d)$. This last formula is the real time analogue of the Feynman-Kac-Nelson formula for a 1-dimensional field theory

$$(\phi_1, e^{-tH}\phi_2) = \int e^{-\int_0^t V(\gamma(\tau))d\tau} \bar{\phi}_1(\gamma(t))\, \phi_2(\gamma(0))\, d\omega(\gamma)$$

$t > 0$ $d\omega$ being Wiener's measure.

Example 2 (Free relativistic scalar boson quantum fields)

We define the Fresnel fourtuple $(\mathcal{H}, B, \Delta, D)$ which is appropriate for expressing the relevant expectations in terms of Fresnel integrals.

$\mathcal{H} = H_1(\mathbb{R}^{S+1})$ = real Sobolev space of distributions defined as the Hilbert space closure of the real Schwartz functions $\phi \in S(\mathbb{R}^{S+1})$ in the norm

$$|\phi|^2 \equiv \int_{\mathbb{R}^{S+1}} [(\tfrac{\partial\phi}{\partial t})^2 + (\nabla\phi)^2 + m^2\phi^2]\ dx\, dt$$

where $m \geq 0$ is a constant, the mass of the field, $x \in \mathbb{R}^S, t \in \mathbb{R}$. The scalar product in $\mathcal{H}$ is accordingly defined in terms of Fourier transforms by

$$(\phi_1, \phi_2) = \int_{\mathbb{R}^{S+1}} \bar{\tilde{\phi}}_1(p)\ (p^2+m^2)\ \tilde{\phi}_2(p)\ dp$$

with $\tilde{\phi}(p) = (2\pi)^{-\frac{S+1}{2}} \int e^{i\,p\cdot x}\phi(x)\ dx$ and $p^2 = p\cdot p$ is the square of the Euclidean norm of p.

Moreover we define

$$D = \{\phi \in \mid \tilde{\phi} \in C^1(\mathbb{R}^{S+1})\quad \|\phi\|_D = |\phi| + \sup_{j,p} \left|\frac{\partial\phi}{\partial p_j}\right| < \infty\}$$

i.e. D is the Banach space whose elements are all functions of $\mathcal{H}$ which have continuous differentiable Fourier transforms and have finite norm $\|\cdot\|_D$.

B is the operator with domain the set of all functions $\phi(x)$ in $\mathcal{H}$ which have compact support in $\mathbb{R}^{S+1}$. For such functions one has by definition

$$(\phi, B\phi) = 2\, S_o(\phi) \equiv \int_{\mathbb{R}^{S+1}} [(\frac{\partial\phi}{\partial t})^2 - (\nabla\phi)^2 - m^2\phi^2]\, dx\, dt$$

where $S_o(\phi)$ is the "classical action along ϕ".

Δ is the symmetric bilinear form Δ_F defined on $D \times D$ by

$$\Delta_F(\phi,\psi) = \int_{\mathbb{R}^S} P \int_{\mathbb{R}} \bar{\tilde{\phi}}(p)(p_o^2 - \vec{p}^2 - m^2)(p^2+m^2)^{-1}\, \tilde{\psi}(p)(p^2+m^2)^2\, dp$$

$$- i\pi \int_{\mathbb{R}^{S+1}} \bar{\tilde{\phi}}(p)\,\delta(p_o^2 - \vec{p}^2 - m^2)(p^2+m^2)^2 \tilde{\psi}(p)\, dp$$

where P means principal value and δ is the Dirac measure. For the verification that ($\mathcal{H}$,D,B,Δ_F) is a Fresnel fourtuple see [1], Section 9. Using now this Fresnel fourtuple we can define for any function $f \in F(D')$ the Fresnel integral relative to Δ_F

$$F_{\Delta_F}(f) \equiv \int_{\mathcal{H}}^{\Delta_F} e^{\frac{i}{2}(\phi,B\phi)} f(\phi)\, d\phi = \int_{\mathcal{H}}^{\Delta_F} e^{i\, S_o(\phi)} f(\phi)\, d\phi \, .$$

3. Asymptotic Expansion around $\hbar = 0$

We consider oscillatory integrals of the form

$$I^\Delta(\hbar) = \int_{\mathcal{H}}^{\Delta} e^{\frac{i}{2\hbar}(x,Bx)}\, e^{-\frac{i}{\hbar}W(x)}\, g(x)\, dx \qquad (3.1)$$

where ($\mathcal{H}$,B,Δ,D) is a Fresnel fourtuple in the sense of the section 2 and W, $g \in F(D')$. We shall discuss the asymptotic expansion of $I^\Delta(\hbar)$ around $\hbar = 0$, along similar lines as in the case where $B = \mathbb{1}$ treated in [2]. From the assumption $W \in F(D')$ and the fact that $F(D')$ is a Banach algebra we have

$$e^{-\frac{i}{\hbar}W} = \sum_{n=o}^{\infty} (\frac{-i}{\hbar})^n \frac{1}{n!} W^n \qquad (3.2)$$

with the sum on the right hand side convergent in the F(D') convergent in the F(D')-norm. Since $f \to \int^{\Delta} e^{\frac{i}{\hbar}(x,Bx)} f(x)\, dx$ is a continuous linear functional on F(D') we get interchanging the normalized integral with the sum and using the definition of the normalized integral for each term in the sum

$$I^{\Delta}(\hbar) = \sum_{n=o}^{\infty} (-\frac{i}{\hbar})^n \frac{1}{n!} \int_D \cdots \int_D e^{-\frac{i\hbar}{2}\Delta(\sum_{j=1}^{n} \alpha_j+\beta, \sum_{k=1}^{n} \alpha_K+\beta)}$$

$$\prod_{j=1}^{n} d\mu(\alpha_j)\ d\nu(\beta) \qquad (3.3)$$

with $\quad W(x) = \int_D e^{i<x,\alpha>} d\mu(\alpha)$

$$g(x) = \int_D e^{i<x,\beta>} d\nu(\beta).$$

Now for $h \neq 0$ such that $\mathrm{Im}\, h\, \Delta \leq 0$ we have that the right hand side is bounded by

$$\sum_{n=o}^{\infty} (\frac{1}{|h|})^n \frac{1}{n!} \int_D \cdots \int_D \prod_{j=1}^{n} d|\mu|(\alpha_j) \quad d|\nu|(\beta)$$

$$= e^{\frac{1}{|h|}\ \|\mu\|} \cdot \|\nu\| \qquad (3.4)$$

where $\|\cdot\|$ means total variation. Hence for such $h \in \Gamma = \{h \in \mathbb{C}\ h \neq 0\ \mathrm{Im}\, h \leq 0\}$ we have uniform convergence of the series (3.3). Moreover, each term in the series is itself an analytic function of h in Γ.

Consider now the n-th term in the series (3.3). Let

$$\Phi_n(\hbar) = \int_D \cdots \int_D e^{-i\frac{\hbar}{2}\Delta(\sum_{j=1}^{n}\alpha_j+\beta, \sum_{k=1}^{n}\alpha_k+\beta)} \prod_{j=1}^{n} d\mu(\alpha_j) d\nu(\nu).$$

We have

$$(\frac{\partial}{\partial\hbar})^k \phi_n(\hbar)\Big|_{\hbar=0} = k!\ (\frac{-i}{2})^k \int_D \cdots \int_D (\sum_{j=1}^{n} \sum_{k=1}^{n} \Delta(\alpha_j+\beta,\alpha_k+\beta))^k \prod_{j=1}^{n} d\mu(\alpha_j) d\nu(\beta).$$

$$(3.5)$$

We first assume that W is such that

$$\int_D d\mu(\alpha) = 0$$

$$\int_D \Delta(\beta,\alpha)\ d\mu(\alpha) = 0$$

for all $\beta \in D$. In this case we have that the derivatives (3.5) vanish for all $0 \le 1 \le n-1$, hence the n-th term of the series (2.3) i.e. $(\frac{-i}{\hbar})^n \frac{1}{n!} \Phi_n(h)$ is regular for $h \to 0$ $\hbar \in \partial\Gamma = \{\hbar \in \mathbb{C} \mid \mathrm{Im}\, \hbar\, \Delta = 0\}$ and all its derivatives at $\hbar = 0$ exist, i.e. each term in the series (2.3) is C^∞ on the whole axis. It can be shown (see [3]) that (3.3) yields in fact an asymptotic expansion of $I^\Delta(h)$ for $\hbar \in \partial\Gamma$. We have namely the following Theorem

<u>Theorem 3.1 [3]</u>

Let $W(x) = \int_D e^{i<x,\alpha>} d\mu(\alpha)$ and $g(x) = \int_D e^{i<x,\beta>} d\nu(\beta)$,

where μ and ν are bounded complex measures, such that

$$\int_D d\mu(\alpha) = \int_D \Delta(\gamma,\alpha)\, d\mu(\alpha) = 0 \quad \text{for all } \gamma \in D,$$

$$\int_D e^{\sqrt{2c}\, N(\alpha)}\, d|\mu|(\alpha) < +1$$

and $\int_D e^{\sqrt{2c}\, N(\beta)}\, d|\nu|(\beta) < +\infty$

for some seminorn $N(\cdot)$ such that $|\Delta(\alpha,\alpha)| \le c[N(\alpha)]^2$ for all $\alpha \in \mathrm{supp}\, d\mu \cup \mathrm{supp}\, d\nu$. Then

$$I^\Delta(\hbar) \equiv \int_{\mathcal{H}}^\Delta e^{\frac{i}{2\hbar}(\gamma,Bx)}\, e^{\frac{-i}{\hbar}W(x)}\, g(x)\, dx$$

as defined by (2.4) is analytic for $\hbar \in \Gamma$ it is a C^∞-function in $\hbar$ for $\hbar \in \partial\Gamma$.

Moreover we have

$$I^\Delta(0) = \sum_{n=0}^{\infty} (\frac{-1}{2})^n (\frac{1}{n!})^2 \int_D \cdots \int_D [\sum_{j,k=1}^{n} \Delta(\alpha_j+\beta,\alpha_k+\beta)]^n \prod_{j=1}^{n} d\mu(\alpha_j)\, d\nu(\beta)$$

and

$$I^\Delta(\hbar) = \sum_{n=0}^{N} \hbar^m (\frac{-i}{2})^m \sum_{n=0}^{\infty} (\frac{-1}{2})^n \frac{1}{n!\,(n+m)!}$$

$$\int_D \cdots \int_D [\Delta(\sum_{j=1}^{n} \alpha_j+\beta, \sum_{k=1}^{n} \alpha_k+\beta)^{n+m} \prod_{j=1}^{n} d\mu(\alpha_j)\, d\nu(\beta) + R_N$$

for any arbitrary integer N. The remainder R_N satisfies the estimate

$$|R_N| \leq |\hbar|^{N+1}(N+1)!\ [1-\int_D e^{\sqrt{2c}\ N(\alpha)}d|\mu|(\alpha)]^{-N-2}\int_D e^{\sqrt{2c}\ N(\beta)}d|\nu|(\beta).$$

In the case where supp $d\mu$ is contained in the range of B then we can replace in the expansions of the above theorem every term of the form $\Delta(\alpha_j,\alpha_k)$ by $(\alpha_j,\ B^{-1}\alpha_k)$ $j,k = 1\ldots n$. Hence the theorem holds with Δ replaced by B^{-1} in this sense.

Moreover if B^{-1} is a bounded operator on $\mathcal{H}$ then $\mathcal{H} = D$ with equivalent norms and $\Delta = B^{-1}$ and in this case

$$|\Delta(\gamma\ \gamma)| = |(\gamma,\ B^{-1}\gamma) \leq |\gamma|^2\ \|B^{-1}\|$$

hence $N(\gamma)$ can be taken to be $|\gamma|$ and the conditions of theorem (3.1) reduce to

$$\int e^{\sqrt{2}\ \|B^{-1}\|^{\frac{1}{2}}|\alpha|}\ d|\mu|(\alpha) < +1 \qquad (3.6)$$

$$\int e^{\sqrt{2}\ \|B^{-1}\|^{\frac{1}{2}}|\beta|}\ d|\mu|(\beta) < +\infty\ . \qquad (3.7)$$

In this case the phase function

$$\Phi(x) \equiv \frac{1}{2}(x,Bx) - W(x) \qquad (3.8)$$

considered for $x \in D = \mathcal{H}$ has $x = 0$ as the unique stationary point, in the sense that the Fréchet derivative exists (for this is actually sufficient that $\int |\alpha|\ d|\mu|(\alpha) < +\infty$), satisfies $d\Phi(x) = 0$ (this follows simply from

$$d\Phi(x) = Bx - dW(x)$$

and the assumption that $dW(0) = 0$).

Moreover under the assumption

$$\int_D |B^{-1}\alpha|\ |\alpha|\ d|\mu|\ (\alpha) < \frac{1}{a^2} \quad \text{for some } \partial \qquad (3.9)$$

which is satisfied when (3.6) holds we get that the mapping

$$x \to B^{-1}dW(x)$$

from D(B) into $\mathcal{H}$ is a contraction. By the contraction mapping principle we then get that the equation

$$Bx = dW(x)$$

has the only solution x = 0. In fact computing $d^2\Phi(x)$ (which exists already under the assumption $\int |\alpha|^2 \, d|\mu|(\alpha) < +\infty$) we see that, if (3.9) is satisfied, that x = 0 is a regular stationary point of $\Phi(x)$ in the sense that $d^2\Phi(0)\, y \neq 0$ for all $y \in D(B)$ $y \neq 0$ (in fact one has even $d^2\Phi(x)\, y \neq 0$ for all $y \in D(B)$ $y \neq 0$ and all $x \in \mathcal{H}$).

Thus we have the following

Theorem 3.2 [3]

Under the same assumptions as in Theorem 3.2 and in addition the assumption that B^{-1} is a bounded linear operator on $\mathcal{H}$, then we have $N(\gamma)^2 = |\gamma|^2$ $c = \|B^{-1}\|$ in Theorem 3.1. Moreover the phase function $\Phi(x) = \frac{1}{2}(x,Bx) - W(x)$ has zero as unique stationary point and this point is regular in the sense that $d^2\Phi(0)y \neq 0$ for all $y \in D(B)$ $y \neq 0$. Zero is the fixed point of the mapping $x \to B^{-1}dW(x)$.

Remark

In this case we have

$$I^\Delta(\hbar) = \sum_{m=0}^{\infty} \hbar^m \left(\frac{i}{2}\right)^m \sum_{n=0}^{\infty} \left(\frac{1}{2}\right)^n \frac{1}{n!\,(n+m)!}$$

$$\left[\left(\sum_{j=1}^{n} \nabla_{x_j} + \nabla_y ,\; B^{-1}\left(\sum_{k=1}^{n} \nabla_{x_k} + \nabla_y \right)\right)^{n+m} W(x_1)\dots W(x_n) g(y) \right]$$

where the value of [ ] is to be taken at the critical point. Then all the terms in the asymptotic expansion of $I_\Delta(\hbar)$ are given by the derivatives of W and g at the critical point.

Under the assumption of Theorem 3.2 we have that there is one and only one stationary point of the phase function

$$\Phi(x) = \frac{1}{2}(x,Bx) - W(x)$$

this point is $\neq 0$ whenever $dW(0) \neq 0$. This stationary point is, under the same assumptions, a regular one. We can prove [3] the following Theorem

Theorem 3.2

Let $W(x) = \int e^{i(\alpha,x)} d\mu(\alpha)$ where μ is a finite complex measure such that $\int e^{\sqrt{2}\,\| B^{-1} \|^{\frac{1}{2}} |\alpha|} d|\mu|(\alpha) < +1$ and $g(x) = \int e^{i(x,\beta)} d\nu(\beta)$ with $\int e^{\sqrt{2}\,\| B^{-1} \|^{\frac{1}{2}} |\beta|} d|\nu|(\beta) < +\infty$. Then there is a unique point $a \in \mathcal{H}$ such that $B\,a = dW(a)$. Let

$$I(h) = \tilde{\int} e^{\frac{i}{2h}(x,Bx)} e^{-\frac{i}{h}W(x)} g(x)\, dx.$$

Then $I(h)$ is analytic in $\Gamma = \{h \in \mathbb{C} \quad \operatorname{Im} h\, B^{-1} \leq 0 \quad h \neq 0\}$

$$I^{*}(h) = e^{-\frac{i}{h}[\frac{1}{2}(a,Ba) - W(a)]} I(h)$$

is a C^{∞}-function of h on the real axis. Moreover its value at zero is given by

$$I^{*}(0) = |\mathbb{1} - B^{-1}d^{2}W(a)|^{-1/2} g(a)$$

where $|\mathbb{1} - B^{-1}d^{2}W(a)|$ is the Fredholm determinant of the operator $\mathbb{1} - B^{-1}d^{2}W(a)$.

As application of the general theory developed in [3] we study in [4] Schrödinger operators of the form $H = -\frac{\hbar^{2}}{2}\Delta + \frac{1}{2}(x,A^{2}x)d + V(x)$, where A^{2} is a strictly positive symmetric dxd matrix and $V \in F(\mathbb{R}^{d})$ is a continuous bounded potential. We prove that the theta-function $\theta(t) = \operatorname{Tr} e^{\frac{itH}{\hbar}}$, the response function $g_{s}(\lambda) = \operatorname{Tr}(\lambda - H)^{-s}$, the zeta-function $\zeta(s) = \Sigma \frac{1}{\lambda_{n}^{s}}$ as well as the number

$N(\lambda)$ of eigenvalues λ_n of H which are $\leq \hbar \lambda$ are well defined. More precisely we give formulae for all these quantities in terms of Fresnel integrals relative to a quadratic form ("Feynman path integrals") over a Hilbert space of closed paths and we give their asymptotic behaviour as $\hbar \to 0$ by the methods developed in [3] in terms of classical periodic orbits. This then yields a connection between quantum mechanical spectrum and the classical mechanical periodic motions. These results give in particular a natural extension of those obtained in the literature for compact manifolds (Duistermaat, Guillemin, A. Huber, McKean, Colin de Verdière, Charazin, Seeley, Sjöstrand...) to the case of Schrödinger operators on $\mathbb{R}^d$.

References

[1] S. Albeverio, R. Høegh-Krohn: "Mathematical theory of Feynman path integrals", Lecture Notes in Math. 523. Berlin-Heidelberg-New York: Springer (1976).

[2] S. Albeverio, R. Høegh-Krohn: "Oscillatory integrals and the method of stationary phase in infinitely many dimensions, with applications to the classical limit of quantum mechanics", Inventiones math. 40, 59-106 (1977).

[3] S. Albeverio, Ph. Blanchard, R. Høegh-Krohn: "Oscillatory integrals and the method of stationary phase in infinitely many dimensions II", Preprint 1980.

[4] S. Albeverio, Ph. Blanchard, R. Høegh-Krohn: "The Poisson formula and the ζ-function for the Schrödinger operators." Preprint 1980.

Scaling Limits of Generalized Random Processes*)

Volker Enss

Fakultät für Physik, Universität Bielefeld

D-4800 Bielefeld, Federal Republic of Germany

Abstract

In Dobrushin's setup of scale transformations with multiplicative renormalization for generalized random processes we study the short- and long-distance behaviour of various processes, in particular of Wick-polynomials of Gaussian fields and of $P(\Phi)_2$-models from constructive quantum field theory.

---

*) Talk presented at T. Hida's Workshop on White Noise Approach to Quantum Dynamics.

I. Introduction and Setup

We will reverse the title of this workshop and study some cases where quantum dynamics approach white noise (or other simple processes). By quantum dynamics we mean generalized random processes Ψ which map real test functions $\phi \in \mathcal{Z}$ into the measurable functions $\Psi(\phi)$ on some measure space, e.g. $(\mathcal{Z}', \mathcal{B}, d\eta)$ where $\mathcal{Z}'$ is the dual of $\mathcal{Z}$, $\mathcal{B}$ the σ-algebra generated by the cylinder sets, η a probability measure. Of primary interest for physics are not the specific representations themselves but only the joint probability distributions of finite collections of random variables $\Psi(\phi_i)$ or equivalently the characteristic functional

$$L[\phi] = \int \exp(i\Psi(\phi))d\eta. \tag{1}$$

Later we will pick the representing process for convenience. The processes we consider are stationary (= translation invariant) Gaussian processes

$$L[\phi] = \exp(-\frac{1}{2}\langle\phi,\phi\rangle), \tag{2}$$

Wick polynomials thereof, and the $P(\Phi)_2$-processes as constructed in Euclidean quantum field theory.

If the quadratic form $\langle\cdot,\cdot\rangle$ in (2) is

$$\langle\phi,\phi\rangle_w = \int d^\nu x \; |\phi(\vec{x})|^2 \tag{3}$$

the process is called white noise over ν-dimensional Euclidean space,

$$\langle\phi,\phi\rangle_m = \int d^\nu p \; (\vec{p}^2 + m^2)^{-1} \; |\tilde{\phi}(\vec{p})|^2 \tag{4}$$

gives rise to the ν-dimensional free Euclidean quantum field of mass $m \geq 0$. We will restrict ourselves mainly to the typical example of a two dimensional free field, our methods and results apply immediatly to a much wider class of processes.

For $0 < \lambda < \infty$ and a given function $b(\lambda) > 0$ the scale transformations on test functions are given by

$$\phi \to \phi_\lambda \; : \; \tilde{\phi}(\vec{k}) \to b(\lambda)\; \tilde{\phi}(\lambda\vec{k}) \; =: \; \tilde{\phi}_\lambda(\vec{k}) . \tag{5}$$

By duality they induce transformations in the set of processes

$$\Psi \to \Psi_\lambda : \; \Psi_\lambda(\phi) = \Psi(\phi_\lambda) . \tag{6}$$

In fact they are transformations of equivalence classes of random processes having the same characteristic functional. There the transformation is

$$L \to L_\lambda \; : \; L_\lambda[\phi] = L[\phi_\lambda] . \tag{7}$$

A subset of the scaling transformations are those where $b(\lambda)$ is a power:

$$b(\lambda) = \lambda^\kappa , \tag{8}$$

for each value of κ they form a group, the renormalization group (RG) transformations. White noise is an RG-fixed point for $\kappa = {}^{\nu}/_{2}$, the massless free Euclidean field for $\kappa = {}^{(\nu-2)}/_{2}$. RG-fixed points are also called "automodel processes".

The block-spin-formation of statistical mechanics can be seen to be a special case with $\lambda = 2^n$.

One might consider the limits

$\lambda \to 0$: *"infrared", "long distance", or "large scale" limit;*

$\lambda \to \infty$: *"short distance", "ultraviolet", or "high energy" limit.*

Convergence is given by the weak topology, i.e. the characteristic functionals converge pointwise to a characteristic functional.

In the sequel we will exclude the trivial limit $L[\phi] \equiv 1$ which occurs if $b(\lambda)$ is chosen small enough.

There arise two natural questions:

a) Does there exist a $b(\lambda)$ such that Ψ_λ converges to a nontrivial limit as $\lambda \to 0$ and/or $\lambda \to \infty$?

b) If the limit exists, is it unique and what are its properties?

The first question is hard in general, we will give some partial answers below, whereas the second has a satisfactory answer. The limit is unique up to a multiplicative constant: if the process converges for both $b(\lambda)$ and $b'(\lambda)$ then $b(\lambda)[b'(\lambda)]^{-1} \to$ const. Furthermore $b(\lambda)$ is regularly varying near 0 or ∞, i.e.

$$b(\lambda) = \lambda^{\kappa} b_1(\lambda) \tag{9}$$

for some κ, b_1 a slowly varying function [8]. The limit process is an RG fixed point with parameter κ. The values of κ may be different as $\lambda \to 0$ or ∞ if both limits exist (e.g. the massive free Euclidean field in any dimension, it converges to the massless field as $\lambda \to 0$ and to white noise as $\lambda \to \infty$).

This setup is due to Dobrushin [2], [9], his work has set the stage to combine the methods, results, and folk lore from probability theory, Euclidean field theory and statistical mechanics for the investigation of scaling behaviour in these areas.

Returning to question a) about convergence we start with a simple model: If the process is Gaussian (2), or more general: if the characteristic functional depends only on a quadratic form in ϕ:

$$L[\phi] = F(\langle\phi,\phi\rangle) \tag{10}$$

then it is sufficient to check whether

$$\langle\cdot,\cdot\rangle \to \langle\cdot,\cdot\rangle_\lambda, \quad \langle\phi,\phi\rangle_\lambda = \langle\phi_\lambda,\phi_\lambda\rangle \tag{11}$$

converges to a continuous nontrivial limit.

We can see here that the limit may depend on the test function space Z. Take for example the massive free field in $\nu = 2$ dimensions (4). On the real Schwartz space $Z = S(\mathbb{R}^2)$ with $b(\lambda) = [-2\pi \ln(\lambda m)]^{-1/2}$ we get as $\lambda \to 0$ (with $p = |\vec{p}|$):

$$\langle\phi,\phi\rangle_\lambda = [-2\pi \ln(\lambda m)]^{-1} \int d^2p \, (p^2 + (\lambda m)^2)^{-1} \, |\tilde{\phi}(\vec{p})|^2 \to |\tilde{\phi}(\vec{0})|^2. \tag{12}$$

This form annihilates a subspace, the reduced Schwartz space $S_r(\mathbb{R}^2)$ of test functions vanishing at the origin in momentum space. If one takes $Z = S_r(\mathbb{R}^2)$ instead with $b(\lambda) = 1$ then we obtain as $\lambda \to 0$ a richer structure:

$$\langle\phi,\phi\rangle_\lambda = \int d^2p \ (p^2+(\lambda m)^2)^{-1} \ |\tilde{\phi}(\vec{p})|^2 \qquad (13)$$

$$\to \int d^2p \ p^{-2} \ |\tilde{\phi}(\vec{p})|^2,$$

the massless free field. This is a special case of an "infrared singular" process: the covariance $\tilde{C}(\vec{p}) = p^{-2}$ is not locally integrable at the origin of momentum space.

If in the quadratic form

$$\langle\phi,\phi\rangle = \int d^\nu p \quad \tilde{C}(\vec{p}) \quad |\tilde{\phi}(\vec{p})|^2 \qquad (14)$$

the covariance $\tilde{C}(p) > 0$ for $p > 0$ is regularly varying with exponent a in $p = |\vec{p}|$ as $p \to \infty$ (0) then on a suitable test function space with $b^2(\lambda) = \lambda^\nu \ \tilde{C}(1/\lambda)^{-1}$

$$\langle\phi,\phi\rangle_\lambda \longrightarrow \int d^\nu p \quad p^a \ |\tilde{\phi}(\vec{p})|^2 \quad \text{as} \quad \lambda \to 0 \ (\infty). \qquad (15)$$

Now let $L_i[\phi]$ be a finite collection of characteristic functionals. If $b(\lambda)$ can be chosen such that all $L_{i,\lambda}$ converge (to possibly trivial limits) and at least one of the limits is nontrivial, then the product converges to a nontrivial limit. A similar statement is true for convex combinations.

For processes with independent values at each point as given in [5] the above remark on factorization reduces the problem to the study of the scaling behaviour of finite dimensional measures. We won't give further details here.

There are not too many more generalized random processes explicitly known. Our main results are about Wick powers (= multiple Ito integrals) of Gaussian processes and the models of Euclidean quantum field theory. We will turn first to the (local) Wick powers.

II. Limits of Wick Powers

a) A Simple Case

Let Φ be a stationary Gaussian process with covariance $\tilde{C}(\vec{p})$. $\tilde{\Phi}(d^\nu p)$ denotes the random variable-valued measure on the Borel sets of $\mathbb{R}^\nu$ such that ([5], Section III, 3.4)

$$\int \overline{\tilde{\phi}(\vec{p})} \quad \tilde{\Phi}(d^\nu p) = \Phi(\phi), \tag{16}$$

$$<\tilde{\Phi}(d^\nu k) \quad \overline{\tilde{\Phi}(d^\nu p)}> = \delta(\vec{p}-\vec{k}) \; \tilde{C}(\vec{p}) \quad d^\nu k \; d^\nu p. \tag{17}$$

If for suitable test functions the following limit exists as a random variable it defines the n-th Wick popwer $:\Phi^n:$ of the process Φ:

$$:\Phi^n: \; (\phi) = (2\pi)^{(1-n)\nu/2} \int \overline{\tilde{\phi}}(\vec{p}_1+\ldots+\vec{p}_n) :\tilde{\Phi}(d^\nu p_1)\ldots\tilde{\Phi}(d^\nu p_n):$$

$$= \lim_{k\to\infty} (2\pi)^{(1-n)\nu/2} \int \phi_k(\vec{p}_1,\ldots,\vec{p}_n) \; \tilde{\Phi}(d^\nu p_1)\ldots\tilde{\Phi}(d^\nu p_n), \tag{18}$$

where $\lim\limits_{k\to\infty} \phi_k(\vec{p}_1,\ldots,\vec{p}_n) = \overline{\tilde{\phi}}(\vec{p}_1+\ldots+\vec{p}_n)$ in $L^2(\mathbb{R}^{n\nu}, \Pi\tilde{C}(\vec{p}_i)d^\nu p_i)$ and $\phi_k(\vec{p}_1,\ldots,\vec{p}_n) = 0$ in an open neighbourhood of the points $\pm\vec{p}_i = \vec{p}_j$, $i \neq j$, $\forall$ k.

The probability distribution of $:\Phi^n:(\phi)$ and thus the characteristic functional depend only on the probability distributions of the process Φ, thus the Wick powers of the various representations equivalent to Φ lie all within one equivalence class of processes.

The limit exists iff for some $\ell \in \mathbb{N}$

$$(1 + |\vec{p}_1+\ldots+\vec{p}_n|)^{-\ell} \quad \tilde{C}(\vec{p}_1)d^\nu p_1\ldots \; \tilde{C}(\vec{p}_n) \; d^\nu p_n < \infty. \tag{19}$$

This integrability condition is fulfilled e.g. for the free Euclidean field in $\nu = 2$ dimensions with mass $m > 0$, any n, and for $\nu = 3$, $m \geq 0$, $n = 2$. It is violated for infrared singular fields like $\nu = 2$, $m = 0$ ($\tilde{C}(\vec{p})$ not locally integrable) and for white noise ($\tilde{C}(\vec{p}) \equiv 1$, it does not have sufficient decay at infinity).[1] Although a finite integral (19) remains finite for $0 < \lambda < \infty$ it may diverge as $\lambda \to 0$ or ∞. Thus we have to be aware of sudden changes of the type of the process in the scaling limit.

A case where the transition is smooth and the result as expected is the Wick square of the three-dimensional Euclidean massive field in the short distance limit $\lambda \to 0$.

Let Φ_o be any representation (e.g. the canonical one guaranteed by Minlos' theorem) of the massless Euclidean free field in 3 dimensions, i.e. the Gaussian process with correlation function

$$\langle \Phi_o(\phi)\ \Phi_o(\psi)\rangle = \int d^3p\ \ p^{-2}\ \ \bar{\tilde{\phi}}(\vec{p})\ \tilde{\psi}(\vec{p}). \tag{20}$$

The massive process Φ_m can then be represented conveniently as

$$\Phi_m\colon\ \phi \to \Phi_m(\phi) = \int \bar{\tilde{\phi}}(\vec{p})\ [p(p^2+m^2)^{-1/2}]\Phi_o(d^3p). \tag{21}$$

Applying the scaling transformations with $b(\lambda) = \sqrt{\lambda}$: $\Phi_m \to \Phi_{m,\lambda}$ we will get in general new processes, but the process $\Phi_{m,\lambda}$ is in the same equivalence class of representations (has the same characteristic functional) as the process $\Phi_{\lambda m}$ for any $m \geq 0$. Thus weak convergence of $\Phi_{o,\lambda}$ and of $:\Phi^2_{o,\lambda}:$ are trivially true because the characteristic functionals do not depend on λ at all.

With $b(\lambda) = \lambda$ we obtain

$$:\Phi^2_m:_\lambda = :\Phi^2_{m,\lambda}: \ . \tag{22}$$

It would be sufficient to show that $:\Phi^2_{m,\lambda}:(\phi) \to :\Phi^2_{o,\lambda}:(\phi)$ in probability as $\lambda \to 0$ but we will show the stronger result

$$\lim_{\lambda\to 0}\ ||\ :\Phi^2_{m,\lambda}:(\phi) - :\Phi^2_{o,\lambda}:(\phi)||_2 = 0:$$

$$\tag{23}$$

$$||\ :\Phi^2_{m,\lambda}:(\phi)\ -:\Phi^2_{o,\lambda}:(\phi)\ ||^2_2$$

$$= 2\int|\tilde{\phi}(\vec{q}_1+\vec{q}_2)|^2\ \{q_1q_2[(q_1^2+\lambda^2m^2)(q_2^2+\lambda^2m^2)]^{-1/2} - 1\}^2 \quad \times$$

---
[1] Recently T. Hida [7] gave a generalization of these notions which allows to define Wick polynomials of Brownian motion.

$$\times \quad \lambda(q_1/\lambda)^{-2}\, d^3(q_1/\lambda) \qquad \lambda(q_2/\lambda)^{-2}\, d^3(q_2/\lambda) \to 0$$

because the curly brackets tend to 0 as $\lambda \to 0$ and the remainder is a bounded measure on $\mathbb{R}^6$ independent of λ.

Thus the massive field converges to the massless field, both admit formation of the Wick square, then the scaling limit of the Wick square of the massive field converges to the Wick-square of the massless field. We will find a different behaviour if the limit is infrared singular or white noise. Nevertheless our method of proof applies there as well: We approximate the scaled random variables by another sequence in L^2-norm. The approximating sequence does not converge in L^2 but its probability distribution is constant or has a simple (Gaussian) limit. Using L^2 allows to estimate test functions instead of random variables which is simpler.

b) The Infrared Singular Case

As a typical example we consider the massive free Euclidean field in $\nu = 2$ dimensions (4) which converges in the short distance limit $\lambda \to 0$ to the infrared singular massless field. The latter is defined only on test functions $\phi \in S_r$, i.e. $\tilde{\phi}(\vec{o}) = 0$. For this field we studied above two limits, (12) and (13) which both play a role here. Similarly we will study the limit of the Wick powers on $S(\mathbb{R}^2)$ and $S_r(\mathbb{R}^2)$. To simplify notation we omit λ and study $m \to 0$. Define

$$[f(m,\delta)]^2 = \int_{p\leq\delta} (p^2+m^2)^{-1}\, d^2p \sim -\ln m, \tag{24}$$

$$\tilde{g}_{m,\delta}(\vec{p}) = [f(m,\delta)]^{-1}\, p(p^2+m^2)^{-1/2}\, \Theta(\delta-p). \tag{25}$$

Then with the representation analogous to (21)

$$\Phi_o(g_{m,\delta}) = [f(m,\delta)]^{-1} \int_{p\leq\delta} \tilde{\Phi}_m(d^2p) \tag{26}$$

are centered normalized Gaussian random variables for any $m,\delta>0$, their probability distribution is

$$d\mu_1(y) = (2\pi)^{-1/2} \exp(-\tfrac{1}{2}\, y^2)\, dy. \tag{27}$$

Let $He_k(x)$ be the Hermite polynomial where the k-th power has coefficient 1. The Wick powers of the random variables $\Phi_o(g_{m,\delta})$

$$:\Phi_o(g_{m,\delta})^k : = He_k(\Phi_o(g_{m,\delta})) \tag{28}$$

have the probability distributions μ_k independent of $m,\delta > 0$, which are the probability distributions of $He_k(x)$ w.r.t. the measure μ_1.

One easily sees that for any $\delta > 0$

$$\lim_{m\to 0} \{[f(m,\delta)]^{-1} - [f(m,1)]^{-1}\} \int_{p\leq\delta} \tilde{\Phi}_m(d^2p) = 0 \tag{29}$$

in L^2, and also in any L^p, $p < \infty$, because it is Gaussian. Now we are prepared to treat the limit on $S(\mathbb{R}^2)$ with $b(m) = [f(m,1)]^{-n}$ for the n-th Wick power (as $\phi(\vec{x})$ is real so is $\tilde{\phi}(\vec{0})$).

$$(2\pi)^{n-1} [f(m,1)]^{-n} : \Phi_m^n : (\phi)$$

$$= [f(m,1)]^{-n} \int \bar{\tilde{\phi}}(\vec{p}_1+\ldots+\vec{p}_n) : \Phi_m(d^2p_1)\ldots\Phi_m(d^2p_n) :$$

$$= \tilde{\phi}(\vec{0}) : \{[f(m,1)]^{-1} \int_{p\leq\delta} \tilde{\Phi}_m(d^2p)\}^n : \tag{30}$$

$$+ \int_{p_i\leq\delta} \{\bar{\tilde{\phi}}(\vec{p}_1+\ldots+\vec{p}_n) - \tilde{\phi}(\vec{0})\}[f(m,1)]^{-n} : \tilde{\Phi}_m(d^2p_1)\ldots\tilde{\Phi}_m(d^2p_n):$$

$$+ [f(m,1)]^{-1} \int_{\bigcup_j p_j\geq\delta} \bar{\tilde{\phi}}(\vec{p}_1+\ldots+\vec{p}_n)[f(m,1)]^{1-n}:\tilde{\Phi}_m(d^2p_1)\ldots\tilde{\Phi}_m(d^2p_n):$$

In the second term on the r.h.s. of (30) the curly bracket tends to 0 as $\delta \to 0$ whereas the remainder is uniformly bounded in $\delta \leq 1$ and $m > 0$. In the third term the integral is bounded for any $\delta > 0$ uniformly in $m > 0$ because at least one of the p_i's is bigger then δ, thus the integral over p_i remains finite as $m \to 0$ even without an f (the detailed estimate is analogous to Lemma 6.3. in [3]). The factor $[f(m,1)]^{-1}$ in front of the integral annihilates this contribution as $m \to 0$ for any $\delta > 0$. The surviving term is thus the first one, by (29) it can be approximated in L^2 as

$$\lim_{m\to 0} \left\| (2\pi)^{n-1} [f(m,1)]^{-n} : \Phi_m^n : (\phi) - \tilde{\phi}(\vec{0}) : \Phi_o(g_{m,\delta})^n : \right\|_2 \leq \varepsilon(\delta), \tag{31}$$

and $\varepsilon(\delta) \to 0$ as $\delta \to 0$. The characteristic functional of the approximating process does not depend on m and δ:

$$L[\phi] = \int \exp\,(i\ \tilde{\phi}(\vec{0})\ x)\ d\mu_n(x). \tag{32}$$

Obviously the result is a characteristic functional.

Starting with Wick polynomials instead of Wick powers yields a dense set of probability measures instead of μ_n in (30). $L[\phi] = \int \exp\,(i\ \ell(\phi)\ x)\ d\nu(x)$ is the general case of a process which depends on ϕ through a continuous real linear functional only.

We would have obtained the same result if we considered a sequence of Gaussian processes admitting n-th Wick powers, which converges in L^2 on a reduced test function space to an infrared singular process, no matter which one.

If we study the limit on the reduced test function space $S_r(\mathbb{R}^2)$ the limits will have a richer structure which depends on the massless field, but it is still independent of the approximating sequence. It is sufficient to find the limit on the dense set of test functions ϕ with $\tilde{\phi}(\vec{p}) = 0$ if $p \leq n\delta_o$ for some $\delta_o > 0$. This time one of the factors remains finite because of the zero of $\tilde{\phi}$, thus we will show convergence of the following sequence of processes:

$$\Psi_m^n(\phi) = \frac{1}{n}\left[\frac{2\pi}{f(m,1)}\right]^{n-1} : \Phi_m^n : (\phi)$$

$$= \frac{1}{n}[f(m,1)]^{1-n} \int_{p_i \leq \delta} \tilde{\tilde{\phi}}(\vec{p}_1+\ldots+\vec{p}_n) : \tilde{\Phi}_m(d^2p_1)\ldots\tilde{\Phi}_m(d^2p_n) :$$

$$+ : \Phi_m(\phi)\{[f(m,1)]^{-1} \int_{p\leq\delta} \tilde{\Phi}(d^2p)\}^{n-1} : \tag{33}$$

$$+ \frac{1}{n}\sum_{i=1}^{n} \int_{\substack{p_i \geq \delta \\ p_j \leq \delta,\, j\neq i}} \{\tilde{\tilde{\phi}}(\vec{p}_1+..+\vec{p}_n) - \tilde{\tilde{\phi}}(\vec{p}_i)\} : \tilde{\Phi}_m(d^2p_i) \times$$

$$\times \prod_{j \neq i} [f(m,1)]^{-1} \tilde{\Phi}_m(d^2p_j):$$

$$+ \frac{1}{n}[f(m,1)]^{-1} \int_{\bigcup_{i \neq j} p_i \geq \delta, p_j \geq \delta} \tilde{\tilde{\phi}}(\vec{p}_1+..+\vec{p}_n)[f(m,1)]^{2-n}:\tilde{\Phi}_m(d^2p_1)..\tilde{\Phi}_m(d^2p_n):$$

This time the first term which was leading in the previous example vanishes identically in m for any $\delta \leq \delta_o$. In the third term the curly bracket vanishes for $p_i \leq \delta_o$ and tends to 0 as $\delta \to 0$; the remainder is bounded uniformly in $m > 0$, $0 < \delta \leq \delta_o$. The integral in the fourth summand is finite for any $\delta > 0$ uniform in m, with the factor in front of it it tends to zero as $m \to 0$. The leading second term can be approximated using (29), the error vanishes in L^2 as $m \to 0$, by

$$: \Phi_o(\phi) \; \Phi_o(g_{m,\delta})^{n-1} :$$

Finally $\Phi_o(\phi)$ can be taken outside the Wick ordering because it is independent from $\Phi_o(g_{m,\delta})$. Thus

$$\lim_{m\to 0} \left\| \Psi_m^n(\phi) - \Phi_o(\phi) : \Phi_o(g_{m,\delta})^{n-1}: \right\| < \varepsilon(\delta), \tag{34}$$

$\varepsilon(\delta) \to 0$ as $\delta \to 0$. The characteristic functional of the approximating process is easy to compute because of the independence of the factors. It is independent of m and $\delta \leq \delta_o$:

$$L[\phi] = \int \exp\{-\frac{1}{2} y^2 <\phi,\phi>\} \; d\mu_{n-1}(y), \tag{35}$$

where $<\phi,\psi>$ is the two point function of the process Φ_o. As a convex combination of characteristic functionals it is evidently a characteristic functional. Again Wick polynomials yield a dense set among the processes

$$L[\phi] = \int \exp\{-\frac{1}{2} y^2 <\phi,\phi>\} \; d\nu(y), \tag{36}$$

ν a probability measure. The 2N-point functions are expressed through the 2-point function but with different coefficients than in the Gaussian case. A physicist would call that a degenerate ground state. A special case of this limit had been obtained previously by Karwowski and Streit [10]. For further details, more general processes and full proofs see [3].

To show convergence of the sequence of processes we first had to guess a renormalization function $b(\lambda)$. In all cases treated here we have chosen it such that the two point function converges, this can easily be calculated. Certainly this works only if the sequence and the limit process all have finite second moments. In the general case (e.g. stable distributions) one will need more fantasy.

c) White Noise

In the previous section we have seen that the limiting process is concentrated near the origin or close to the coordinate axes in $\nu\cdot n$-dimensional momentum space. This time the opposite is true, the process is almost evenly spread over a slice around a hyperplane of momentum space such that we can sum up the pieces using the Central Limit Theorem.

We consider Wick powers of Gaussian processes where the latter tend to white noise in the long distance ($\lambda \to \infty$) scaling limit, e.g. the massive free Euclidean fields.

If the n-th Wick power exists then by (19) the renormalization constants

$$b^{-2}(\lambda,\vec{q}) := (2\pi)^{(1-n)\nu} n! \int \tilde{C}(\vec{p}_1/\lambda)\, d^\nu p_1 \ldots \tilde{C}(\vec{p}_n/\lambda)\, d^\nu p_n\, \delta(\vec{p}_1+\ldots+\vec{p}_n-\vec{q}) \tag{37}$$

exist and they will diverge as $\lambda^{\nu(n-1)}$ for $\lambda \to \infty$.

Let $\tilde{\Phi}(d^\nu p)$ be the random variable valued measure of the white noise process in some representation. The Gaussian field Φ_1 to start with can be represented conveniently as

$$\tilde{\Phi}_1(d^\nu p) = h(\vec{p})\, \tilde{\Phi}(d^\nu p) \tag{38}$$

where $h^2(\vec{p}) = \tilde{C}(\vec{p}) = (p^2+m^2)^{-1}$. Let the scaled field Φ_λ be:

$$\tilde{\Phi}_\lambda(d^\nu p) = h(\vec{p}/\lambda)\, \tilde{\Phi}(d^\nu p). \tag{39}$$

We shall consider the sequence of fields

$$\Psi_\lambda(\phi) := b(\lambda) \int : \tilde{\Phi}_\lambda(d^\nu p_1)\ldots\tilde{\Phi}_\lambda(d^\nu p_n) : \bar{\tilde{\phi}}(\vec{p}_1+\ldots+\vec{p}_n), \tag{40}$$

$b(\lambda) = b(\lambda,\vec{0})$ as given in (37).

For convenience we take the test functions ϕ with $\tilde{\phi}(\vec{p}) \in \mathcal{D}(\mathbb{R}^\nu)$, later one can extend the result by continuity uniform in λ. It is easy to check that for any bounded set $B \subset \mathbb{R}^\nu$

$$\lim_{\lambda\to\infty} \sup_{\vec{q}\in B} \left| b^2(\lambda,\vec{0})\, b^{-2}(\lambda,\vec{q}) - 1 \right| = 0, \tag{41}$$

this implies

$$\| \Psi_\lambda(\phi) \|_2 \longrightarrow \| \phi \|_2 \text{ as } \lambda \to \infty. \tag{42}$$

Thus the variance behaves as it should. It remains to show that the random variables become Gaussian. We will indicate the proof for n = 2, for higher powers it requires some more effort. The general results and full proofs will be given in [4].

A similar calculation like that leading to (41) yields

$$\lim_{\lambda\to\infty} b(\lambda) \left\| h(p_1/\lambda)[h(p_2/\lambda) - h(p_1/\lambda)]\, \bar{\tilde{\phi}}(\vec{p}_1+\vec{p}_2) \right\|_2 = 0. \tag{43}$$

Denote $\mathbb{Z}_+^\nu = \{i \in \mathbb{Z}^\nu \mid i_1 \geq 0\}$,

$$\Delta_i = \{\vec{p} \in \mathbb{R}^\nu \mid i_j < p_j \leq i_j+1,\ i \in \mathbb{Z}_+^\nu\}. \tag{44}$$

The step function approximation

$$\bar{h}(\vec{p}/\lambda) = \bar{h}(-\vec{p}/\lambda) = h(i/\lambda) \quad \text{if } \vec{p} \in \pm\Delta_i \tag{45}$$

is asymptotically good:

$$\lim_{\lambda\to\infty} b(\lambda) \left\| \bar{h}^2(\vec{p}/\lambda) - h^2(\vec{p}/\lambda) \right\|_2 = 0. \tag{46}$$

Define for $i \in \mathbb{Z}_+^\nu$:

$$X_i := \iint_{\vec{p}_1\in\Delta_i\cup-\Delta_i} \bar{\tilde{\phi}}(\vec{p}_1+\vec{p}_2) : \tilde{\Phi}(d^\nu p_1)\, \tilde{\Phi}(d^\nu p_2): \quad . \tag{47}$$

They are real, centered, identically distributed, pairwise orthogonal (in L^2) random variables, $X_i \in L^p \;\; \forall\, p < \infty$. Furthermore X_i and X_j are independent if $|i-j| > m$ for some $m < \infty$ (which depends on the size of supp $\tilde{\phi}$).

From (43) and (46) we know that

$$\lim_{\lambda\to\infty} \left\| \Psi_\lambda(\phi) - b(\lambda) \sum_{i\in\mathbb{Z}_+^\nu} h^2(i/\lambda)\, X_i \right\|_2 = 0. \tag{48}$$

Thus we can approximate $\Psi_\lambda(\phi)$ by a weighted sum of weakly dependent random variables, the latter can be handled using the following Lemma which is a variant of the Central Limit Theorem:

Lemma: Let $X_i \in L^p$, $p \geq 3$, $i \in \mathbb{Z}^\nu$ be a set of m-dependent, centered, identically distributed, orthonormal (in L^2) random variables; let $f \in L^2(\mathbb{R}^\nu)$ be a bounded, continuous, nonnegative function. Then the probability distribution of

$$S_n = (n^{\nu/2} \, \|f\|_2)^{-1} \sum_{i\in\mathbb{Z}^\nu} f(i/n)\, X_i \tag{49}$$

converges weakly to the normal distribution.

Various proofs of the CLT can be modified to show this Lemma (cf. chapter 7 in [1]). f could have finitely many points of discontinuity, for $\nu = 1$, f the characteristic function of the unit interval, the Lemma coincides with the usual CLT. In general the sharp cut-off at n is replaced by a smooth one.

This finishes the sketch of our proof that the Wick square of massive free fields converges weakly as a process to white noise.

III. Long Distance Limit of $P(\Phi)_2$-Models

In this section we show by a different method that some models from constructive quantum field theory converge to white noise in the long distance scaling limit (see [11] as a general reference for these models).

The n-point Schwinger functions S_n are the distributions which determine the n-th moment of the random variables $\Phi(\phi)$:

$$E[(\Phi(\phi))^n] = \int d^\nu x_1 \ldots d^\nu x_n \quad S_n(\vec{x}_1,\ldots,\vec{x}_n)\ \phi(\vec{x}_1)\ldots\phi(\vec{x}_n). \quad (50)$$

The truncated Schwinger functions S_n^T are defined recursively by

$$S_n(\vec{x}_1,\ldots,\vec{x}_n) = \Sigma\, S^T_{r_1}\ (\vec{x}_{\ell_1,1},\ldots,x_{\ell_1,r_1}) \times \ldots$$
$$\times\, S^T_{r_s}\ (\vec{x}_{\ell_s,1},\ldots x_{\ell_s,r_s}) \quad (51)$$

where the sum extends over all splittings of $(\vec{x}_1,\ldots,\vec{x}_n)$ into subsets, $r_1+\ldots+r_s = n$.

S_n and S_n^T are invariant under joint translations of all arguments.

Below we have to assume the following cluster property:

$$S_n^T(\vec{0},\vec{x}_2,\ldots,\vec{x}_n) \in L^1(\mathbb{R}^{\nu(n-1)},\ \prod_2^n d^\nu x_i). \quad (52)$$

Due to Euclidean invariance the 2-point Schwinger function $S_2(\vec{x}_1,\vec{x}_2) = S_2(\vec{x}_1-\vec{x}_2)$ can be represented as the Fourier transform of

$$\tilde{S}_2(\vec{p}) = \int (p^2+\sigma^2)^{-1}\, d\rho(\sigma^2).$$

Our second assumptions will be on the measure ρ:

$$\text{supp } d\rho(\sigma^2) \subset [m^2,\infty)\ ,\ m > 0\ ; \quad (53)$$

$$\int \sigma^{-2}\, d\rho(\sigma^2) < \infty\ . \quad (54)$$

Our assumptions should be fulfilled for most non-critical, massive fields. They are known to hold if the cluster expansion converges: for weakly coupled $P(\Phi)_2$-models and for very strong coupling [12]. In the latter case, e.g. for $\lambda\Phi_2^4$, there are two pure phases, then the assumptions presumably hold in each single phase separately, but not for mixtures. For these fields we show that they converge to white noise, which is what one would have expected anyway: all these models have nonzero mass, i.e. finite correlation length. From very far away they should look like a Gaussian field with zero correlation length, i.e. independent values at each point : white noise.

Theorem: Let a Euclidean quantum field theory obey assumptions (52), (53), and (54), then it converges in the long distance scaling limit to white noise if $b(\lambda)$ is chosen as

$$b(\lambda) = \lambda^{\nu/2} \{\int \sigma^{-2} \, d\rho(\sigma^2)\}^{-\frac{1}{2}} . \tag{55}$$

Proof: We will show that the interacting fields can be approximated by a sequence of generalized free fields and that the latter tend to white noise. First observe

$$\langle\phi,\phi\rangle_\lambda = S_2(\phi_\lambda \otimes \phi_\lambda) =$$

$$= b^2(\lambda) \int d\rho(\sigma^2) \int d^\nu p (p^2+\sigma^2)^{-1} |\tilde{\phi}(\lambda \vec{p})|^2$$

$$= \{\int \sigma^{-2} \, d\rho(\sigma^2)\}^{-1} \int d\rho(\sigma^2) \int d^\nu p \, (p^2/\lambda^2 + \sigma^2)^{-1} |\tilde{\phi}(\vec{p})|^2$$

$$\longrightarrow \| \phi \|_2^2 \quad \text{as } \lambda \to \infty.$$

Thus the generalized free field (i.e. the Gaussian process) with the same 2-point function $\langle\phi,\phi\rangle_\lambda$ as the interacting field converges to white noise. It remains to show that the deviation of the interacting field from the Gaussian counterpart disappears as $\lambda \to \infty$. This can be easily done using the truncated functions because the Gaussian fields have $S_n^T = 0$ for $n \geq 3$.

Now for any $f \in L^1(\mathbb{R}^k, d^k u)$

$$\lim_{\lambda\to\infty} \lambda^k \, |f(\lambda \vec{u})| \, d^k u = \delta(\vec{u}) \, d^k u \cdot \|f\|_1$$

as a linear functional on bounded continuous functions. Thus by translation invariance and (52)

$$S_n^T(\phi_\lambda^{\otimes n}) = b^n(\lambda) \lambda^{-n\nu} \int d^\nu x_1 \ldots d^\nu x_n \; S_n^T(\vec{x}_1,\ldots,\vec{x}_n) \; \phi(\vec{x}_1/\lambda) \ldots \phi(\vec{x}_n/\lambda)$$

$$\leq \text{const.} \; \lambda^{-\nu(n-2)/2} \int d^\nu y_1 \ldots d^\nu y_n \left| \phi(\vec{y}_1) \ldots \phi(\vec{y}_n) \right| \quad \times$$

$$\times \quad \lambda^{\nu(n-1)} \left| S_n^T(\vec{0}, \lambda(\vec{y}_2 - \vec{y}_1), \ldots, \lambda(\vec{y}_n - \vec{y}_1)) \right|$$

$$\leq \text{const. } \lambda^{-\nu(n-2)/2} \int d^\nu y \; |\phi(\vec{y})|^n + \varepsilon(\lambda)$$

$$\longrightarrow 0 \text{ as } \lambda \to \infty \text{ for } n \geq 3.$$

This estimate shows that the moments of the random variables $\Phi(\phi_\lambda)$ converge to the Gaussian moments. But the Gaussian probability distribution is uniquely determined by its moments, then convergence of the moments implies weak convergence of the probability distribution of $\Phi(\phi_\lambda)$ to the corresponding one of the generalized free field (Theorems 4.5.5 and 4.4.2 in [1]). This finishes the proof of the theorem.

For a different approach to this problem see [6].

References

[1] Chung, K.L.: A course in probability theory, New York, Academic Press 1974.

[2] Dobrushin, R.L.: a) Automodel generalized random fields and their renorm-group. In: Multicomponent Random Systems. Dekker, New York, to appear. b) Ann. Prob. 7, 1-28 (1979)

[3] Enss, V.: Rep. Math. Phys. 13, 87-99 (1978).

[4] Enss, V.: Random processes converging to white noise, in preparation.

[5] Gel'fand, I.M., Vilenkin, N.Ya.: Generalized functions Vol. IV, New York, Academic Press 1964.

[6] Haba, Z.: Scale limit and low momentum behaviour of Euclidean fields II., preprint Bielefeld BI-TP 78/25.

[7] Hida, T.: Proc. Japan Acad. 54, SerA, 55-58 (1978).

[8] Ibragimov I.A., Linnik, Yu.V.: Independent and stationary sequences of random variables, Groningen, Wolters-Noordhoff 1971.

[9] Jona-Lasinio, G.: in "New developments in quantum field theory and statistical mechanics - Cargèse 1976", M. Levy and P. Mitter ed., New York, Plenum 1977.

[10] Karwowski, W., Streit, L.: Rep. Math. Phys. 13, 1-12 (1978).

[11] Velo, G., Wightman, A. ed.: Constructive quantum field theory, Lecture Notes in Physics 25, Berlin, Springer 1973.

[12] Eckmann, J.P., Magnen, J., Sénéor, R.: Commun. Math. Phys. 39, 251-271 (1975); Glimm, J., Jaffe, A., Spencer, T.: Ann. Phys. (N.Y.) 101, 610-630, 631-669 (1976); Fröhlich, J.: Adv. Math. 23, 119-180 (1977).

Renormalization Group Analysis of Some Highly Bifurcated Families

P. Collet, Harvard University, Cambridge, Mass., USA

J.-P. Eckmann, University of Geneva, Switzerland

O.E. Lanford, University of California, Berkeley, USA

Abstract

Using a renormalization group technique, we investigate the sequence of bifurcations occurring in a one parameter family of maps on an interval. This analysis provides a proof of the conjectures of Feigenbaum [F].

I. The Work of Feigenbaum

We first briefly describe the numerical results of Feigenbaum [F], using notations which are slightly different from his.

Let ψ be a function belonging to $C^2[-1,1]$ which is even, concave, has its maximum at $x = 0$ and is such that $\psi(0) = 1$ and $\psi(1) \geq -1$. For each value of a parameter μ in $[0,1]$, we consider a transformation T_μ from the interval $[-1,1]$ into itself, given by

$$x \longrightarrow T_\mu(x) = \psi(\mu x).$$

(Feigenbaum considered a transformation given by $x \to \mu\psi(x)$ which is of course conjugated to T_μ).

Following the ideas of Metropolis, Stein, and Stein [MSS], one can look for the values of μ for which T_μ has a stable periodic orbit. Starting at $\mu = 0$, one finds a sequence of contiguous intervals $[0,b_1]$, $[b_1,b_2]$, such that if μ belongs to $]b_n,b_{n+1}[$, T_μ has a periodic orbit of period 2^n. A simple way to localize these intervals is to consider the value of μ for which $x = 0$ (the point where ψ is maximum) belongs to a stable periodic orbit. According to [MSS] this value exists for every n and is unique. This result will also be a consequence of our results. Such a value of μ will be called a superstable value, and denoted by μ_n if the period is 2^n. We have therefore the following equation

$$T_{\mu_n}^{2^n}(0) = 0, \quad \text{and } T_{\mu_n}^{j}(0) \neq 0 \quad \text{if} \quad 0 < j < 2^n.$$

We can now state the first conjecture of Feigenbaum

<u>Conjecture (1)</u>

<u>For every ψ with a quadratic maximum, one has</u>

$$\operatorname{Log}(\mu_\infty - \mu_n)/n \longrightarrow -\operatorname{Log}\delta ,$$

where δ *is independent of* ψ.

This conjecture generalizes in the following way

Conjecture (1')

If ψ *behaves like* $1-a|x|^{1+\varepsilon}$ *around its maximum for some positive number* ε, *then*

$$\mathrm{Log}(\mu_\infty - \mu_n)/n \longrightarrow -\mathrm{Log}\ \delta(\varepsilon)$$

where $\delta(\varepsilon)$ *depends on* ε *but not on the particular function considered*.

This conjecture seems to be true in much more general cases. One can for example verify it numerically for the Hénon map [D.G.P.], and for some hydrodynamical models [B.F.].

The second conjecture is obtained by comparing the graphs of $T^{2^n}_{\mu_{n+1}}$ and $T^{2^{n+1}}_{\mu_{n+1}}$ around $x = 0$. These functions look very similar if one performs a scaling transformation and a reflection with respect to the x-axis. This remark is the origin of Feigenbaum's second conjecture.

Conjecture (2)

Let $\phi_n = |T^{2^{n-1}}_{\mu_n}(0)|$, *then*

$$(-1)^n\ \phi_n^{-1}\ T^{2^{n-1}}_{\mu_n}(\phi_n \cdot) \xrightarrow[n\to+\infty]{} \psi_*(\cdot)$$

where ψ_* *is only a function of* ε.

The third conjecture describes the asymptotic behavior of the sequence $(\phi_n)_{n\in\mathbb{N}}$.

Conjecture (3)

$$\frac{\phi_{n+1}}{\phi_n} \longrightarrow \lambda(\varepsilon)$$

where $\lambda(\varepsilon)$ *is only a function of* ε.

All these conjectures were numerically verified with great accuracy in [F], where the numbers $\delta(1)$ and $\lambda(1)$ for example are given with twelve digits precision.

II. Renormalization Group Analysis of the Problem [C.E.L.]

As we have already said, the graphs of $T^{2^n}_{\mu_n}$ and $T^{2^{n+1}}_{\mu_{n+1}}$ look very similar after a scaling and a symmetry with respect to the x-axis. From the relation

$$T^{2^{n+1}}_{\mu_{n+1}} = T^{2^n}_{\mu_{n+1}} \circ T^{2^n}_{\mu_{n+1}} ,$$

one is tempted to study the functions which are invariant under composition followed by scaling and change of sign. Namely, one can look for a function ϕ and a positive number λ such that

$$- \frac{1}{\lambda} \phi \circ \phi (\lambda x) = \phi(x) \quad \forall\ x \in [-1,1], \quad \text{and } \phi(o) = 1. \tag{1}$$

Theorem 1 For sufficiently small and positive λ, equation (1) has a solution ϕ_λ of the form $\phi_\lambda(x) = f_\lambda(|x|^{1+\varepsilon(\lambda)})$. This is the unique solution for which

i) $t \to f_\lambda(t)$ is in $C^2([0,1])$

ii) $f_\lambda(t) = 1 - (1+\lambda)t + O(\lambda^2)$ in the $C^2([0,1])$ topology

iii) $\varepsilon(\lambda) = -\lambda/(1+\text{Log}\lambda) + O(\lambda^2)$.

Moreover, f_λ extends to a function which is analytic in the disc $|t| < 2$, and $\lambda \to (f_\lambda, \varepsilon(\lambda))$ is real analytic in $[0,\varepsilon_o]$.

Remark The natural parameter is ε, however it is easier to state and prove theorem 1 using λ as a parameter. In any case, one can invert the function $\lambda \to \varepsilon(\lambda)$ on some interval $[0,\varepsilon_o]$ with ε_o strictly positive, and reexpress the conclusions of theorem 1 in terms of ε.

We now consider a transformation N given by

$$\psi \to N(\psi) , \text{ with } N(\psi)(x) = \frac{1}{\psi(1)} \psi \circ \psi \quad (-\psi(1)x).$$

Let (λ,ϕ_λ) be a solution of (1), putting $x = 0$ in (1) and using $\phi_\lambda(0) = 1$, one easily obtains

$$\phi_\lambda(1) = -\lambda.$$

Hence

$$N(\phi_\lambda) = \frac{1}{\phi_\lambda(1)} \phi_\lambda \circ \phi_\lambda(-\phi_\lambda(1)\cdot) = \phi_\lambda$$

since ϕ_λ is even.

From now on, λ will be fixed according to the hypothesis of theorem 1. Moreover, we shall consider the action of N on functions ϕ which are of the form $\phi(x) = f(|x|^{1+\varepsilon(\lambda)})$ where f is analytic. It will be more convenient to look at the equivalent transformation T given by

$$g \to T(g) \ , \text{ with}$$

$$T(g)(t) = \frac{1}{f_\lambda(1)+g(1)} (f_\lambda+g)([(f_\lambda+g)((-f_\lambda(1)-g(1))^{1+\varepsilon(\lambda)}t)]^{1+\varepsilon(\lambda)}) - f_\lambda(t).$$

In the following, T will be considered as a map in the Banach space E defined by

$$E = \{g| \ \ g \text{ is analytic in the disc } |z| < 2, \text{ and } g(0) = 0\}$$

and equipped with the sup norm.

Notice that from $f_\lambda(1) = -\lambda$, T is not defined on the manifold W_o given by

$$W_o = \{g|g(1) = \lambda\}.$$

We now investigate the spectrum of the tangent map of T at the fixed point g = 0. This tangent map will be denoted by A.

<u>Proposition 2</u> <u>In the Banach space</u> E, <u>we have</u>

$$Sp(A) = \{S\} + \Sigma$$

<u>where</u> $S = 2 + \varepsilon/\lambda + \mathcal{O}(\lambda)$ <u>is a simple eigenvalue</u>, <u>and</u> $\operatorname{Sup}_{z\in\Sigma} |z| = \mathcal{O}(\lambda)$.
<u>The eigenspace associated with</u> δ <u>is called</u> E_u <u>and is generated by an eigenvector</u> e_1 <u>given by</u>

$$e_1(t) = t(1+\mathcal{O}(\lambda)) \ .$$

<u>The spectral subspace associated with</u> Σ <u>is called</u> E_S <u>and is a supplement of</u> E_u.

We can now apply a general theorem [H.P.S.] to obtain stable and unstable manifolds W_S and W_u respectively. W_S is of codimension 1 and tangent to E_S at the origin, while W_u is of dimension 1 and tangent to E_u at the origin.

In order to apply the Renormalization Group ideas, we need a local C^1 linearization of T around its fixed point [C.E.]. This is not possible in general. However a weaker result will be enough for our purposes. We now use the spectral decomposition $E = E_u \oplus E_S$ and denote by A' the operator $A|_{E_S}$. $\mathcal{L}(E_S)$ will be the space of bounded linear operators in E_S.

<u>Theorem 3</u> <u>There is a</u> C^1 <u>diffeomorphism</u> S <u>of a neighborhood of zero in</u> E <u>such that</u>

$$STS^{-1} = t$$

<u>where</u> t <u>is given by</u>

$$t\binom{x}{y} = \binom{\delta x}{A'y+L(x,y)y}$$

<u>and</u> $L(\cdot)$ <u>is a</u> C^1 <u>function from</u> E <u>to</u> $\mathcal{L}(E_S)$ <u>such that</u>

$$\| L(x,y) \|_{\mathcal{L}(E_S)} \leq \mathcal{O}(1) \, \| (x,y) \| \quad .$$

We notice that $A' + L(x,y)$ is a contraction of E_S if $\|(x,y)\|$ is small enough. With our notations, the family of functions $x \to \psi(\mu x)$ is given by

$$t \to f_\lambda(\mu^{1+\varepsilon(\lambda)}t) - f_\lambda(t) + g(\mu^{1+\varepsilon(\lambda)}t) = g_\mu(t)$$

$\mu \to g_\mu$ is a C^1 curve in E, and it can be easily proven that this curve is transverse to the stable manifold for some value μ_∞ of the parameter. Consider now a function g on W_o. The associated ψ satisfies $\psi(0) = 1$ and $\psi(1) = 0$. W_o is therefore the space of representatives of transformations having superstable period 2. From this remark we deduce that μ_n satisfies the following relation:

$$N^{n-1}(\psi(\mu_n \cdot))(1) = o$$

which is equivalent to

$$T^{n-1}(g_{\mu_n}) \in W_o$$

or

$$S\, t^{n-1}\, S^{-1}(g_{\mu_n}) \in W_o \; .$$

$\mu \to S^{-1} g_\mu$ is again a C^1 curve in E which is transversal to E_S at the intersection point $\mu = \mu_\infty$.

Therefore we have

$$S^{-1} g_\mu = (\mu_\infty - \mu)(1+O\,(\mu-\mu_\infty))e_1 + \Omega(\mu)$$

with $\Omega(\mu)$ in E_S.

We now apply theorem 3 and obtain

$$t^{n-1}\, S^{-1} g_\mu = (\mu_\infty - \mu)\; \delta^{n-1}\; a(1+O(\mu-\mu_\infty))e_1 + \Omega_n(\mu)$$

with $||\Omega_n(\mu)|| \to o$ if $n \to \infty$ and if $(\mu_\infty-\mu)\delta^{n-1}$ remains small. Due to transversality, a is a non zero constant. Finally we obtain

$$S\, t^{n-1}\, S^{-1} g_\mu = (\mu_\infty - \mu)\delta^{n-1}\; a(1+O(\mu-\mu_\infty))e_1 + \sigma_n(\mu)$$

where $||\sigma_n(\mu)|| \to o$ if $n \to \infty$. Now $S\, t^{n-1}\, S^{-1} g_{\mu_n} \in W_o$ reads by definition

$$(S\; t^{n-1}\, S^{-1}\; g_{\mu_n})(1) = \lambda$$

which is

$$(\mu_\infty - \mu_n)\delta^{n-1}\; a(1+O(\mu_n-\mu_\infty))\; e_1(1) + \sigma_n(\mu_n)(1) = \lambda .$$

Using $a \neq o$ and $e_1(1) \neq o$, conjecture (1) is proven for those functions ψ which are of the form

$$\psi(x) = f_\lambda(|x|^{1+\varepsilon(\lambda)}) + g(|x|^{1+\varepsilon(\lambda)})$$

where λ is small (and hence $\varepsilon(\lambda)$ also) and g is a small analytic function. Moreover the number δ appearing in conjecture (1) is the unstable eigenvalue of A. From the definition of N, we derive

$$N^{n-1}(\psi(\mu\cdot)) = (-1)^n \phi_n^{-1} \psi^{2^{n-1}}(\phi_n \cdot)$$

with

$$\phi_n = \prod_{i=1}^{n-1} |N^i(\psi(\mu\cdot))(1)|.$$

From the definition of N, we have $N^{n-1}(\psi(\mu\cdot))(o) = 1$. Using the value of μ_n which we have already determined, and theorem 3, one can easily prove that

$$N^{n-1}(\psi(\mu_n\cdot)) \to \psi_*$$

with

$$\psi_*(x) = f_\lambda(|x|^{1+\varepsilon(\lambda)}) + g_*(|x|^{1+\varepsilon(\lambda)})$$

where g_* is the intersection of W_o and W_u.

This proves conjecture (2) if the hypothesis on ψ are the same as in our proof of conjecture (1).

The proof of (3) is now similarly easy. We have

$$N^i\ (\psi(\mu_n\cdot)(1) = \phi_\lambda(1) + (\mu_\infty-\mu_n)\delta^i\ e_1(1) + O(\mu_\infty-\mu_n)\delta^i$$

$$= (-\lambda)(1-\delta^{i-n} + o(\delta^{i-n})).$$

Therefore

$$\prod_{i=1}^{n-1} N^i\ (\psi(\mu\cdot))(1) = (-\lambda)^{n-1}\ s_n.$$

where $s_n \to s_\infty \neq o$ if $n \to +\infty$, which proves (3) under the same restrictions as in the proof of (1). Here again we have an interpretation of the number $\varepsilon\ \lambda$ appearing in the conjecture.

III. Properties Of the Transformation T_{μ_∞} [C.E.L.].

We briefly describe some properties of the transformation T_{μ_∞}, assuming that the function ψ satisfies the hypotheses which were used in the proof of the conjectures.

Theorem 4 The non wandering set $\tilde{\Omega}$ of T_{μ_∞} is given by

$$\tilde{\Omega} = \Omega \cup \Omega_1$$

where Ω_1 is countable, and Ω is a Cantor set [N].

Ω_1 is the set of all the unstable periodic points and of all their predecessors. Therefore we shall only investigate the ergodic properties of the map T_{μ_∞} on Ω.

Theorem 5

i) $T_{\mu_\infty}|_\Omega$ is topologically transitive

ii) The uniform measure ν on the Cantor set Ω is invariant by T_{μ_∞}

iii) The ergodic theorem is true in the sense that for dx almost all x we have

$$(N+1)^{-1} \sum_{n=o}^{N} g(T^n_{\mu_\infty})(x) \xrightarrow{N\to+\infty} \int_\Omega g(y)d\nu(y),$$

for any g in $C^o([-1,1])$.

iv) However $T_{\mu_\infty}|_\Omega$ is not weakly mixing .

Similar results were obtained by Misiurewicz [M].

References

[B.F.] C. Boldrighini, V. Franceschini, A five-dimensional truncation of the plane incompressible Navier-Stokes equations. Preprint University of Modena (1978)

[C.E.] P. Collet, J.-P. Eckmann, A Renormalization Group Analysis of the Hierarchical Model in Statistical Mechanics. Lecture Notes in Physics No. 74, Springer, Berlin-Heidelberg-New York 1978.

[C.E.L.] P. Collet, J.-P. Eckmann, O.E. Lanford III., Universal properties of maps on an interval. To appear

[D.G.P.] B. Derrida, A. Gervois, Y. Pomeau, Universal metric properties of bifurcations of endomorphisms. Preprint CEN SACLAY 1978

[F] M.J. Feigenbaum, Quantitative universality for a class of nonlinear transformations. Journal of Stat.Phys. 19, 25-52, (1978)

[H.P.S.] M.W. Hirsch, C.C. Pugh, M. Shub, Invariant Manifolds. Lecture Notes in Mathematics No. 583, Springer, Berlin-Heidelberg-New York 1977

[M.S.S.] N. Metropolis, M.L. Stein, P.R. Stein, On finite limit sets for transformations on the unit interval. Journal of Combinatorial Theory (A), 15, 25-44, (1973)

[M] M. Misiurewicz, Invariant measures for continuous transformations of [0,1] with zero topological entropy. Preprint, Warsaw University (1978)

[N] S. Newhouse, The abundance of wild hyperbolic sets and smooth stable sets for diffeomorphisms, Publ.Math. IHES, 50, 101-152 (1979)

Anticommutative Integration and Fermi Fields

P. Krée

Département de Mathématiques

Université de Paris VI, France

Contents:

Anticommutative Integration and Fermi Fields

Let Z be a complex and separable Hilbert space. Norbert Wiener has constructed an isometry of the symmetric Fock Space $\mathrm{Fock}^{-}(Z)$ onto some L^2-space. I. Segal and L. Gross have proposed an L^2-picture of the antisymmetric Fock space $F^{+}(Z)$ using non-commutative integration theory. The scope of this lecture is the introduction of a new L^2-picture of $\mathrm{Fock}^{+}(Z)$. An explicit formula is given for the intertwining operator; and the result is similar to the corresponding result in the commutative case [3]. Hence a free index $\varepsilon = \pm$ is introduced: $\varepsilon = -$ means "symmetric and $\varepsilon = +$ means "antisymmetric". Hence the case of mixed fields with bosons and fermions can be treated by tensor product.

1. Symmetric Algebras and Grassmann Algebras

For more details, we refer to [2]. All vector spaces are defined on a fixed field $\mathbb{K}$ with characteristic zero. We work in the category of pairs of vector spaces X - X', Y - Y'... in duality, the morphisms are the transposable maps. The algebra of all ε-symmetric tensors on X is denoted by $T^{\varepsilon}(X) = \oplus_{k=o}^{\infty} T_k^{\varepsilon}(X)$; an element of this algebra is denoted by $t = \Sigma_o^{\infty} t_k$ or $t(\zeta) = \Sigma\, t_k(\zeta)$. The algebraic dual of $T^{\varepsilon}(X)$ is the space $F^{\varepsilon}(X) = \Pi_{k=o}^{\infty} F_k^{\varepsilon}(X)$ of all ε-symmetric forms $f(x) = f = \Sigma_{k=o}^{\infty} f_k(x)$. The twisted duality

$$\langle f,t\rangle = \Sigma_{k=o}^{\infty} k!\, \langle f_k, t_k\rangle_k \tag{1.1}$$

is used between the vector spaces, where $\langle\ ,\ \rangle_k$ denotes the bilinear form on $F_k^{\varepsilon}(X) \times T_k^{\varepsilon}(X)$ such that

$$\langle f_k, x_1\, x_2 \ldots x_k\rangle_k = f_k(x_k, x_{k-1}, \ldots, x_1). \tag{1.2}$$

An involution of the type $t = x_1 x_2 \ldots x_k \longrightarrow t^{v} = x_k \ldots x_1$ is called twist. Because $T^{\varepsilon}(X)$ and $T^{\varepsilon}(X')$ are also in a twisted duality, $T^{\varepsilon}(X)$ is imbedded in $F^{\varepsilon}(X')$. The image of this map is the space $F^{\varepsilon}_{cyl}(X')$ of cylindrical forms on X'. If $\ell: X \longrightarrow Y$ is a morhphism, the transpose of $L = T^{\varepsilon}(\ell) : T^{\varepsilon}(X) \longrightarrow T^{\varepsilon}(Y)$ associates with any $g = g(y) \in F^{\varepsilon}(Y)$, the inverse image $g(\ell x) \in F^{\varepsilon}(X)$ of g by the linear map ℓ.

For any fixed $t \in T^{\varepsilon}(X)$, the directional left derivation is defined by transposition of the linear operator $t' \to t't$ of right product by t in $T^{\varepsilon}(X)$. In the same way, $f \to f_{\partial t}$ is defined by transposition of $t' \to tt'$.

Forms on a product (1.3)

Let X and Y be two vector spaces and $Z = X\times Y$. Introducing the canonical maps $\Pi_1: Z \to X$ and $i_1: X \to Z$, any $t \in T^{\varepsilon}(X)$ (resp. any $f \in F^{\varepsilon}(X)$) is identified with its image $(T^{\varepsilon}i_1)(t)$ (resp. with its inverse image $f(\Pi_1 x)$. The bilinear map $(t;t') \to t't = Sym_{\varepsilon}(t'\otimes t)$, defines by the universal property of tensor products, a linear isomorphism

$$T^{\varepsilon}(X) \otimes T^{\varepsilon}(Y) \xrightarrow{\simeq} T^{\varepsilon}(X \times Y). \tag{1.4}$$

The transposition of this isomorphism gives the structure of $F^{\varepsilon}(X \times Y)$. Moreover, for $f \in F^{\varepsilon}(X)$, $g \in F^{\varepsilon}(Y)$, $t \in T^{\varepsilon}(X)$, $t' \in T^{\varepsilon}(Y)$

$$\langle f(x)g(y),\ t'(\eta)t(\xi)\rangle = \langle f,t\rangle\langle g,t'\rangle. \tag{1.5}$$

If X and X' are two vector spaces in duality, and $\varepsilon = \pm$, the canonical two-form on $X \times X' = Y$ is

$$\begin{aligned} &Y \times Y \xrightarrow{\quad x.\xi \quad} \mathbb{K} \\ &y = (x;\xi),\ y_1 = (x_1;\xi_1) \to 2^{-1}(\langle x,\xi_1\rangle - \varepsilon\langle x_1\xi\rangle). \end{aligned} \tag{1.6}$$

The exponential of this form, in the algebra $F^{\varepsilon}(Y)$, is $\exp(x.\xi) = \sum_{k=o}^{\infty} k!^{-1}(x.\xi)^k$. In particular, if $\varepsilon = -$, the corresponding formal series on Y is $(x;\xi) \to \exp\langle x,\xi\rangle$. The form $\xi.x$ on $X' \times X$ is defined in the same way.

Basis and Coordinates (1.7)

Let $I = \{1,2,\ldots\}$ be a totally ordered set, with cardinal $d = |I|$. We suppose that their exists bases $\{\xi_i \ i \in I\}$ and $\{x_1', i' \in I\}$ in X and X', resp., such that $\langle \xi_i, x_{i'} \rangle = \delta_{i,i'}$. The set $T(+,d)$ of multi-indices is the set of all finite subsets j of I. If $j \neq \emptyset$, j is characterized by the ordered list of its elements: $\{j = j_1, j_2, \ldots, j_k\}$; and we set $\xi_j = \xi_{j_1} \wedge \xi_{j_2} \wedge \ldots \wedge \xi_{j_k}$; $|j| = k$; $j! = 1$. The set $T(-,d)$ is the set of families $j = \{j_1, j_2 \ldots\}$ of integers $j_n \geq 0$, vanishing for n sufficiently large. With the convention $(\xi_j)^0 = 1$; $0! = 1$, we set:

$$\xi^j = (\xi_1^{\otimes j_1}) \otimes (\xi_2^{\otimes j_2}) \otimes \ldots = (\xi_1)^{j_1} (\xi_2)^{j_2} \ldots$$

$$|j| = j_1 + j_2 + \ldots; \ j! = j_1! \, j_2! \ldots . \tag{1.8}$$

For arbitrary p integer ≥ 0, $T^p(\varepsilon,d)$ is the set of indices $j \in T(\varepsilon,d)$, with length $|j| = p$. For $\varepsilon = \pm$, the tensors $(\xi^j)_j$ and $(x^{j'})_{j'}$ are dual algebraic bases of $T^\varepsilon(X)$ and $T^\varepsilon(X')$ resp. and for j and $j' \in T(\varepsilon,d)$:

$$\langle (x^j)^v , \xi^{j'} \rangle = \delta_{j,j'} = \begin{cases} 1 & \text{if } j = j' \\ 0 & \text{if not} \end{cases} . \tag{1.9}$$

For example, if X has finite dimension d:

$$\exp(x.\xi) = \sum_{\alpha \in T(\varepsilon,d)} \alpha!^{-1} (x^\alpha)^v \, \xi^\alpha . \tag{1.10}$$

This holds also for $d = \infty$. in fact $E^* = F^\varepsilon(Y)$ is the algebraic dual of $E = T^\varepsilon(Y)$. The weak$_d$ topology on E^* is the weak topology $\sigma(E^*,E)$, if the field $\mathbb{K}$ has the discrete topology; and the R.H.S. of (1.10) converges to $\exp(x.\xi)$ for the weak$_d$ topology. For $\varepsilon = \pm$, any $f \in F^\varepsilon(X)$, any $j \in T^\varepsilon(\varepsilon,d)$, we set $\partial^j f = \partial_{\xi^j} f$; this agrees with the familiar convention of multi-indices in distribution theory (in the particular case $\varepsilon = -$ and $X = \mathbb{R}^d$, d finite).

The Taylor formula is useful in conventional analysis. In fact, for $\varepsilon = \pm$ and $f \in F^\varepsilon(X)$, the inverse image of f by the addition map $(x;x') \to x + x'$, is given by [2]

$$f(x+x') = f(x'+x) = \sum_{\alpha \in T(\varepsilon,d)} \alpha!^{-1} (x'^{\alpha})^{v} (\partial^{\alpha} f)(x). \quad (1.11)$$

A more intrinsic formula can also be given [2].

2. Notations Concerning the Symbolic Calculus

If E and F are vector spaces, $L(F,E^*) \sim (E \otimes F)^*$ denotes the vector space of all linear operators $F \to E^*$; the space $L(F,E^*)$ is equipped with the weak_d topology. If X and Y are two vector spaces, this can be combined with (1.4) and applied to $E = T^{\varepsilon}(X)$, $F = T^{\varepsilon}(Y)$, if any $h = h(x,y) \in F^{\varepsilon}(X \times Y)$ is identified with the following bilinear form on $E \times F$:

$$(t;u) \to \langle h(x,y), u(\eta)t(\xi)\rangle. \quad (2.1)$$

The corresponding homeomorphism

$$L(T^{\varepsilon}(Y), F^{\varepsilon}(X)) \to F^{\varepsilon}(X \times Y) \quad (2.2)$$

$$\hat{Q} \longmapsto \tilde{Q} = \tilde{Q}(x,y)$$

is called the kernel map; $\tilde{Q}$ is the kernel of $\hat{Q}$. The contraction of $u \in T^{\varepsilon}(Y)$ with the right argument of $\tilde{Q}$ is defined by

$$(\hat{Q}\, u)(x) = \langle \tilde{Q}(x,y), u(\eta)\rangle . \quad (2.3)$$

This contraction of $\tilde{Q}$ and u on their dual arguments y and η has natural properties. For example any $t \in T^{\varepsilon}(X)$

$$\langle\langle \tilde{Q}(x,y), u(\eta)\rangle, t(\xi)\rangle \equiv \langle \tilde{Q}(x,y), u(\eta)\, t(\xi)\rangle \quad . \quad (2.4)$$

Let T be the linear isomorphism $F^{\varepsilon}(X \times Y) \to F^{\varepsilon}(Y \times X)$ s.t. $\forall f(x;y)$, $\forall t \in T^{\varepsilon}(X)$ and $u \in T^{\varepsilon}(Y)$: $\langle Tf, t't\rangle = \langle f, t\, t'\rangle$. Then the kernel of the transposed operator $\hat{Q}'$: $T^{\varepsilon}(X) \to F^{\varepsilon}(Y)$ is $T\tilde{Q}$. But if the following new definition:

$$\langle t(\xi), \tilde{Q}(x,y)\rangle = (\hat{Q}'t)(ty) \quad (2.5)$$

of contraction of t with the left argument of $\tilde{Q}$ is introduced; then $\tilde{Q}$ is also the "right kernel" of $\hat{Q}'$; and (2.4) is replaced by

$$\langle u(\eta),\ \langle t(\xi),\ \tilde{Q}(x,y)\rangle\rangle = \langle u(\eta)\ t(\xi),\ \tilde{Q}(x,y)\rangle\ . \tag{2.6}$$

Let X - X', Y - Y' and Z - Z' be three pairs of vector spaces in duality and let $\hat{L}$ and $\hat{Q}$ be two linear operators

$$T^{\varepsilon}(Y) \xrightarrow{\hat{L}} F^{\varepsilon}(X);\ T^{\varepsilon}(X') \xrightarrow{\hat{Q}} F^{\varepsilon}(Z). \tag{2.7}$$

The product of $\hat{L}$ and $\hat{Q}$ is defined if the image of $\hat{L}$ is contained in $T^{\varepsilon}(X')$ or if $\hat{Q}$ is canonically extended to a linear operator $F^{\varepsilon}(X) \to F^{\varepsilon}(Z)$. The operator $\hat{L}$ is called regular if $\mathrm{Im}(\hat{L}) \subset T^{\varepsilon}(X')$; then $\hat{L}'$ can be extended to a linear operator $F^{\varepsilon}(X') \to F^{\varepsilon}(Y)$. Conversely, if $\hat{L}'$ is regular, $\hat{L}$ is extended by bitransposition. If $\hat{R} = \hat{Q}\ \hat{L}$ is defined, the contraction of their kernels is defined by

$$\hat{R}(z,y) = \langle\hat{Q}(z,\xi),\ \hat{L}(x,y)\rangle\ . \tag{2.8}$$

<u>Extension of a linear map $\hat{Q}: T^{\varepsilon}(Y) \to F^{\varepsilon}(X)$</u> (2.9)

If I(E) denotes the identity map of a vector space, the map $\hat{Q} \otimes I(E): T^{\varepsilon}(Y) \otimes E \to F^{\varepsilon}(X) \otimes E$, is also denoted $\hat{Q}$. A transposition argument permit sometimes the extension of $\hat{Q}$ to some spaces of vectorial forms on Y', with values in an algebraic dual E*. In fact, let $g \in F^{\varepsilon}(Y',E^*)$ be such that the transpose of

$$\begin{aligned} &T^{\varepsilon}(X' \times Y') \xrightarrow{A(g)} F^{\varepsilon}(X,E^*) \\ &\tilde{Q}(x,y) \to \langle\tilde{Q}(x,y)\ ,\ g(\eta)\rangle \end{aligned} \tag{2.10}$$

is regular. Then A(g) is extended by the bitransposed operator $A(g)'': F^{\varepsilon}(X \times Y) \to F^{\varepsilon}(X,E^*)$; and we set $\hat{Q}(g) = A(g)''\tilde{Q}$.

<u>Example:</u> $E = T^{\varepsilon}(X')$, $E^* = F^{\varepsilon}(X')$. The form $\exp(x.\xi')$ on $X \times X'$ is considered as a vectorial form $g_v(x) = \exp(x.\xi') \in F^{\varepsilon}(X,E^*)$. Then

$$\hat{Q}_x(\exp x.\xi') = \tilde{Q}(x,\xi'). \tag{2.11}$$

Symbol (2.12)

Let Ω be an open subset of $X = \mathbb{R}^d$. The symbol of a linear differential operator $\hat{P}$ on Ω is defined as the function $P(x,\xi)$ on $\Omega \times X'$ such that

$$\hat{P}(e^{i<x,\xi>}) = P(x,\xi)e^{i<x,\xi>}. \tag{2.13}$$

In our case, if the symbol of $\hat{Q}$ is defined by

$$Q(x,\zeta) = \tilde{Q}(x,\xi)(\exp-x.\xi), \tag{2.14}$$

then (2.11) gives a formula similar to (2.13):

$$\hat{Q}_x(\exp x.\xi') = Q(x,\xi')\exp x.\xi' . \tag{2.15}$$

For example $f \in F^{\varepsilon}(X)$ and $g \in F^{\varepsilon}(X')$:

$$\text{Symbol of } (u \to f\,\partial_g u) = f(x)g(\xi). \tag{2.16}$$

Hence for g and $g' \in F^{\varepsilon}(X')$

$$\text{Symbol of } (u \to (\partial_g)(\partial_{g'})u) = gg'. \tag{2.17}$$

The space of left derivations $u \to \partial_g u$ is a subalgebra of $\mathrm{End}\,(T^{\varepsilon}(X'))$; and this subalgebra is isomorphic with $F^{\varepsilon}(X')$. In particular $u \to \partial_g u$ is invertible iff $g_o \neq 0$ with $g = \Sigma_{k=o}^{\infty} g_k$. We have also

$$(\partial_g u)(x) = <g(\xi'),\ u(x+x')> . \tag{2.18}$$

This algebra is called the algebra of left convolution operators in $T^{\varepsilon}(X')$.

N.B. For $\varepsilon = +$, the following formula is wrong in general

$$(\partial_g u)(x) = <u(x+x')>, g(\xi')> .$$

Hence the introduction of the definition (2.5) is necessary. For similar reasons left integrals and right integrals will be used in the paragraph 4.

3. The Phase Space of a Dirac Field

Let G be a complex and separable prehilbertian space. The anti-space $\bar{G}$ of G is defined by the additive group of G equipped with the product by scalars $(\lambda;z) \to \bar{\lambda}z$. The identity map of G defines a semi-linear bijection $G \to \bar{G}$ denoted $z \to \bar{z}$; with this convention, the element of $\bar{G}$ denoted before $\bar{\lambda}z$ is written now $\overline{\bar{\lambda}z} = \lambda\bar{z}$. The injection $z \to (\bar{z};z)$ in $\bar{G}\times G$ of the <u>real</u> vector space G_r defined by G is denoted α; and the pair $(\alpha,\bar{G}\times G)$ is a complexification of G_r. The scalar product of G is denoted $(z,z') = \langle\bar{z},z'\rangle$ or shortly $\bar{z}z' = z'\bar{z}$. The generic point of $X = \bar{G}\times G$ is denoted $x = (\bar{z};z')$. A second copy $X_1 = \bar{G}\times G$ of X is introduced. If the generic point of X_1 is denoted $\xi = (\bar{\zeta},\zeta')$, the identity map ℓ: $X_1 \to X$ is denoted $(\bar{\xi},\zeta') \to (\bar{\zeta},\zeta')$. we suppose $\varepsilon = +$. In order to obtain short notations, the canonical symplectic two-form $b = b(\xi) = \bar{\zeta}.\zeta'$ on X_1, associated by the formula (1.6) to the duality $\langle\bar{\zeta},\zeta'\rangle$ on $\bar{G}\times G$, is sometimes denoted

$$[\xi = (\bar{\zeta};\zeta);\ \xi_1 = (\bar{\zeta}_1,\ \zeta_1)] \xrightarrow{\bar{\zeta}.\zeta} (\bar{\zeta}.\zeta_1 - \bar{\zeta}_1.\zeta)/2 \qquad (3.1)$$

i.e. the index "prime" is suppressed. This simplification is possible, because the form b defined on the complexification $\bar{G}\times G$ of G_r, is characterized by the corresponding form (3.1) on the real space G_r. The following antisymmetric duality will be systematically used between the complex vector spaces X and X_1.

$$\begin{aligned} X \times X_1 &\longrightarrow \mathbb{C} \\ (x;\xi) \to \langle x,\xi\rangle &= \bar{z}\zeta - \bar{\zeta}z \end{aligned} \qquad (3.2)$$

<u>Proposition</u> (3.3)

By the formula (1.6) the bilinear form $\langle x,\xi\rangle$ on $X\times X_1$ defines the following antisymmetric forms:

$$x.\xi = \bar{z}.\zeta + \bar{\zeta}.z \qquad \text{on } X\times X_1 \qquad (3.4)$$

and $\xi.x = z.\bar{\zeta} + \zeta.\bar{z}$ on $X_1 \times X$. (3.5)

Moreover the inverse image of b by the linear map $(x;\xi) \to \ell^{-1}x+\xi$ is the following form on $X\times X_1$

$$b((\ell^{-1}x) + \xi_1) = b(\ell^{-1}x) + b(\xi_1) + x.\xi_1. \qquad (3.6)$$

Proof

The generic point of $Y = X \times X_1$ is denoted $y = (x;\xi) = (\bar{z};z;\bar{\zeta};\zeta)$, or $y_1 = (x_1;\xi_1) = (\bar{z}_1;z_1;\bar{\zeta}_1;\zeta_1)$. Hence the two-form $II = \bar{z}.\zeta+\bar{\zeta}.z$ on Y satisfies

$$II(y;y_1) = 2^{-1}(\bar{z}\ \zeta_1 - \bar{z}_1\ \zeta + \bar{\zeta}\ z_1 - \bar{\zeta}_1\ z).$$

But in view of (1.6)

$$(x.\zeta)(y;y_1) = 2^{-1}(<x,\xi_1> - <x_1,\xi>)$$

$$= 2^{-1}(\bar{z}\ \zeta_1 - \bar{\zeta}_1\ z - (\bar{z}_1\ \zeta - \bar{\zeta}\ z_1))$$

$$= II(y;y_1).$$

Hence $II = x.\zeta$ and (3.4) is proven. The same procedure gives (3.5) and (3.6).

Basis (3.7)

We suppose that there exists an ordered basis $(e_i,\ i \in I') = (e_i, e_2 \ldots)$ of the vector space G such that $(e_i, e_j) = \delta_{i,j}$ and we set $d' = |I'|$. For any $z \in G$ there exists complex numbers z_1 s.t. $z = \Sigma\, z_1\, e_i$; hence $\bar{z} = \Sigma\, \bar{z}_i\, \bar{e}_i$. The element $e_i \in G$ (resp. $\bar{e}_i \in \bar{G}$) is identified with the linear form $-\bar{\zeta}_1 : \xi \to -\bar{\zeta}_i$ on $\bar{G}$ (resp. $\zeta_i : \xi \to \zeta_i$ on G). If the generic point of another copy of G is denoted $\zeta = \Sigma\, \zeta_1\, \phi_1$, we set $\phi_i = \bar{z}_i$, $\bar{\phi}_i = z_i$. Hence for arbitrary $(i;j) \in I^2$ the isomorphism $\ell : X_1 \to X$ maps $-z_i$ on ζ_i and $\bar{z}_j$ on $-\bar{\zeta}_j$. For the duality $<x,\xi>$ between X and X_1:

$$<\zeta_i, \bar{z}_j> = <-\zeta_i,\ z_j> = \delta_{i.j} \tag{3.8}$$

$$<\zeta_i, z_j> = <\zeta_i,\ \bar{z}_j> \quad = 0. \tag{3.9}$$

If G has finite dimension d, it is well known that the following 2d-form on X

$$\bar{z}_1 \wedge z_1 \wedge \bar{z}_2 \wedge z_2 \ldots \bar{z}_d \wedge z_d = \bar{z}_1 \wedge \bar{z}_2 \ldots \wedge \bar{z}_d \wedge z_d \ldots \wedge z_2$$

is independent of the choice of an orthonormal basis $(\bar{z}_i)_i$ in G. This form is denoted $\bar{z}z^v = \bar{z}^v z$. By application of (1.9)

$$<\bar{z}\, z^{v},\ (-1)^{d}\, \bar{z}\, z^{v}> = 1. \tag{3.10}$$

In the applications, G will be the linear subspace generated by the elements of an orthonormal basis of a complex and separable Hilbert space Z, where $\bar{Z} \times Z$ is the phase space of a Dirac field.

<u>The form picture of $Fock^{+}(\bar{Z} \times Z)$</u> (3.11)

This is a space [2] of antisymmetric forms on $Z \times \bar{Z}$, isometric with the hilbertian sum:

$$Fock^{+}(\bar{Z} \times Z) = \oplus_{k=o}^{\infty} (\bar{Z} \times Z)_{k}^{+} \tag{3.12}$$

where $(\bar{Z} \times Z)_{k}^{+}$ is the closure of $T^{+}(\bar{Z} \times Z)$ in the hilbertian tensor product ${}_{H}\hat{\otimes}_{k}(\bar{Z} \times Z)$. The scalar product of $\bar{Z} \times Z$ is

$$((\bar{z};z');\ (\bar{z}_1;z_1')) = \overline{(z,z_1)} + (z',z_1'). \tag{3.13}$$

Any $f_k \in (\bar{Z} \times Z)_{k}^{+}$ is identified with the following homogeneous form of degree k on $\overline{(\bar{Z} \times Z)} = Z \times \bar{Z}$:

$$(x_1;\ x_2 \ldots;\ x_k) \rightarrow (x_1\ x_2 \ldots x_k,\ f_k) \tag{3.14}$$

and using this identification for k = 0,1... the form picture of (3.12) is defined as the space $FH^{+}(Z \times \bar{Z})$ of antisymmetric forms $f(z,\bar{z}) = \Sigma_{k=o}^{\infty} f_k(z;\bar{z})$ on $Z \times \bar{Z}$ such that

$$\| f \|^{2} = \Sigma_{k=o}^{\infty}\ k!\ \| f_k \|_{k}^{2} < \infty. \tag{3.15}$$

The corresponding scalar product is

$$(f,g) = \Sigma_{k=o}^{\infty}\ k!\ (f_k,\ g_k)_k.$$

The set of tensors

$$t_{\beta,\alpha} = \{z^{\beta}\ \bar{z}^{\alpha}\ \ ;\ \alpha \text{ and } \beta \in T(o,d)\} \tag{3.16}$$

is an orthonormal basis in $FH^{+}(Z \times \bar{Z})$. Let X_1 be a second copy of $\bar{Z} \times Z$. Let $L' = Z^{+}(\ell')$ be the map $T^{+}(X_1) \rightarrow T^{+}(X)$ naturally associated to the isomorphism $\ell' = -\ell : X_1 \to X$. There exists a semi-linear map T of $T^+(X_1)$ such that for any $t \in T^+(X)$, $t' \in T^+(X_1)$

$$(L' t', t) = \langle T^{-1} t', t \rangle . \tag{3.17}$$

In particular

$$t' = z^{\alpha} \bar{z}^{\beta} \Rightarrow T^{-1}(t') = (-1)^{|\beta|} z^{\beta\nu} \bar{z}^{\alpha\nu} . \tag{3.18}$$

4. Coforms

The general formalism, valid for an arbitrary vector space X, and $\varepsilon = \pm$ is applied systematically to the particular case $\varepsilon = +$ and $X = \bar{Z} \times Z$.

Definition (4.1)

Let X' be a vector space. For any vector space X in duality with X', a cylindrical coform on X is a linear form on $F^{\varepsilon}_{cyl}(X) \simeq T^{\varepsilon}(X')$.

Hence the vector space $F^{\varepsilon'}_{cyl}(X)$ of all cylindrical coforms M on X is isomorphic with $F^{\varepsilon}(X')$; this isomorphisms $M \to \hat{M}$ is called the Laplace transform; and the action of M on $\Psi \in F^{\varepsilon}_{cyl}(X)$ is written

$$\langle \Psi, M \rangle = \int \Psi(x) M(x) \quad \text{or} \quad \langle M, \Psi \rangle = \int M(x) \Psi(x). \tag{4.2}$$

Hence

$$\langle \Psi, M \rangle = \langle \Psi, \hat{M} \rangle . \tag{4.3}$$

A normal space of ε-symmetric forms on X is a locally convex space $\Gamma(X)$ of forms on X such that the injection $T^{\varepsilon}(X') \to \Gamma(X)$ is continuous with dense range. Hence by transposition $\Gamma'(X) \subset F^{\varepsilon}_{cyl}(X)$; and an element of $\Gamma'(X)$ is called a coform.

The image by a linear map (4.4)

Let X' and Y' be vector spaces, and let $\ell : Y' \to X'$ be a linear map. The transpose of $L = T^{\varepsilon}(\ell) : T^{\varepsilon}(Y') \to T^{\varepsilon}(X')$ is a linear map

$$M \to L'(M) : F^{\varepsilon'}_{cyl}(X) \to F^{\varepsilon'}_{cyl}(Y) \text{ and}$$
$$\widehat{L'(M)}(\eta) = \hat{M}(\ell\eta). \tag{4.5}$$

Projective representation of a coform (4.6)

The notation $X_\alpha \subset\subset X$ means that X_α is a finite dimensional subspace of X. Application of (4.4) to the canonical injection $i_\alpha : X'_\alpha \to X'$, gives a coform $s_\alpha(M) = M_\alpha$ on $X_\alpha = X/(X'^{\perp}_\alpha)$, where s_α denotes the canonical surjection $X \to X_\alpha$ and $\hat{M}_\alpha(\xi) = \hat{M}(i_\alpha \xi)$.

Partial Integrations (4.7)

Let X - X' and Y - Y' be two pairs of vector spaces in duality. The inverse of the Laplace transform $F^{\varepsilon'}_{cyl}(Y) \to F^{\varepsilon}(Y')$ is denoted $\Psi \to \delta\Psi$. A form $f(x,y) \in F^{\varepsilon}(X \times Y)$ is the kernel of a map $\hat{Q}: T^{\varepsilon}(Y) \to F^{\varepsilon}(X)$. If the transposed map is regular, $\hat{Q}$ admits an extension: $F^{\varepsilon}(Y') \to F^{\varepsilon}(X)$. And for any $\Psi \in F^{\varepsilon}(Y')$ we set

$$\int f(x,y)\delta\Psi(y) = \langle f(x,y), \Psi(\eta)\rangle = \hat{Q}(\Psi). \tag{4.8}$$

A symmetrical convention is possible, because f is the right kernel of $\hat{Q}': T^{\varepsilon}(X) \to F^{\varepsilon}(Y)$. If the transposed map is regular we set for any $\Phi \in F^{\varepsilon}(X')$

$$\int (\delta\phi(x))\; f(x,y) = \langle \Phi(\xi),\; f(x,y)\rangle = \hat{Q}'(\Phi). \tag{4.9}$$

Convolution. Left and right product (4.10)

Let M and N be two cylindrical coforms on X in duality with X'. We set $M * N = s(M \otimes N)$, where s is the addition map. Hence $\hat{M}\,\hat{N} = \widehat{M * N}$. for $f \in T^{e}(X')$ the coforms fM and Mf are defined by

$$\langle g, f\, M\rangle = \langle g\; f,\; M\rangle \;;\; \langle M\; f, g\rangle = \langle M,\; f\; g\rangle \tag{4.11}$$

for arbitrary $g \in T^{\varepsilon}(X')$. Hence

$$\widehat{f\,M} = \partial_f \hat{M} \;;\quad \widehat{Mf} = \hat{M}\partial_f . \tag{4.12}$$

Let $\varepsilon = +$, and let f be a homogeneous antisymmetric form of maximum degree on the vector space X of finite dimension d. The Fermi coform $\mu = df(x)$ on X is the antisymmetric coform on X with Laplace transform $\hat{\mu}$ homogeneous of degree d on X' such that $\langle\hat{\mu}, f\rangle = 1$. The following proposition eliminates an ambiguity of sign indicated in [1], page 55.

Proposition (4.13)

For arbitrary $j \in T(+,d)$ and $g \in F^+(X)$

$$\int \mu(\partial^i g)(x) e^{x.\xi} = (-1)^{(d+1)|i|} \xi^i \int \mu g(x) e^{x.\xi} .$$

Proof

For arbitrary $\varphi \in F^+(X)$

$$<\widehat{\mu(\partial^i g)}, \varphi> = <\mu(\partial^i g), \varphi> = <\mu, (\partial^i g)\varphi>$$

$$= (-1)^{(d+1)|i|} <\mu, g(\varphi\partial^i)> = <\mu g, \varphi\partial^i>\sigma$$

$$= <\hat{\mu g}, \varphi\partial^i>\sigma = <\zeta^i \hat{\mu g}, \varphi>\sigma \; ; \text{ with } \sigma = (-1)^{(d+1)|i|} .$$

Hence $\widehat{\mu(\partial^i g)} = \xi^i \hat{\mu g}$, and the proposition is proven.

Definition (4.14)

With the notations of §3, the canonical cylindrical normal coform ν on $X = \bar{Z} \times Z$ denotes the anticommutative cylindrical coform with Laplace transform

$$\nu(\bar{\zeta}, \zeta') = \exp(\bar{\zeta}.\zeta') .$$

Proposition (4.15)

For arbitrary $Z_\alpha \subset\subset Z$, σ_α denotes the canonical surjection $\bar{Z} \times Z \to \bar{Z}_\alpha \times Z_\alpha$. Then for any orthonormal basis $\{\bar{z}_1, \ldots \bar{z}_d\}$ of Z, the coform $\nu_\alpha = \sigma_\alpha(\nu)$ is

$$\nu'(\bar{z}, z) = (-1)^d e^{-\bar{z}.z} d(\bar{z}z^V) = (-1)^d d(\bar{z}z^V) e^{-\bar{z}.z} . \quad (4.16)$$

Proof

By (4.4) the Laplace transform of the coform ν_α on $\bar{Z}_\alpha \times Z_\alpha$ is the form $\exp \bar{\zeta}.\zeta$. In general, the Laplace transform of a cylindrical coform M on a space X is

$$\hat{M}(\xi) = \int M(x) e^{x.\xi} = \int e^{\xi.x} M(x) . \quad (4.17)$$

Hence the first relation (4.16) is equivalent to

$$\exp \bar{\zeta}.\zeta = \int (\exp(z\bar{\zeta} + \bar{\zeta}z - \bar{z}z))(-1)^d d(\bar{z}z^V)$$

and this can be proven for example by a computation using coordinates.

5. Hermite Forms and Normalized Laplace Transform

With the notations of §3, $\hat{W}$ denotes the left convolution operator in $T^+(X_1)$ with symbol $W(\bar{\zeta},\zeta) = \exp(\bar{\zeta}.\zeta)$. The inverse of this operator has symbol $\exp(-\bar{\zeta}.\zeta)$.

Definition (5.1)

For α and $\beta \in T(+,d)$, the corresponding Hermite form is defined by

$$H_{\alpha,\beta}(\bar{z},z) = \hat{W}^{-1}(\bar{z}^{\alpha}z^{\beta}). \tag{5.2}$$

Hence by (2.18)

$$H_{\alpha,\beta}(\bar{z},z) = \langle e^{\bar{\zeta}'.\zeta'}, (\bar{z}+\bar{z}')^{\alpha}(z+z')^{\beta}\rangle. \tag{5.3}$$

For example $G = \mathbb{C}$ and $d = 1$:

$$H_{0,0}(\bar{z},z) = 1 \qquad H_{0,1}(\bar{z},z) = z$$

$$H_{1,0}(\bar{z},z) = \bar{z} \qquad H_{1,1}(\bar{z},z) = (1+\bar{z}z).$$

Proposition (5.4)

$$H_{\alpha,\beta}(\bar{z},z) = (-1)^{|\beta|}(\partial_{\bar{z}}^{\beta}\ \partial_{z}^{\alpha}\ e^{-\bar{z}z})\ e^{\bar{z}.z} \tag{5.6}$$

$$\Sigma_{\alpha,\beta}(-1)^{|\beta|}\ \bar{\zeta}^{\beta v}\ \zeta^{\alpha v}\ H_{\alpha,\beta}(\bar{z},z) = e^{z.\bar{\zeta}+\zeta.\bar{z}\ -\ \bar{\zeta}.\zeta}\ . \tag{5.7}$$

Proof

Both members of (5.3) are multiplied on the left by $[(\bar{\zeta})^{\beta}\zeta^{\alpha}]^{v}$. Hence by addition, if I denotes the left hand side of (5.7)

$$I = \langle e^{-\bar{\zeta}.\zeta'}, \exp - \bar{\zeta}(z'+z) + \zeta(\bar{z}'+\bar{z})\rangle$$

$$= (\exp(z\bar{\zeta} + \zeta\bar{z}))\langle\exp - \bar{\zeta}'.\zeta', \exp z'\bar{\zeta} + \zeta\bar{z}'\rangle$$

$$= \exp(z\bar{\zeta} + \zeta\bar{z} - \bar{\zeta}\zeta)$$

and this proves (5.7). Then, as in the familiar commutative, (5.6) is deduced from (5.7) using the Taylor formula (1.11).

Definition

The normalized Laplace transform θ on $\bar{Z}\times Z$ associate to any cylindrical coform M on $\bar{Z}\times Z$, the following form on X_1:

$$(\theta M)(\bar{\zeta},\zeta) = \hat{M}(\bar{\zeta},\zeta).e^{-\bar{\zeta}.\zeta} \; . \tag{5.8}$$

Hence

$$(\theta M)(\bar{\zeta},\zeta) = \int M(\bar{z},z)e^{\bar{z}.\zeta+\bar{\zeta}.z - \bar{\zeta}.\zeta} \; .$$

Theorem (5.9)

For arbitrary t and t' $\in T^+(\bar{G}\times G)$ we set $u = \hat{W}^{-1}(t)$, $u' = \hat{W}^{-1}(t')$. Then

$$\theta(\nu u) = L'(t) = (T^+\ell)(t) \tag{5.10}$$

$$e^{\bar{\zeta}.\zeta}\partial_u = e^{\bar{\zeta}.\zeta}\; L'(t) \tag{5.11}$$

$$\int u'(x)u(x)\nu(x) = \langle t,L'(t')\rangle \; . \tag{5.12}$$

Proof

The formula (5.6) can be written

$$H_{\alpha,\beta}(\bar{z},z) = (-1)^{\beta}(\partial_{(-\bar{\zeta})^{\alpha}\zeta^{\beta}}e^{-\bar{z}.z})e^{\bar{z}.z}. \tag{5.13}$$

Hence, the application of the formula (4.13) to the Fermi-coform $\mu = (-1)^d\, d\bar{z}z^V$, gives

$$\mu(\partial_{(-\bar{\zeta})^{\alpha}\;\zeta^{\beta}}\; e^{-\bar{z}.z}) = (-1)^{|\alpha|+|\beta|}(-\bar{\zeta})^{\alpha}\;\zeta^{\beta}\; e^{\bar{\zeta}.\zeta} \; .$$

Hence, combining this result with (5.13) and (4.16)

$$\widehat{\nu H_{\alpha\beta}} = \bar{\zeta}^{\alpha}\;\zeta^{\beta}\; e^{\bar{\zeta}.\zeta} \; .$$

In view of the definition (5.8) of the θ transform

$$\theta(\nu H_{\alpha\beta}) = \bar{\zeta}^{\alpha}\zeta^{\beta} = L'(\bar{z}^{\alpha}z^{\beta}).$$

By linearity, this gives (5.10).

Finally

$$\langle t;\ L'(t')\rangle = \langle \partial_W u(x),\ e^{b(\zeta)}(T_\zeta t')(\zeta)e^{-b(\zeta)}\rangle$$

$$= \langle u(x),\ e^{b(\zeta)}(T_\zeta t')(\zeta)\rangle = \langle u(x),\ e^{b(\zeta)}\partial_{u'}\rangle$$

$$= \langle u'(x)u(x),\ \exp b(\zeta)\rangle = \int u'(x)u(x)\nu(x)$$

<u>Corollary</u> (5.14)

For arbitrary $v \in T^+(x_1)$ we set

$$v^+ = W^{-1}\ T^{-1}Wv\ .$$

Hence for arbitrary $f,\ g \in T^+(X_1)$

$$\int gf^+\nu = (L'Wf,\ L'Wg) = (\theta(nf),\theta(\nu g)). \qquad (5.15)$$

Hence the cylindrical normal coform ν on $X = \bar{Z} \times Z$ is positive in the following sense

$$\forall f \text{ and } g \in F^+_{cyl}(X) \qquad \int \qquad \nu \geq 0. \qquad (5.16)$$

The bilinear map

$$(f\nu, g\nu) = \int \qquad \nu \qquad (5.17)$$

is a scalar product on the following space of cylindrical coforms on X

$$\nu F^+_{cyl}(X) = \{\nu\ f\ ;\quad f \in F^+_{cyl}(X)\ . \qquad (5.18)$$

Finally we have the

<u>Theorem</u>

The normalized Laplace transform θ is an isometry of $\nu F^+_{cyl}(X)$ on a dense subspace of the form picture $FH^+(Z \times \bar{Z})$ of the Fock space $\bar{Z} \times Z$. The completion of $\nu F^+_{cyl}(X)$ is in fact the space of coforms $M = (M_\alpha)$ on $\bar{Z} \times Z$ such that $\forall Z_\alpha \subset\subset Z$, $M_\alpha \in \nu_\alpha F^+_{cyl}(\bar{Z}_\alpha \times Z_\alpha)$ and

$$||M|| = \sup_\alpha ||M_\alpha|| < \infty.$$

References

[1] F.A. Berezin, The method of second quantization. Academic Press 1966

[2] P. Krée, Séminaire sur les équations aux dérivées partielles en dimension infinie 3e année 1976-79. Secrétariat mathématique de l'Institut H. Poincaré

[3] P. Krée, Solutions faibles d'équations aux dérivées fonctionnelles. Séminaire Lelong Analyse. Lecture notes in mathematics
I n° 410 - 1972/73, p. 143-181
II 1973/74

Homogeneous Self-Dual Cones and Jordan Algebras

J. Bellissard and B. Iochum
Université de Provence, Marseille, and
Centre de Physique Théorique
CNRS Marseille, France

Abstract

It is shown that the category of homogeneous self-dual cones in Hilbert spaces is isomorphic to the category of Jordan Banach algebras with predual (JBW-algebras).

I. The Algebras $M_n(\mathbb{R})$ and $M_n(\mathbb{C})$

The aim of this talk is to give a characterization of the ordered vector space given by some important algebras, like von Neumann and Jordan algebras.

In order to illustrate this fact let us begin with the simple case of matrix algebras.

Let K be the real or complex field, and let $M_n(K)$ be the space of n×n matrices with entries in K. $M_n(K)$ has a natural structure of associative algebra and a *-operation given by:

$$(x.y)_{ij} = \sum_{k=1}^{n} x_{ik} y_{kj}$$
$$(x^*)_{ij} = \bar{x}_{ji}. \tag{1}$$

We denote by $M_n^h(K)$ the real subspace of self adjoint matrices (i.e., $x = x^*$) in $M_n(K)$; a Jordan product is defined on it by:

$$x \circ y = \tfrac{1}{2}(x.y + y.x) \quad x,y \in M_n^h(K) \tag{2}$$

It is bilinear and symmetric with respect to x, and y and it satisfies the identities:

$$\text{i)} \quad x \circ x = x^2$$
$$\text{ii)} \quad x \circ (x^2 \circ y) = x^2 \circ (x \circ y). \tag{3}$$

Let $\mathfrak{A}$ be the Jordan algebra $(M_n^h(K), \circ)$ [1,2]. The set $\mathfrak{A}_+$ of positive matrices in $M_n(K)$ is a closed convex cone included in $\mathfrak{A}$, containing the identity $\mathbb{1}$; the set $(\mathfrak{A}, \mathfrak{A}_+, \mathbb{1})$ becomes an order-unit space [3]. For x in $M_n(K)$ we put:

$$tr(x) = \sum_{i=1}^{n} x_{ii}. \tag{4}$$

This trace operation allows us to define a real hilbert structure on $M_n^h(K)$ by mean of the inner product

$$\langle x|y\rangle = tr(x\circ y) = tr(x.y). \tag{5}$$

Let $\mathfrak{H}_K$ (or $\mathfrak{H}$ if there is no confusion) be the real hilbert space $(M_n^h(K), \langle|\rangle)$ and $\mathfrak{H}_+$ the set of positive matrices in $\mathfrak{H}_K$. As sets $\mathfrak{H}_K$, $\mathfrak{H}_+$ coincide with $\mathfrak{A}$, $\mathfrak{A}_+$ but we will distinguish them from structural point of view.

The first main property of $\mathfrak{H}_+$ is the self-duality: x is a positive matrix if and only if $tr(xy) \geq 0$ for any positive y. This means:

$$\mathfrak{H}_+ = \{x \in \mathfrak{H};\ \langle x|y\rangle \geq 0 \quad \forall y \in \mathfrak{H}_+\}. \tag{6}$$

In the sequel we will denote by $\leq$ the order relation defined by $\mathfrak{H}_+$ in :

$$x \leq y \iff y-x \in \mathfrak{H}_+ \qquad x,y \in \mathfrak{H}. \tag{7}$$

A face of $\mathfrak{H}_+$ is a convex cone F in $\mathfrak{H}_+$ such that

$$0 \leq y \leq x \in F \quad \text{implies} \quad y \in F. \tag{8}$$

In our case one can check that F is a face if and only if there exists an idempotent e_F such that

$$F = e_F \mathfrak{H}_+ e_F = \{e_F x e_F;\quad x \geq 0\}. \tag{9}$$

It is then natural to define P_F as the following projection map (called the <u>facial projection</u> associated to F):

$$P_F(x) = e_F x e_F \qquad x \in \mathfrak{H} \tag{10}$$

P_F is self andjoint and

$$F = P_F \mathfrak{H}_+. \tag{11}$$

The Set $F^{\perp} = \{x \geq 0;\ \langle x|y\rangle = 0\ y \in F\}$ is a face called the orthogonal face of F. One can see that

$$e_{F^\perp} = \mathbb{1} - e_F \tag{12}$$

and therefore

$$F^{\perp\perp} = F. \tag{13}$$

Following A. Connes [4] we are led to introduce:

$$\delta_F = \frac{1}{2}\,(\mathbb{1} \quad + P_F - P_{F^\perp}). \tag{14}$$

The main property of δ_F comes form

$$\delta_F(x) = e_F \circ x. \quad x \in \mathfrak{h} \tag{15}$$

Thus δ_F connects the geometric structure of $\mathfrak{h}_+$, by means of the face F, and the algebraic structure of $\mathfrak{A}$.

Clearly we get

$$e^{t\delta_F} x = e^{\frac{t}{2}e_F}.x.e^{\frac{t}{2}e_F} \qquad x \in \mathfrak{h} \tag{16}$$

which shows that the one-parameter group $(e^{t\delta_F})_{t\in\mathbb{R}}$ leaves $\mathfrak{h}_+$ invariant. This is what A. Connes called "facial homogeneity" [4].

<u>Proposition 1:</u> The positive cone in $M_n^h(K)$ is facially homogeneous.

Now we want to show on this example, how to reconstruct $\mathfrak{A}$ form $\mathfrak{h}_+$.

We will call "derivations of $\mathfrak{h}_+$" any linear operator δ in $\mathfrak{h}$ such that $e^{t\delta}$ leaves $\mathfrak{h}_+$ invariant for real "t". The set of derivations of $\mathfrak{h}_+$ is a real self-adjoint Lie algebra. We denote by $\mathfrak{M}$ its self-adjoint part (i.e. $\delta = \delta^*$). It can be shown that any element δ of $\mathfrak{M}$ has the form δ_a

$$\text{where} \quad \delta_a(x) = a \circ x \qquad \text{for some } a \in \mathfrak{A}. \tag{17}$$

Clearly δ_a is self-adjoint and

$$e^{t\delta_a}(x) = e^{\frac{ta}{2}}.x.e^{\frac{ta}{2}} \qquad x \in \mathfrak{h}. \tag{18}$$

Since δ_a is linear with respect to a, the linear map $a \to \delta_a$ is a linear isomorphism between $\mathfrak{A}$ and $\mathfrak{M}$.

Let $\mathfrak{M}_+$ be the set of positive elements of $\mathfrak{M}$. It is a closed convex cone containing the identity $\mathbb{1}_{\mathfrak{h}}$. It contains also any facial derivation by (14), and therefore

$$0 \leq \delta_F \leq \mathbb{1}_{\mathfrak{h}} . \tag{19}$$

Now, if $a \in \mathfrak{A}$, the spectral theorem shows that "a" can be written as:

$$a = \sum_{i=1}^{k} \alpha_i \, e_{F_i} \tag{20}$$

where $k \leq n$, the α_i's are real numbers, and e_{F_i} are mutually orthogonal idempotents. Moreover $a \in \mathfrak{A}_+$ if and only if $\alpha_i \geq 0 \ \forall i$. It follows that:

$$\delta_a = \sum_{i=1}^{k} \alpha_i \, \delta_{F_i} \tag{21}$$

and since $P_F \delta_G P_F = 0$ if $G \subset F^\perp$, we get:

$$a \in \mathfrak{A}_+ \Leftrightarrow \delta_a \in \mathfrak{M}_+ . \tag{22}$$

Therefore the map $a \to \delta_a$ becomes on order-isomorphism between the order-unit spaces $(\mathfrak{A}, \mathfrak{A}_+, \mathbb{1})$ and $(\mathfrak{M}, \mathfrak{M}_+, \mathbb{1}_{\mathfrak{h}})$.

Now the question is: Can we define "geometrically" a Jordan product in $\mathfrak{M}$, such that this map becomes a Jordan isomorphism?

To answer this question, we proceed as follows. Let F be a face and $\mathcal{P}_F$ be the projector on $\mathfrak{M}$ corresponding to P_F by the isomorphism:

$$\mathcal{P}_F(\delta_a) = \delta_{(P_F(a))} . \tag{23}$$

It is not difficult to see that

$$P_{F^\perp} \mathcal{P}_F(\delta_a) = 0 \quad \text{and} \quad P_F \, \mathcal{P}_F(\delta_a) = P_F \delta_a P_F . \tag{24}$$

These two properties actually characterize $\mathcal{P}_F(\delta_a)$. For if δ is a self-adjoint derivation which is zero both on F and $F^{\perp}$, then it is zero every where in $\mathfrak{h}$. Therefore a Jordan product can be defined in $\mathfrak{M}$ by using the spectral theorem and the formula:

$$\delta_F \circ \delta_a = \frac{1}{2}(\mathbb{1} + P_F - P_{F^{\perp}})(\delta_a). \tag{25}$$

A simple calculation shows that this product is bilinear and symmetric, and gives the answer to the question since

$$\delta_F \circ \delta_a = \delta_{(e_F \circ a)}. \tag{26}$$

<u>Proposition 2:</u> The set $\mathfrak{M}$ of self-adjoint derivations of $\mathfrak{h}_+$ has a natural structure of Jordan algebra such that it becomes isomorphic to $\mathfrak{A}$.

II. Problems in the Infinite Dimensional Case

On this example we have described the essential steps to reconstruct the Jordan algebra $\mathfrak{A}$ from the geometrical properties of the cone $\mathfrak{h}_+$. Conversely starting from $\mathfrak{A}$, the construction of $\mathfrak{h}$ and $\mathfrak{h}_+$ is, in this case, very easy since $\mathfrak{A}$ has a trace given by the formula (4). The main property used implicitly is [5]

$$\mathrm{tr}(x\circ(y\circ z)) = \mathrm{tr}((x\circ y)\circ z). \tag{27}$$

Unfortunately there are infinite dimensional Jordan algebras without trace. A simple example is given by the hermitian part of a type-III W*-algebra with the product defined in (2). In this case it is possible to construct the cone $\mathfrak{h}_+$ by mean of the Tomita-Takesaki's theory [4, 9, 10]. But not all Jordan algebra is the real part of a von Neumann algebra. Very recently G. Effros and E. Størmer [6] have shown that a class of Jordan algebras (namely J.W.-algebras) are image under a positive projection of the self-adjoint part of a W*-algebra. Let us explain the main idea.

Let M be the W*-algebra $M_n(\mathbb{C})$, and α be the antiautomorphism given by the transposition map:

$$[\alpha(x)]_{ij} = x_{ji} \qquad x \in M. \tag{28}$$

It commutes with the involution. Denoting by $\mathfrak{A}$ and $\mathfrak{B}$ the Jordan algebras defined respectively by $M_n^h(\mathbb{C})$ and $M_n^h(\mathbb{R})$ we remark the following:

$$M = \mathfrak{A} \oplus i\, \mathfrak{A}$$
$$\mathfrak{B} = \{a \in \mathfrak{A}\ ;\ \alpha(a) = a\}. \tag{29}$$

The real hilbert structure $\mathfrak{h}_{\mathbb{C}}$ on $\mathfrak{A}$ can be extended to a complex hilbert structure on M by:

$$(x|y) = \mathrm{tr}(x^*y) \qquad x,y \in M. \tag{30}$$

Denoting by T_α the map $x \to \alpha(x)$, T_α becomes a unitary operator on $\tilde{\mathfrak{h}}_{\mathbb{C}} = \mathfrak{h}_{\mathbb{C}} \oplus i\, \mathfrak{h}_{\mathbb{C}}$, such that $T_\alpha^2 = \mathbb{1}_{\tilde{\mathfrak{h}}_{\mathbb{C}}}$. The real hilbert space $\mathfrak{h}_{\mathbb{R}}$ associated to $\mathfrak{B}$ is nothing but the fixed-point of T_α inside $\mathfrak{h}_{\mathbb{C}}$. Moreover:

$$\mathfrak{h}_{\mathbb{R}}^+ = \left(\frac{\mathbb{1}+T_\alpha}{2}\right)\mathfrak{h}_{\mathbb{C}}^+ . \tag{31}$$

On the other hand let $\mathcal{R}(\mathfrak{A})$ (resp. $\mathcal{R}(\mathfrak{B})$) be the real associative closed algebra generated by $\mathfrak{A}$ (resp. $\mathfrak{B}$) in M. We get:

$$\text{i)}\quad M = \mathcal{R}(\mathfrak{A}) \cap i\,\mathcal{R}(\mathfrak{A})$$
$$\text{ii)}\quad \mathcal{R}(\mathfrak{B}) \cap i\,\mathcal{R}(\mathfrak{B}) = \{0\} \tag{32}$$
$$\text{iii)}\quad M = \mathcal{R}(\mathfrak{B}) \oplus i\,\mathcal{R}(\mathfrak{B}) .$$

The map "α" is then defined by:

$$\alpha(x+iy) = x^* + i\,y^* \qquad \text{if } x,y \in \mathcal{R}(\mathfrak{B}). \tag{33}$$

Although the three previous cases (i.e., M: Jordan algebra with trace, $\mathfrak{A}$: self-adjoint part of a W*-algebra, $\mathfrak{B}$: fixed point Jordan algebra of a W*-algebra under antiautomorphism) are not the most general examples, they are generic: any Jordan algebra $\mathfrak{A}$ we will consider (namely J.B.W. algebra), can be decomposed into a direct sum [7].

$$\mathfrak{A} = \mathfrak{A}_{tr} \oplus \mathfrak{A}_{W^*} \oplus \mathfrak{B} \tag{34}$$

where $\mathfrak{A}_{tr}$ is a Jordan algebra with a faithful trace, $\mathfrak{A}_{W*}$ is the self-adjoint part of a von Neumann algebra, and $\mathfrak{B}$ is the fixed point of the hermitian part of a W*-algebra M under an antiautomorphism. In this last case, we get

$$\mathcal{R}(\mathfrak{B}) \cap i\, \mathcal{R}(\mathfrak{B}) = \{0\} \qquad \text{and } M = \mathcal{R}(\mathfrak{B}) \oplus i\, \mathcal{R}(\mathfrak{B}). \tag{35}$$

Using the Tomita-Takesaki's theory for M [8], we construct its natural cone $P_M^\natural$ [4, 9, 10], and the antiautomorphism α is represented by a unitary operator leaving $P_M^\natural$ invariant. The fixed points of T_α in $P_M^\natural$ give a subcone $P_{\mathfrak{B}}^\natural$ canonically associated to $\mathfrak{B}$, which is self-dual and facially homogeneous.

For a Jordan algebra with a finite faithful trace we use the method described in Section I [11,12].

III. Definitions and Results

In the previous sections we have shown formally the method to describe the isomorphism between facially homogeneous self-dual cones and Jordan algebras. We want to give now, in a more precise form, the results.

Definition 1: Let $\mathfrak{h}$ be a real hilbert space. A cone $\mathfrak{h}_+$ is self dual if

$$\mathfrak{h}_+ = \{x \in \mathfrak{h}; \quad \langle x|y\rangle \geq 0 \quad \forall y \in \mathfrak{h}_+\}. \tag{36}$$

As in the Section I, we define a face of $\mathfrak{h}_+$ as a convex subcone F of $\mathfrak{h}_+$ such that

$$\text{if } y \in \mathfrak{h}_+ \text{ and } x \in F \text{ satisfies } x-y \in \mathfrak{h}_+, \text{ then } y \in F. \tag{37}$$

We denote by P_F the orthogonal projection on the closed linear hull of F, and by $F^\perp$ its orthogonal face:

$$F^\perp = \{x \in \mathfrak{h}_+ ; \quad \langle x|y\rangle = 0 \quad y \in F\} . \tag{38}$$

Definition 2 [4]: Let $\mathfrak{h}_+$ be a self dual cone in $\mathfrak{h}$, $\mathfrak{h}_+$ is called facially homogeneous if for any face F and for any real number t, $\exp t(P_F - P_{F^\perp})$ leaves $\mathfrak{h}_+$ invariant.

<u>Definition 3:</u> A derivation of $\mathfrak{h}_+$ is a bounded linear operator δ on $\mathfrak{h}$, such that for any $t \in \mathbb{R}$, $\exp t\delta$ leaves $\mathfrak{h}_+$ invariant. The set of self-adjoint derivations of $\mathfrak{h}_+$ is denoted by $\mathfrak{M}(\mathfrak{h}_+)$.

On the other hand we recall that a Jordan algebra $\mathfrak{A}$ is a real vector space, equipped with a bilinear product $(x,y) \in \mathfrak{A} \times \mathfrak{A} \to x \circ y \in \mathfrak{A}$, which is symmetric and satisfies the Jordan identity:

$$x \circ ((x \circ x) \circ y) = (x \circ x) \circ (x \circ y). \qquad (39)$$

<u>Definition 4 [13]:</u> A J.B.-algebra is a Jordan algebra which is a Banach algebra with a norm such that:

i) $\| a^2 \| = \| a \|^2$

ii) $\| a^2 - b^2 \| \leq \max(\| a \|^2, \| b \|^2)$

$\mathfrak{A}$ is a JBW algebra if moreover it is the dual of some Banach space $\mathfrak{A}_*$.

The main results in this respect are summarized in the following theorems.

<u>Theorem A [14]:</u> Let $\mathfrak{h}_+$ be a facially homogeneous self dual cone. Then $\mathfrak{M}(\mathfrak{h}_+)$ has a natural structure of JBW algebra, and the correspondance $\mathfrak{h}_+ \to \mathfrak{M}(\mathfrak{h}_+)$ is functorial.

<u>Theorem B [15]:</u> ("Standard form of a JBW algebra"): Let $\mathfrak{A}$ be a JBW algebra. Then there is a real hilbert space $\mathfrak{h}$ and a facially homogeneous self-dual cone $\mathfrak{h}_+$ in $\mathfrak{h}$ such that $\mathfrak{A}$ is canonically isomorphic to $\mathfrak{M}(\mathfrak{h}_+)$. Moreover, if ω is a normal state on $\mathfrak{A}$ there is a vector ξ_ω in $\mathfrak{h}_+$ such that ω coincides with the state $\delta \in \mathfrak{M}(\mathfrak{h}_+) \to \langle \xi_\omega | \delta \xi_\omega \rangle$ by this isomorphism.

Let us give some indication about the proof and the literature.

To prove the Theorem A the crucial point concerns the spectral theorem in $\mathfrak{M}(\mathfrak{h}_+)$ [16]. Once this result is given, we can construct the projection P_F in $\mathfrak{M}(\mathfrak{h}_+)$ by the equations (24), and the Jordan product is defined as in the equation (25)(see [14]. The functorial character of the correspondance $\mathcal{V}_+ \to \mathfrak{M}(\mathcal{V}_+)$ is only a question of calculation.

In order to prove the theorem B, we use the existence of the decomposition (34) for a JBW-algebra $\mathfrak{A}$ (see [7]). If $\mathfrak{A}$ has a finite faithful trace, the proof was given in [11] (see also [5]). The extension to the semi-finite case is only a technical question. If $\mathfrak{A}$ is the real part of a von Neumann algebra the classical results about its natural cone make the job [4,9,10]. If $\mathfrak{A}$ is the set of fixed points of the real part of a von Neumann algebra M, under an antiautomorphism α, then we construct the natural cone $P^\natural_M$ the unitary operator T_α defined by α, leaving $P^\natural_M$ invariant (i.e. $T_\alpha \Delta^{1/4} x\xi = \Delta^{1/4}\alpha(x)\xi$, $x \in M^+$, where ξ is a cyclic vector associated to ω), and we define the cone $P^\natural_{\mathfrak{A}}$ has the set of invariant vectors of $P^\natural_M$. It is not difficult to see that $P^\natural_{\mathfrak{A}}$ is then homogeneous, self-dual and independent of ω, and that $\mathfrak{M}(P^\natural_{\mathfrak{A}})$ can be identified with $\mathfrak{A}$[15]. (After the time this talk was given, the authors have learned that U. Haagerup and H. Hanche-Olsen [17] have developped a Tomita-Takesaki's theory for JBW-algebras.) As a consequence we get a decomposition theory for JBW-algebras. For if $\mathcal{V}_+$ is a separable facially homogeneous self-dual cone it can be decomposed in a unique way into (see [11], th. (1.13) and prop. (7.3))

$$\mathcal{V}_+ \simeq \int_Z^\oplus \mathcal{V}_+(\zeta)\,d\mu(\zeta)$$

where $\mathcal{V}_+(\zeta)$ are facially homogeneous self-dual and irreducible μ-almost everywhere, Z is a standard Borel space, μ a Borel measure on Z. It corresponds to this formula a decomposition for $\mathfrak{M}(\mathcal{V}_+)$ as direct integral of JBW factors.

Another consequence concerns the classification of finite dimensional facially homogeneous self-dual cones. By theorem A and B, it is equivalent to classify finite dimensional JBW algebras [18,19]. The P. Jordan - J. von Neumann - E. Wigner theorem gives the answer for factors. There are five classes of JBW factors. $M_n^h(K)$ for $K = \mathbb{R}$, $\mathbb{C}$ and $\mathbb{H}$ the quaternionic field, the spin factors generated by $\mathbb{1}$ and $(b_i)_{i\in I}$ (I finite or not) such that $b_i = b_i^*$ and $b_i \quad b_j = 2\delta_{ij}\,\mathbb{1}$ and the exceptional algebra M_3^8 of symmetric 3×3 matrices with entries in the Cayley algebra $\mathbb{K}$ [1,2].

As a last remark, the construction of G. Effros and E. Størmer suggests that the classification of JBW-algebras is connected with the classification of involutive antiautomorphisms of W*-algebra up to automorphisms. For type I factors, there are at most two such classes of antiautomorphisms. For the II_1 and II_∞ injective factors there is only one [21, 22] and it seems that for the III_λ injective factors $(0 < \lambda < 1)$ there are exactly two classes, while for the III_1 injective factors there is only one [22].

References

[1] H. Braun and M. Koecher, Jordan algebras, Springer-Verlag, Berlin-Heidelberg-New York, 1966

[2] N. Jacobson, "Structure and representations of Jordan algebras", Amer. Math. Soc. Pub, 39, Providence, 1968

[3] E.M. Alfsen, "Compact convex sets and boundary integrals", Ergebnisse der Mathematik und ihrer Grenzgebiete, Vol. 57, Springer-Verlag, Berlin - Heidelberg - New York, 1971

[4] A. Connes, "Caracterisation des espaces vectoriels ordonnés sous-jacents aux algèbres de von Neumann". Ann. Inst. Fourier, Grenoble, 24, 4 (1974), 121-155

[5] G. Janssen, "Formal-reelle Jordanalgebren unendlicher Dimension und verallgemeinerte Positivitätsbereiche". Journ. reine und angew. Math. 249 (1971), 143-200

[6] G. Effros and E. Størmer, "Positive projections and Jordan Structure in operator algebras". Preprint Series, Universitetet i Oslo, Mathematics no. 14, Nov. 20 (1978)

[7] E. Størmer, "Jordan algebras of type I", Acta Math. 115 (1966), 165-184

[8] M. Takesaki, "Tomita's theory of modular Hilbert Algebras and its applications", Springer Lecture notes in Mathematics no. 128 (1970)

[9] H. Araki, "Some properties of modular conjugation operator of a von Neumann Algebra and a non-commutative Radon-Nykodim theorem with a chain-rule", Pacif. Journ. Math. 50 no. 2 (1974), 309-354

[10] U. Haagerup, "The Standard form of Von Neumann Algebras", Math. Scand. 37 (1975), 271-283

[11] J. Bellissard and B. Iochum, "Homogeneous self-dual cones versus Jordan algebras. The theory revisited", Ann. Inst. Fourier, Grenoble, 28 no. 1 (1978), 27-67

[12] W. Bös, "The structure of finite homogeneous cones and Jordan algebras", Osnabrück preprint, July 1976

[13] E.M. Alfsen, F.W. Schultz and E. Størmer, "A Gelfand-Neumark theorem for Jordan algebras", Advances in Math. 28 (1978), 11-56

[14] J. Bellissard and B. Iochum, "L'algèbre de Jordan d'un cône autopolaire facialement homogène", Comptes Rend. Acad. Sciences Paris, 288 (1979), 229-232

[15] cf. reference [11] and B. Iochum "These d'Etat", to appear.

[16] J. Bellissard and B. Iochum, "Spectral theory for facially homogeneous self-dual cones", to appear in Math. Scand.

[17] U. Haagerup and H. Hanche-Olsen, private communication

[18] E.B. Vinberg, "The theory of convex homogeneous cones", Trans. Moscow Math. Soc. 12 (1963), 340-403

[19] E.B. Vinberg, "The structure of groups of automorphisms of a homogeneous convex cone", Trans. Moscow Math. Soc. 14 (1965), 63-93

[20] P. Jordan, J. von Neumann and E. Wigner, "On an algebraic generalization of the quantum mechanical formalism", Ann. of Math. 36 (1934), 29-64

[21] E. Størmer, "Real structure in the hyperfinite factor", Oslo preprint Mathematic No. 13, Sept. 1979

[22] T. Giordano, to appear (private communication)

Generators of One-Parameter Groups of *-Automorphisms on UHF-Algebras

Ola Bratteli

Institute of Mathematics

University of Oslo

Norway

Abstract

Some results and conjectures on unbounded derivations on C*-algebras are reviewed, with particular emphasis on UHF-algebras. Several conditions ensuring that a derivation is the generator of a one-parameter group of *-automorphisms is exposed, involving analytic elements, linearly increasing surface terms, and approximate commutation.

I. Introduction

A C*-algebra $\mathfrak{A}$ is called a UHF-algebra if it contains an increasing sequence $\{\mathfrak{A}_n\}$ of C*-subalgebras such that $\bigcup_n \mathfrak{A}_n$ is dense in $\mathfrak{A}$, each $\mathfrak{A}_n$ is a full complex matrix algebra, and all the $\mathfrak{A}_n$'s has a common identity element 1, [3], [13], [23]. In mathematical physics, these algebras occur in two contexts.

1. The algebra of the canonical anticommutation relations (CAR) (over a separable Hilbert space) is a UHF algebra. This follows from the fact that the CAR-algebra over a n-dimensional Hilbert space is isomorphic with the algebra of all $2^n \times 2^n$ complex matrices [3], [23].

2. The C*-algebras of certain quantum lattice systems are UHF-algebras. The lattice is a countable set L (usually $L = \mathbb{Z}^\nu$, $\nu = 1,2,\ldots$), and to each point $x \in L$ is associated a system with C*-algebra $\mathfrak{A}_{\{x\}}$ isomorphic with the $N \times N$ matrix algebra, where N is the number of "degrees of freedom" of the elementary system. Then $\mathfrak{A} = \bigotimes_{x\in L} \mathfrak{A}_{\{x\}}$ is a CAR algebra. We have $\mathfrak{A} = \overline{\bigcup_\Lambda \mathfrak{A}_\Lambda}$, where Λ ranges over all finite subsets of L, and $\mathfrak{A}_\Lambda = \bigotimes_{x\in\Lambda} \mathfrak{A}_{\{x\}}$, [2], [27].

If $\mathfrak{A}$ represents a reversible physical system which is homogeneous in time, one would expect the time development to be represented by a one-parameter group $t \in R \to \tau_t^*$ of affine automorphism of the state space $E_{\mathfrak{A}}$ of $\mathfrak{A}$ continuous in the sense that for each $A \in \mathfrak{A}$ there exists a weakly compact subset $K_{A,t}$ of $\mathfrak{A}$ such that

$$\begin{aligned} &|(\tau_t^*(\omega_1) - \tau_t^*(\omega_2))(A) \leq \sup_{B\in K_{A,t}} |(\omega_1-\omega_2)(B)| \\ &|(\tau_{-t}^*(\omega_1) - \tau_{-t}^*(\omega_2))(A)| \leq \sup_{B\in K_{A,t}} |(\omega_1-\omega_2)(B)| \end{aligned} \qquad (*)$$

for all $\omega_1, \omega_2 \in E_{\mathfrak{A}}$, and assume also that

$$t \to \tau_t^*(\omega)(A)$$

is continuous for each $\omega \in E_{\mathfrak{A}}$, $A \in \mathfrak{A}$. It follows that there exists a one-parameter group $t \to \tau_t$ of *-automorphisms of $\mathfrak{A}$ such that

$$\tau_t^*(\omega)(A) = \omega(\tau_t(A))$$

and $t \to \tau_t$ is strongly continuous i.e. $t \to \tau(A)$ is continuous in norm for each $A \in \mathfrak{A}$, [17] or [2] , Theorem 3.2.11 and Corollary 3.2.12. Although the condition (*) is not always fullfilled in realistic models, we will consider only strongly continuous one-parameter groups τ of *-automorphisms from now on.

Recall that a state ω of $\mathfrak{A}$ is called a (τ,β)-KMS state on $\mathfrak{A}$ if

$$\omega(A\tau_{i\beta}(B)) = \omega(BA)$$

for all $A,B \in \mathfrak{A}_\tau$ = the entire analytic elements for τ. Here $\beta \in \mathbb{R}$ and ω is interpreted as an equilibrium state at temperature β^{-1} (here we adopt the physicist's convention with respect to the sign of β, converse to the convention in Takesakis talk). Several alternative characterizations of the KMS-condition can be given. One useful one is in terms of the infinitesimal generator δ of τ, defined as the operator from $D(\delta) \subseteq \mathfrak{A}$ into $\mathfrak{A}$ given by

$$D(\delta) = \{A \in \mathfrak{A},\ \lim_{t\to o}\frac{1}{t}(\tau_t(A)-A) \text{ exists in norm } (\Leftrightarrow \text{ exists weakly})\}$$

$$\delta(A) = \lim_{t\to o}\frac{1}{t}(\tau_t(A)-A)\ ,\ A \in D(\delta).$$

Note that δ is a *-derivation, i.e.

$$\delta(AB) = \delta(A)B + A\delta(B)$$

$$\delta(A^*) = \delta(A)^*$$

for all $A, B \in D(\delta)$. An equivalent version of the KMS condition, known as the autocorrelation lower bounds, is then

$$-i\ \beta\ \omega(A^*\delta(A)) \geq \omega(A^*A)\ \ \log\left(\frac{\omega(A^*A)}{\omega(AA^*)}\right)$$

for all $A, B \in \mathfrak{A}$, [30], [2].

A state ω is called a ground state, or a $(\tau, +\infty)$-KMS state if

$$-i\omega(A^* \delta(A)) > 0$$

for $A \in D(\delta)$. Similarly ceiling states, or $(\tau,-\infty)$-KMS states, are defined by reversal of the last inequality.

Consider the following three conjectures for a UHF-algebra $\mathfrak{A}$, [28], [25]

Conjecture 1: Each one-parameter group τ admits KMS states at all values $\beta \in \mathbb{R} \cup \{\pm \infty\}$.

Conjecture 2: For each τ there exists a sequence $H_n = H_n^* \in \mathfrak{A}$ such that

$$\tau_t(A) = \lim_{n\to\infty} e^{itH_n} A e^{-itH_n}$$

for each $A \in \mathfrak{A}$, $t \in \mathbb{R}$ (uniformly for t in compacts).

Conjecture 3: For each τ there exists an increasing sequence $\{\mathfrak{A}_n\}$ of full matrix subalgebras of $\mathfrak{A}$ such that $\bigcup_n \mathfrak{A}_n \subseteq D(\delta)$ and $\bigcup_n \mathfrak{A}_n$ is a core for δ (i.e. δ is the closure of its restriction to $\bigcup_n \mathfrak{A}_n$ where δ is the infinitesimal generator of τ).

We will show in a moment that these conjectures are listed in order of increasing strength, but first we will mention two results that seems to corroborate the conjectures and one result that points to some cohomological problems which may indicate a negative result.

Theorem 4 [28]: For each τ there exists an increasing sequence $\{\mathfrak{A}_n\}$ of full matrix subalgebras of $\mathfrak{A}$ such that $\bigcup_n \mathfrak{A}_n$ is dense in $\mathfrak{A}$, and $\bigcup_n \mathfrak{A}_n$ consists of analytic elements for δ.

This theorem does not imply conjecture 3 since $\bigcup_n \mathfrak{A}_n$ is not necessarily mapped into itself by δ.

Proposition 5 [24]: If α is an automorphism of $\mathfrak{A}$, then there exists a sequence $\{U_n\}$ of unitaries in $\mathfrak{A}$ such that

$$\alpha(A) = \lim_{n\to\infty} U_n A U_n^*$$

for all $A \in \mathfrak{A}$.

The last Proposition seemingly reduces the proof of Conjecture 2 to a cohomological problem. This problem might be non-trivial, however, as indicated by the following example.

Example 6 [9]: If $\mathfrak{A}$ is the CAR-algebra, there exists an automorphism α of $\mathfrak{A}$ and a unitary $U \in \mathfrak{A}$ such that

α is outer

$\alpha^2(A) = U A U^*$

$\alpha(U) = -U$

$\alpha^4 = 1$.

Note in this connection that the existence of a unitary $U \in \mathfrak{A}$ such that for example

$$\tau_t(U) = e^{it}U$$

would imply the non-existence of (τ,β)-KMS states for all $\beta \in (\mathbb{R} \cup \{\pm\infty\})\setminus\{0\}$.

Note also that the Conjectures 1-3, if they are true at all, are very specific for UHF-algebras $\mathfrak{A}$, and do not hold for general simple C*-algebras. This is connected with the abundance of projections in UHF-algebras, i.e. UHF algebras are in some sense the nonabelian analogue of the C*-algebra of continuous functions on the Cantor set, and the latter algebra admits no non-trivial strongly continuous oneparameter groups of *-automorphisms, [4]. In his talk, Cuntz gave examples of C*-dynamical systems $(\mathfrak{A}, \tau)$ where $\mathfrak{A}$ is simple and τ has a (τ,β)-KMS state at only one value $\beta \in \mathbb{R} \setminus \{0\}$, and he also gave an example where $(\mathfrak{A}, \tau)$ has a ground state, but no KMS state for finite β, [10], [21]. It is known that the crossed product of the C*-algebra $C(\mathbb{T})$ of continuous functions on the circle by the automorphism defined by an irrational rotation is simple, [11], and if one lifts the oneparameter group given by rotations of $\mathbb{T}$ to this crossed product in the canonical manner, one obtains a C*-dynamical system $(\mathfrak{A},\tau)$ having a trace (= $(\tau,0)$-KMS state) but no KMS state at values $\beta \neq 0$, and no ground or ceiling states. One can even show that given any closed set $F \subseteq \mathbb{R} \cup \{\pm\infty\}$, then there exists a C*-dynamical system $(\mathfrak{A},\tau)$

where $\mathfrak{A}$ is simple, τ is periodic with period 1, and $(\mathfrak{A},\tau)$ has (τ,β)-KMS states if and only if $\beta \in F$, [7].

We now prove the interrelationships between the conjectures.

Conjecture 3 $\Rightarrow$ Conjecture 2 [12], [25]: Let δ be any *-derivation with $\bigcup_n \mathfrak{A}_n \subseteq D(\delta)$. Fix n, and let $\{E_{ij}\}$ be matrix units for $\mathfrak{A}_n$. Define

$$H_n = -i \sum_k \delta(E_{k1}) E_{1K} .$$

One then computes that

$$H_n = H_n^*$$

and

$$\delta(A) = [iH_n, A]$$

for $A \in \mathfrak{A}_n$. Define $\delta_n(A) = [iH_n, A]$ for $A \in \mathfrak{A}$. If δ is a generator, and $\bigcup_n \mathfrak{A}_n$ is a core for δ, then δ_n converges to δ in the graph sense. This means that $A \in D(\delta)$ if and only if there exists a sequence $A_n \in \mathfrak{A}$ such that $A_n \to A$ and $\delta_n(A_n)$ converges, and in this case $\delta(A) = \lim_n \delta_n(A_n)$. It follows from a version of the Trotter-Kato theorem that

$$\tau_t(A) = \lim_{n\to\infty} e^{t\delta_n}(A) = \lim_{n\to\infty} e^{itH_n} A e^{-itH_n}$$

uniformly for t in compacts, [2], [19].

Conjecture 2 $\Rightarrow$ Conjecture 1, [25], see also [14]: This proof depends only on the existence of a normalized trace tr and an identity on $\mathfrak{A}$. Define

$$\omega_n(A) = \mathrm{tr}\,(e^{-\beta H_n}A) \,/\, \mathrm{tr}\,(e^{-\beta H_n}).$$

Then ω_n is a $(e^{t\delta_n}, \beta)$-KMS state on $\mathfrak{A}$ (and unique such in the UHF-case due to the uniqueness of tr). If ω is any weak *-limit point of $\{\omega_n\}$, one may exploit graph convergence of $\delta_n \to \delta$ and the auto-correlation lower bounds to show that ω is a (τ,β)-KMS state, [2].

Similarly one gets τ-ground states as weak *-limits of (τ,β)-KMS states as $\beta \to \infty$, or as weak *-limits of $e^{t\delta_n}$ - ground states as $n \to \infty$.

II. Dissipative Operators on Inductive Limits of Banach Spaces

The introduction, and in particular Conjecture 3, motivates the following problem. Given a *-derivation δ on a UHF-algebra $\mathfrak{A} = \bigcup_n \mathfrak{A}_n$, when does δ extend by closure to a generator of a one-parameter group of *-automorphisms? A general criterion for this is given by the following proposition. We use the notion dissipative operator in the sense of Lumer-Phillips, i.e., a linear operator L on a Banach space X with dense domain D(L) is called dissipative if for each nonzero $A \in D(L)$ there exists a nonzero linear functional $f \in X^*$ which is a tangent-functional at A in the sense $f(A) = ||f||\ ||A||$, and $\text{Re}\, f(L(A)) \leq 0$. Note that generators of contraction semigroups are dissipative. If L is dissipative, then L is closeable and its closure $\bar{L}$ is dissipative [20]. If L is dissipative and $A \in D(L)$, $A \neq 0$, then $\text{Re}\, f(L(A)) \leq 0$ for all tangent-functionals f at A, [1].

Proposition 7: Let δ be a *-derivation on a UHF-algebra $\mathfrak{A} = \bigcup_n \mathfrak{A}_n$ with $D(\delta) = \bigcup_n \mathfrak{A}_n$. Then $\pm\delta$ are dissipative, and in particular δ is closeable. Hence the closure $\bar{\delta}$ is a generator of a one-parameter group of automorphisms if one of the following two conditions are fulfilled:

1. $\bigcup_n \mathfrak{A}_n$ consists of analytic elements for $\bar{\delta}$.
2. $(\lambda-\delta)(\bigcup_n \mathfrak{A}_n)$ is dense in $\mathfrak{A}$ for one positive and one negative λ.

Proof: Once the dissipativeness is established. 1. follows from [20] and 2. from [5], see also [2]. A general result in this direction is true: If δ is a *-derivation on a C*-algebra $\mathfrak{A}$ with identity $1 \in D(\delta)$, and $D(\delta)$ is closed under the square root operation of positive elements, then δ is dissipative [18]. In the special case covered by the proposition one may however give the following simple argument: By the proof of Conjecture 3 $\Rightarrow$ Conjecture 2 there exists $H_n = H_n^* \in \mathfrak{A}$ such that if $\delta_n = \text{ad}(iH_n)$ one has $\delta|\mathfrak{A}_n = \delta_n|\mathfrak{A}_n$. But δ_n is clearly the generator of the group $t \to \text{Ad}(e^{itH_n})$

hence δ_n is dissipative and $\delta|\mathfrak{A}_n$ is dissipative. Thus δ is dissipative.

We will give five more manageable conditions ensuring that $\bar{\delta}$ is a generator, in Theorem 8, Corollary 11, Corollary 14 and Proposition 15. The proofs consist in all cases in verifying that one of the conditions 1 or 2 are satisfied. Condition 1 holds under the following by now classical assumption.

<u>Theorem 8 [26], [27]:</u> Let $\mathfrak{A} = \bigcup_\Lambda \mathfrak{A}_\Lambda$ be the C*-algebra of a quantum lattice system, and for each finite subset $\Lambda \subset L$, let $\Phi(\Lambda)$ be a self-adjoint element in $\mathfrak{A}_\Lambda$ (the function $\Lambda \to \Phi(\Lambda)$ is called an interaction). If

$$||\Phi||_\lambda = \sum_{n\geq o} e^{\lambda n} \ (\sup_{x\in L} \Sigma \ \{||\Phi(\Lambda)||; \quad \Lambda \ni x, \quad |\Lambda| = n+1\})$$

is finite for a $\lambda > 0$ (where $|\Lambda|$ = the number of elements in Λ) it follows that the *-derivation given by

$$\delta(A) = \sum_\Lambda [i\Phi(\Lambda), A]$$

is well defined on $\bigcup_n \mathfrak{A}_n$, and $\bigcup_n \mathfrak{A}_n$ consists of analytic elements for $\bar{\delta}$. In particular $\bar{\delta}$ is a generator.

The other conditions are less tied up with the C*-algebraic situation we have described, and it is clearer to state them in a more general framework. Let X be a Banach space containing an increasing sequence $\{V_n\}$ of closed subspaces such that $\bigcup_n V_n$ is dense in X. Let L be a dissipative operator with $D(L) = \bigcup_n V_n$. Then the triple $(X, \{V_n\}, L)$ corresponds to $(\mathfrak{A}, \{\mathfrak{A}_n\}, \delta)$ and the Lumer-Phillips theory implies that $\bar{L}$ is the generator of a strongly continous semi-group of contractions if and only if $(\lambda-L)(\bigcup_n V_n)$ is dense for some $\lambda > 0$.

It is one trivial case where $\bar{L}$ is a generator, namely when $L\, V_n \subseteq V_n$ for all n. The restriction of L to V_n then generates a norm-continous semigroups of contractions on V_n, and $\bar{L}$ is the generator of the semigroup U_t defined by

$$U_t|_{V_n} = \exp\{tL|_{V_n}\}.$$

In quantum lattice systems this corresponds to the existence of an increasing sequence of regions with no surface energy, that is zero interaction between the points inside and the points outside the region. One would expect that $\bar{L}$ still would be a generator if the surface energy does not grow to fast with the region, and the next two theorems establish that this is indeed the case (assume throughout that L is dissipative on $D(L) = \bigcup_n V_n$).

Theorem 9 [1]: Let $Q_n: X \to X/V_n$, be the quotient map onto the quotient Banach space X/V_n. $\bar{L}$ is a generator if

$$\sup_n ||Q_n L|_{V_n}|| < +\infty .$$

Theorem 10 [6]: $\bar{L}$ is a generator if there exists constants M, $\alpha > 0$ and linear operators $L_{n,m}: V_n \to V_{n+m}$, n = 1,2,...; m = 0, 1,2,.... such that

$$||L|_{V_n} - L_{n,m}|| \leq M n e^{-\alpha m}.$$

(Alternatively one can allow a slower decrease in m by sacrificing the rapid growth in n, i.e.

$$||L|_{V_n} - L_{n,m}|| \leq M\log(n+1)\ (m+1)^{-1-\alpha}).$$

Before indicating the proof of the last theorems we state an immediate Corollary of these theorems and Proposition 7. This Corollary is stronger than Theorem 8 in one and two dimensional quantum lattice systems, but we shall argue later that the corresponding strong result is false in three and higher dimensions.

Corollary 11: Let $\mathfrak{A} = \overline{\bigcup_n \mathfrak{A}_n}$ be a UHF algebra, δ a *-derivation ith $D(\delta) = \bigcup_n \mathfrak{A}_n$ and $H_n = H_n^*$ elements in $\mathfrak{A}$ such that $\delta|_{\mathfrak{A}} = ad(iH_n)|_{\mathfrak{A}_n}$. $\bar{\delta}$ is a generator if there exists $K_n = K_n^* \in \mathfrak{A}_n$ such that

$$||H_n - K_n|| = O(1) \qquad \text{as } n \to \infty$$

or

$$||H_n - K_{n+m}|| = O(n)e^{-\alpha m}$$

for some $\alpha > 0$.

Outline of proofs of Theorems 9 and 10. There exists a very simple argument for Theorem 9 in the case that there exist linear operators $L_n: V_n \to V_n$ such that

$$||L|_{V_n} - L_n|| \le M < \infty \qquad , n = 1,2,\ldots$$

(Such operators exist under the hypothesis of the theorem if there are projections E_n of norm one from X onto V_n, set $L_n = E_n L E_n$.) Since L is dissipative, it follows from this estimate that $L_n - M$ is dissipative and hence

$$||(\lambda - L_n)^{-1}|| = ||((\lambda - M) - (L_n - M))^{-1}||$$

$$\le \frac{1}{\lambda - M}$$

for $\lambda > M$. Let $\lambda > 2M$, and let $f \in X^*$ be in the annihilator of the range $R(\lambda - L)$, i.e.

$$\lambda f = L^* f.$$

If $A \in V_n$ one has

$$|f(A)| = |f((\lambda - L_n)(\lambda - L_n)^{-1}A)|$$

$$= |f((L - L_n)(\lambda - L_n)^{-1}A)|$$

$$\le ||f|| M(\lambda - M)^{-1}||A||.$$

Since $\bigcup_n V_n$ is dense, it follows that $||f|| \le \frac{M}{\lambda - M}||f||$ and hence $||f|| = 0$, i.e. $f = 0$, $(\lambda - L)(\bigcup_n V_n)$ is dense and $\bar{L}$ is a generator. The proof of the general version of Theorem 9 is much more complicated, see [1].

The idea of proof of Theorem 10 is to use an expansion reminiscent to the Dyson perturbation expansion. The latter expansion is obtained by iteration from the Duhamels formula

$$e^{t(S+P)} - e^{tS} = \int_o^t ds \; e^{sS} \, P \, e^{(t-s)(S+P)}$$

and has the form

$$e^{t(S+P)} = e^{tS}$$

$$+ \sum_{n\geq 1} \int_{o\leq s_1\leq \ldots \leq s_n\leq t} ds_1 \ldots ds_n \, e^{s_1 S} \, P \, e^{(s_2-s_1)S} \, P \ldots e^{(s_n-s_{n-1})S} \, P \, e^{(t-s_n)S} .$$

It is valid if S is the generator of a contraction semigroup and P is bounded. We will use a version of this expansion where the group e^{tS} and the perturbation P is varied in each step. By subtracting suitable multiples of the identity we may assume that each $L_{n,o}$ is dissipative on V_n, and by a change of scale we may assume

$$||L|_{V_n} - L_{n,m}|| \leq n \, e^{-\alpha m}.$$

Set $L(n,m) = L_{n,m}$. If $A \in V_{n_o}$, we define the terms in the expansion of $e^{tL} A$ inductively as follows

$$\sigma_t(0)(A) = A$$

$$\sigma_t(1;1)(A) = \int_o^t ds \, e^{(t-s)L(n_o+1,o)} L(n_o,1)\sigma_s(0)(A)$$

and

$$\sigma_t(k;\, n_1,\ldots,\, n_k)(A)$$

$$= \int_o^t dse^{(t-s)L(n_o+\ldots+n_k,o)} \{L(n_o+\ldots+n_{k-1},n_k)$$

$$-L(n_o+\ldots+n_{k-1},\, n_k-1)\} \, \sigma_s(k-1;n_1,\ldots,n_{k-1})(A)$$

if $k \geq 2$, or if $k = 1$, $n_1 \geq 2$. Here $n_i = 1,2,\ldots,\; i = 1,2,\ldots,k$. Define $M = \max\{||L(n_o,1)||,1\}$. Then one use the fact that $e^{tL(n,o)}$ are semigroups of contraction to inductively derive the estimates

$$||\sigma_t(0)(A)|| \leq ||A||$$

$$||\sigma_t(1,1)(A)|| \leq |t|M||A||$$

and

$$||\sigma_t(k;n_1,\ldots,n_k)(A)||$$

$$\leq M||A||\frac{n_o(n_o+n_1)\ldots(n_o+\ldots+n_{k-1})}{k!}\, e^{-\alpha(n_1+\ldots+n_k)}\,(2e^{\alpha}\, t)^k$$

$$\leq M||A||e^{\alpha' n_o}\, e^{-(\alpha-\alpha')(n_1+\ldots+n_k)}\,(\frac{2e^{\alpha}}{\alpha'}\, t)^k$$

where $0 < \alpha' < \alpha$. The last inequality follows from the estimate

$$\frac{n_o(n_o+n_1)\ldots(n_o+\ldots+n_{k-1})}{k!}$$

$$\leq \frac{(n_o+\ldots+n_{k-1})^k}{k!}\,\frac{\alpha'^k}{\alpha'^k} \leq e^{\alpha'(n_o+\ldots+n_{k-1})}\,\frac{1}{\alpha'^k}\,.$$

Note that the $O(n)$-part of the estimate

$$||L|V_n - L(n,m)|| \leq O(n)\, e^{-\alpha m}$$

is important for the bounds on the σ_t's we have derived here, and that the argument for the convergence of the perturbation expansion breaks down if $O(n)$ is replaced by $O(n^{1+\varepsilon})$ where $\varepsilon > 0$. Probably the theorem itself is wrong in this case, and we shall later exhibit explicit counterexamples in the case $\varepsilon = 1$.

The convergence of the expansion follows from

$$\sum_{n_1=1}^{\infty}\ldots\sum_{n_k=1}^{\infty} e^{-(\alpha-\alpha')(n_1+\ldots+n_k)}$$

$$= (\sum_{n=1}^{\infty} e^{-(\alpha-\alpha')n})^k = (\frac{e^{-(\alpha-\alpha')}}{1-e^{-(\alpha-\alpha')}})^k$$

and hence

$$\sum_{k=1}^{\infty}\sum_{n_1=1}^{\infty}\ldots\ldots\sum_{n_k=1}^{\infty}||\sigma_t(k;\, n_1,\ldots,n_k)(A)||$$

is dominated by a geometric series convergent for $t \in [0, t_o\rangle$, $t_o = \frac{\alpha'(1-e^{-(\alpha-\alpha')})}{2e^{\alpha} e^{-(\alpha-\alpha')}}$. Thus we may define

$$\tau_t(A) = \sigma_t(0)(A)$$

$$+ \sum_{k=1}^{\infty} \sum_{n_1=1}^{\infty} \cdots \sum_{n_k=1}^{\infty} \sigma_t(k; n_1, \ldots, n_k)(A).$$

Then, using estimates similar to the ones derived above and also using that $\bar{L}$ is a closed operator one may show that

$$\int_s^t du \quad \tau_u(A) \in D(\bar{L})$$

and

$$\tau_t(A) - \tau_s(A) = \bar{L}(\int_s^t du \ \tau_u(A))$$

for $0 \le s \le t < t_o$. Since $\bar{L}$ is dissipative it follows from this relation that $t \to ||\tau_t(A)||$ is non-increasing in t.

Assume ad absurdum that $\bar{L}$ is not a generator, i.e., that $R(\lambda-L)$ is not dense for a $\lambda > 0$, and let $f \in X^*$ be a linear functional with $||f|| = 1$ and $f(R(\lambda-L)) = \{0\}$. Then

$$\lambda f(\int_0^t ds \ \tau_s(A)) = f(\bar{L}(\int_0^t ds \ \tau_s(A)))$$

$$= f(\tau_t(A)) - f(A)$$

where the last equality follows from (*). Hence $t \to f(\tau_t(A))$ satisfies the integral equation

$$f(\tau_t(A)) = f(A) + \lambda \int_0^t ds \ f(\tau_s(A))$$

and the unique solution to this equation is

$$f(\tau_t(A)) = e^{\lambda t} f(A).$$

But by a suitable choice of $A \in \bigcup_n V_n$ this contradicts that

$t \to ||\tau_t(A)||$ is non-increasing. It follows that $R(\lambda - L)$ is dense and $\bar{L}$ is a generator.

A posteriori it also follows that the perturbation expansion

$$e^{t\bar{L}}(A) = \tau_t(A) = \sigma_t(0)(A)$$

$$+ \sum_{k=1}^{\infty} \sum_{n_1=1}^{\infty} \cdots \sum_{n_k=1}^{\infty} \sigma_t(k; n_1, \ldots, n_k)(A)$$

is valid when $t \to [0, t_o>$.

In the Hilbert-space case the $O(n)$-growth of Theorem 10 can be relaxed a little bit, i.e. one can allow $O(n\log n)$-growth, but not $O(n^{1+\varepsilon})$-growth with $\varepsilon > 0$. This is a consequence of

Theorem 11 [15]: Let $\mathfrak{h}$ be a Hilbert space, $\{\mathfrak{h}_n\}$ an increasing sequence of closed subspaces with $\bigcup_n \mathfrak{h}_n$ dense in $\mathfrak{h}$, $E_n : \mathfrak{h} \to \mathfrak{h}_n$ the orthogonal projection from $\mathfrak{h}$ onto $\mathfrak{h}_n$ and H a symmetric operator with domain $\bigcup_n \mathfrak{h}_n$. Assume that

$$H\mathfrak{h}_n \subseteq \mathfrak{h}_{n+1}$$

and

$$\sum_{n\geq 1} ||H E_n - E_n H E_n||^{-1} = +\infty.$$

It follows that $\bar{H}$ is self-adjoint.

If H is positive, the theorem remains true, when the last condition is replaced by

$$\sum_{n\geq 1} ||H E_n - E_n H E_n||^{-\frac{1}{2}} = +\infty$$

i.e. one can allow $O(n^2)$-growth in this case, [16]. One can however not allow $O(n^2)$-growth for symmetric operators in general, and experience from the theory of quasi-analytic vectors, [8] would lead one to believe that the estimate of Theorem 11 is the best possible.

Example 12 [15]: There exists a Hilbert space $\mathfrak{h}$, an increasing sequence $\mathfrak{h}_n$ of finite dimensional subspaces with $\overline{\bigcup_n \mathfrak{h}_n} = \mathfrak{h}$, a symmetric operator H on $\bigcup_n \mathfrak{h}_n$ such that

$$H \mathfrak{h}_n \subseteq \mathfrak{h}_{n+1}$$

and

$$\| H E_n - E_n H E_n \| = O(n^2)$$

but H is not essentially self-adjoint on $\bigcup_n \mathfrak{h}_n$ (E_n is the orthogonal projection from $\mathfrak{h}$ onto $\mathfrak{h}_n$).

One example is $H = -i\frac{d}{dx}$ which has deficiency indices (0,1) on $\mathfrak{h} = L^2(0,\infty)$. Define

$$\psi_n(x) = x^{-n} \exp(-x-x^{-1})$$

for $x > 0$ and $n = 0,1,2,\ldots$ Then one can show that

$$H\psi_n = -i(\psi_{n+2} - n\psi_{n+1} - \psi_n)$$

and, setting $\mathfrak{h}_n$ = linear span of $\psi_o, \psi_1, \cdots \psi_{2n}$, the conditions of the example are fulfilled.

Another interesting example is obtained by considering $H = -\frac{d^2}{dx^2} + x^2 - x^4$ on the linear span of the Hermite functions in $L^2(\mathbb{R})$. Setting $\mathfrak{h}_n$ = the linear span of the 4n first Hermite functions the conditions of the example are fulfilled, but H has deficiency indices (2,2) on the Hermite functions. In contrast $-\frac{d^2}{dx^2} + x^2 + x^4$ is essentially selfadjoint on the Hermite functions, for example by the remark after Theorem 11.

III. Related Results

If a sequence of bounded dissipative operators approximately commute in an appropriate sense, their graph limit is a generator:

Theorem 13 [6]: Let $\{L_n\}$ be a sequence of bounded, dissipative operators on a Banach space X with densely defined graph limit L. If

$$\| L_n L_m - L_m L_n \| \le M < +\infty$$

for all n, m, and $S_n = \lim_{m\to\infty} L_nL_m - L_mL_n$ exists strongly for all n, then L is a generator and

$$e^{tL} A = \lim_{n\to\infty} e^{tL_n} A \; ; \; A \in X, \; t \in \mathbb{R}_+ .$$

A corollary is that quantum lattice systems which has interactions which are close to being classical (in partical which is classical) has a timedevelopment in the thermodynamic limit.

Corollary 14 [6]: Let $\mathfrak{A}$ be a C*-algebra, $\{K_n\}_{n\geq n}$ a sequence of self-adjoint elements in $\mathfrak{A}$, and define $\delta_{iK_n}(A) = [iK_n, A]$, $A \in \mathfrak{A}$. If the graph limit δ of δ_{iK_n} is densely defined, and $K_n \in D(\delta)$ and

$$||K_nK_m - K_mK_n|| \leq M < +\infty$$

then δ is a generator and

$$e^{t\delta}(A) = \lim_{n\to\infty} e^{itK_n} A \, e^{-itK_n} \; ; \; A \in \mathfrak{A}, \; t \subset$$

Proof: Use Theorem 13 and the identity

$$\delta \circ \delta_{iK_n} - \delta_{iK_n} \circ \delta = \delta_{i\delta(K_n)} .$$

A related result is:

Proposition 15 [22]: Let $\mathfrak{A}$ be a C*-algebra and $\{K_n\}_{n\geq 1}$ a sequence of self-adjoint elements in $\mathfrak{A}$ with densely defined graph limit δ. If $[K_n, K_m] \to 0$ as $n,m \to \infty$ it follows that δ is a generator.

It is interesting to note that any derivation δ on a UHF-algebra $\mathfrak{A} = \bigcup_n \mathfrak{A}_n$, with $D(\delta) = \bigcup_n \mathfrak{A}_n$, is a sum of two *-derivations δ_1, δ_2 defined on $\bigcup_n \mathfrak{A}_n$ whose closures are generators, even when δ itself cannot be extended to a generator. This follows from:

Theorem 16 [29]: Let $\mathfrak{A} = \overline{\bigcup_n \mathfrak{A}_n}$ be a UHF algebra, and δ a *-derivation on $\mathfrak{A}$ with domain $\bigcup_n \mathfrak{A}_n$.

1. For all $\varepsilon > 0$ there exists a *-derivation δ_ε with $D(\delta_\varepsilon) = \bigcup_n \mathfrak{A}_n$,

$||\delta-\delta_\varepsilon|| < \varepsilon$ and $\delta_\varepsilon\ (\bigcup_n \mathfrak{A}_n) \subseteq \bigcup_n \mathfrak{A}_n$.

2. If $\delta(\mathfrak{A}_n) \subseteq \mathfrak{A}_{n+1}$, then there exists *-derivations δ_1 and δ_2 with $D(\delta_1) = D(\delta_2) = \bigcup_n \mathfrak{A}_n$; $\delta_1(\mathfrak{A}_{2n}) \subseteq \mathfrak{A}_{2n}$, $\delta_2(\mathfrak{A}_{2n+1}) \subseteq \mathfrak{A}_{2n+1}$ for $n = 1,2,\ldots,$ and $\delta = \delta_1 + \delta_2$.

References

[1] Batty, C.J.K., Dissipative mappings with approximately invariant subspaces, Preprint.

[2] Bratteli, O. and D.W. Robinson, Operator algebras and quantum statistical mechanics, 2 vols., Texts and Monographs in Physics, Springer Verlag Berlin-Heidelberg-New York, to appear.

[3] Bratteli, O., Inductive limits of finite-dimensional C*-algebras, Trans.Americ.Math.Soc. 171, (1972), 195-234.

[4] Bratteli, O. and G.A. Elliott, Structure spaces of approximately finite-dimensional C*-algebras II, Jour.Func.Anal., 30(1978), 74-82.

[5] Bratteli, O. and D.W. Robinson, Unbounded derivations of C*-algebras II, Comm.Math.Phys. 46, (1976), 11-30.

[6] Bratteli, O. and A. Kishimoto, Generation of semigroups, and two-dimensional quantum lattice systems, Jour.Func.Anal., to appear.

[7] Bratteli, O., G.A. Elliott and R. Herman, On the possible temperatures of a dynamical system, to appear.

[8] Chernoff, P.R., Quasi-analytic vectors and quasianalytic functions, Bull.Amer.Math.Soc. 81 (1975), 637-646.

[9] Connes, A., Outer Conjugacy classes of automorphisms of factors, Ann.sci.Ec.Norm.Sup., 4^{e} serie, 8, (1975), 383-420.

[10] Cuntz, J., Simple C*-algebras generated by isometries, Comm.Math.Phys. 57 (1977), 173-185.

[11] Effros, E.C. and F. Hahn, Locally compact transformation groups and C*-algebras, Mem.Amer.Math.Soc. 75, (1967).

[12] Elliot, G.A., Derivations of matroid C*-algebras, Invent. Math. 9, (1970), 253-269.

[13] Glimm, J., On a certain class of operator algebras, Trans. Amer.Math.Soc. 95, (1960), 318-340.

[14] Jørgensen, P.E.T., Trace states and KMS states for approximately inner dynamical one-parameter groups of *-automorphisms, Comm.Math.Phys. 53 (1977), 135-142.

[15] Jørgensen, P.E.T., Approximately reducing subspaces for unbounded linear operators, Jour.Func.Anal. 23, (1976), 392-414.

[16] Jørgensen, P.E.T., Essential self-adjointness of semibounded operators, Math. Ann. 237 (1978), 187-192.

[17] Kadison, R.V., Transformation of states in operator theory and dynamics, Topology 3,(1965), 177-198.

[18] Kishimoto, A., Dissipations and derivations, Comm.Math. Phys. 47,(1976), 25-32.

[19] Kuntz, T.G., Extensions of Trotter's operator semigroup approximation theorems, Jour.Func.Anal. 3,(1969), 354-375.

[20] Lumer, G. and R.S. Phillips, Dissipative operators in a Banach space, Pac.J.Math. 11, (1961), 679-698.

[21] Olesen, D. and G.K. Pedersen, Some C*-dynamical systems with a single KMS-state, Math. Scand. 42 (1978), 111-118.

[22] Olesen, D. and G.K. Pedersen, Groups of automorphisms with spectrum condition and the lifting problem, Comm.Math.Phys. 51, (1976), 85-95.

[23] Powers, R., Representations of the canonical anticommutation relations, Thesis, Princeton University (1967).

[24] Powers, R., Representations of uniformely hyperfinite algebras and their associated von Neumann rings, Ann.Math. 86, (1967), 138-171.

[25] Powers, R. and S. Sakai, Existence of ground states and KMS-states, Comm.Math.Phys. 39, (1975), 273-288.

[26] Robinson, D.W., Statistical mechanics of quantum spin systems II, Comm.Math.Phys. 7, (1968), 337-348.

[27] Ruelle, D., Statistical mechanics, Benjamin, New York (1969).

[28] Sakai, S., On one-parameter groups of *-automorphisms on operator algebras and the corresponding unbounded derivations, Amer.J.Math. 98, (1976), 427-440.

[29] Sakai, S. The theory of unbounded derivations in C*-algebras, Copenhagen and Newcastle lecture notes (1977).

[30] Sewell, G.L., KMS-conditions and local thermodynamic stability II, Comm.Math.Phys. 55, (1977), 53-61.

[31] Yosida, K., Functional analysis, 5th Ed., Springer-Verlag, Berlin-Heidelberg-New York (1978).

[illegible]

[26] Robinson, D.W., [illegible]

[illegible]

[illegible] Math. Phys. 98 (1977) [illegible]

[illegible] theory [illegible] Copenhagen and Newcastle lecture notes (1977).

[30] Sewell, G.L., KMS-conditions and local thermodynamical stability [illegible] Comm. Math. Phys. [illegible]

[31] Yosida, K., Functional analysis, 5th ed., Springer-Verlag, Berlin-Heidelberg-New York (1978).

Automorphisms of Certain Simple C*-algebras

J. Cuntz

Institut für angewandte Mathematik

Universität Heidelberg

Federal Republic of Germany

Abstract

It has been observed by M. Takesaki (private communication) that for the C^*-algebras $\mathcal{O}_n$ there is a bijective correspondence between unitaries in $\mathcal{O}_n$ and endomorphisms that leave the unit fixed. We use this idea to study automorphisms of $\mathcal{O}_n$. The group Aut $\mathcal{O}_n$ is shown to contain a maximal abelian subgroup T (that is canonically associated with the construction of $\mathcal{O}_n$ from a topological Markov chain; cf. the paper by W. Krieger and the author: A class of C^*-algebras and topological Markov chains). This maximal commutative subgroup of Aut $\mathcal{O}_n$ is similar to a maximal torus in a semisimple Lie group in that it is an inductive limit of tori, and its Weyl group, i.e. the quotient of its normalizer in Aut $\mathcal{O}_n$ by T itself is discrete.

Then one-parameter subgroups of Aut $\mathcal{O}_n$ that are contained in T, are studied. In particular, one class of such groups, related to Bernoulli-shifts in $\mathcal{O}_n$ is shown to have unique KMS-states. On the other hand, (unlike to the von Neumann algebra case) any two different one-parameter groups in this class are not exterior equivalent, i.e. do not just differ by a unitary cocycle. Finally, automorphisms in $N(T)$ and their action on a certain class of pure states of $\mathcal{O}_n$, related to T, are studied.

Automorphisms of Certain Simple C*-Algebras

In [2] it has been shown that, among infinite simple C*-algebras, certain algebras $\mathcal{O}_n$ play a kind of universal role, in that these algebras are contained in every simple C*-algebra which contains an infinite projection. $\mathcal{O}_n$ can be constructed as follows. On an infinite-dimensional Hilbert space H take isometries $S_1,\ldots,S_n$ from H onto orthogonal subspaces $H_1,\ldots,H_n$ such that $H = H_1 \oplus \ldots \oplus H_n$. Algebraically, this is expressed by the relations $S_i^* S_i = 1$, $\sum_{i=1}^{n} S_i S_i^* = 1$. In the language of Roberts [4], the S_i span an n-dimensional Hilbert space in L(H). The following is the main result of [1].

Theorem 0: The C*-algebra $C^*(S_1,\ldots,S_n)$ generated by $S_1,\ldots,S_n$ is simple. If $\hat{S}_1,\ldots,\hat{S}_n$ is another family of isometries satisfying the same relations then the map $S_i \to \hat{S}_i\,(i=1,\ldots,n)$ extends to an isomorphism from $C^*(S_1,\ldots,S_n)$ onto $C^*(\hat{S}_1,\ldots,\hat{S}_n)$.

(The isomorphism class of) $C^*(S_1,\ldots,S_n)$ is denoted by $\mathcal{O}_n$. There is a natural one-parameter group $\{\lambda_t\}$ of automorphisms of $\mathcal{O}_n$ defined by $\lambda_t(S_k) = n^{-it} S_k$ $(k = 1,\ldots,n)$. This group has a unique KMS-state ϕ at inverse temperature 1 [3]. The group $\{\lambda_t\}$ may be considered as the modular automorphism group for ϕ. The fixed point algebra F for $\{\lambda_t\}$ is a UHF-algebra of type n^∞ and we have a natural projection $f: \mathcal{O}_n \to F$.

In the following we study the group Aut $\mathcal{O}_n$ of automorphisms of $\mathcal{O}_n$ which looks in many respects like an infinite-dimensional semi-simple Lie group. Also, we show that certain natural one-parameter groups of automorphisms behave like modular automorphism groups for a von Neumann algebra, while on the other hand, there is no unique (modulo inner automorphisms) modular automorphism group for $\mathcal{O}_n$.

§1 Generalities about Automorphisms of $\mathcal{O}_n$

Let α be any endomorphism of $\mathcal{O}_n$ such that $\alpha(1) = 1$. Then there is a unitary $Z_\alpha \in \mathcal{O}_n$ such that $\alpha(S_k) = Z_\alpha^* S_k$ $(k = 1,\dots,n)$. In fact, $Z_\alpha^* = \sum_{i=1}^{n} \alpha(S_i)S_i^*$. On the other hand by Theorem 0 every unitary $U \in \mathcal{O}_n$ determines a unique (isometric) endomorphism λ_U of $\mathcal{O}_n$ such that $\lambda_U(S_i) \subset U^*S_i$ and $\lambda_U(1) = 1$.

Then Z_α satisfies a cocycle identity and λ is a "twisted representation".

1.1 Proposition: Let α,β be endomorphisms and U,V unitaries in $\mathcal{O}_n$.

(a) $Z_{\alpha\beta} = Z_\alpha\, \alpha(Z_\beta)$

(b) $\lambda_U\lambda_V = \lambda_{U\lambda_U(V)}$

(c) λ_U is an automorphism if and only if there is a unitary $V \in \mathcal{O}_n$ such that $\lambda_U(V) = U^*$.

Proof:

(a) $\alpha\beta(S_i) = \alpha(Z_\beta^* S_i) = \alpha(Z_\beta^*)Z_\alpha^* S_i$

(b) $\lambda_U\lambda_V(S_i) = \lambda_U(V^*S_i) = \lambda_U(V^*)U^*S_i = \lambda_{U\lambda_U(V)}(S_i)$

(c) follows from (b).

Note also that λ is bicontinuous if we equip End $\mathcal{O}_n$ with the topology of pointwise norm convergence.

If $U \in \mathcal{O}_n$ is unitary, then $US_iU^* = U\Phi(U^*)S_i$ $(i=1,\dots,n)$, where $\Phi\colon \mathcal{O}_n \to \mathcal{O}_n$ is the endomorphism determined by $\Phi(X) = \sum_{i=1}^{n} S_iXS_i^*$.

Thus λ_V is an inner automorphism if and only if there is $U \in \mathcal{O}_n$ such that $V = \Phi(U)U^*$.

We know that $\mathcal{O}_n \cong M_n \otimes \mathcal{O}_n$ and, in such a representation, Φ can be written as $\phi(X) = 1\otimes X$.

Put $M = M_n \otimes 1$. Then the algebras $\Phi^k(M)$ $(k = 0,1,2\ldots)$ pairwise commute and their union generates F.

1.2 Proposition:

(a) If $U \in \mathcal{O}_n$ is unitary and $X \in F$, then $\lambda_U(X) = \lim_{k\to\infty} \mathrm{Ad}\,(U^*\Phi(U^*) \ldots \Phi^k(U^*))\,(X)$.

(b) $\lambda_U(F) \subset F$ if and only if $U \in F$.

Proof (a): We have $\lambda_U \Phi = \mathrm{Ad}\, U^* \Phi \lambda_U$.

It suffices to prove the assertion for $X \in \Phi^m(M)$. This is done inductively, using the relations $\lambda_U(X) = U^*XU$ $(X \in M)$ and $\lambda_U(\Phi(X)) = (\mathrm{Ad}U^*)(\Phi(\lambda_U(X)))$ $(X \in \mathcal{O}_n$ arbitrary).

Proof (b): If $U \in F$, then $\lambda_U(F) \subset F$ by (1).

If $\lambda_U(F) \subset F$, then $\lambda_t \lambda_U \lambda_{t^{-1}} = \lambda_{\lambda_t(U)}$ $(t \in \mathbb{T})$,

where $\mathbb{T}$ denotes the group of complex numbers of modules one. Since $\lambda_t (t \in \mathbb{T})$ leaves F pointwise fixed, we get $\lambda_{U|F} = \lambda_{\lambda_t(U)|F}$.

Therefore $\lambda_{\lambda_t(U)} = \lambda_U \lambda_{t'} = \lambda_{Ut'}$ with $t' \in \mathbb{T}$ and U is an eigenvector. However, the only unitary eigenvectors for λ_t are the unitaries in F.

Let now $\mathcal{D}_o$ be a maximal commutative C*-subalgebra of M. Define $\mathcal{D}$ to be the commutative C*-algebra generated by $\bigcup_{k=o}^{\infty} \Phi^k(\mathcal{D}_o)$. It is well known that $\mathcal{D}$ is maximal commutative in F. However, it can be shown that $\mathcal{D}$ is also maximal commutative in $\mathcal{O}_n$ and that there is a faithful conditional expectation d: $\mathcal{O}_n \to \mathcal{D}$. We write $\mathcal{D}^u$ for the group of unitaries in $\mathcal{D}_n$ and $\lambda_{\mathcal{D}^u} = \{\lambda_D | D \in \mathcal{D}^u\}$.

1.3 Proposition:

(a) The fixed point algebra for $\lambda_{\mathcal{D}^u}$ is $\mathcal{D}$.

(b) The map $\lambda: \mathcal{D}^u \to \lambda_{\mathcal{D}^u}$ is a isomorphism of groups. In particular, every λ_D $(D \in \mathcal{D}^u)$ is an automorphism.

(c) $\lambda_{\mathcal{D}^u} = \{\alpha \in \mathrm{Aut}\, \mathcal{O}_n \,|\, \alpha|_{\mathcal{D}} = \mathrm{id}\}$.

(d) $\lambda_{\mathcal{D}^u}$ is a maximal commutative subgroup of Aut $\mathcal{O}_n$.

Proof:

(a) $\mathcal{D}$ is contained in the fixed point algebra for $\lambda_{\mathcal{D}^u}$ by Proposition 1.2(a). If, on the other hand, X is in the fixed point algebra for $\lambda_{\mathcal{D}^u}$, then in particular $(\mathrm{Ad}D)X = X$ for all $D \in \mathcal{D}^u$, whence $X \in \mathcal{D}$ by the maximality of $\mathcal{D}$.

(b) By Proposition 1.1(b) we have $\lambda_D\lambda_E = \lambda_{D\lambda_D(E)} = \lambda_{DE}$ for all $D,E \in \mathcal{D}$.

(c) Suppose $\alpha = \lambda_U$ is trivial on $\mathcal{D}$. Then $\alpha(S_iDS_i^*) = U^*S_iDS_i^*U = S_iDS_i^*$ for all $D \in \mathcal{D}$, whence $U \in \mathcal{D}$.

(d) From $\alpha\mathrm{Ad}D = \mathrm{Ad}D\alpha$ $(D \in \mathcal{D})$ it follows $\mathrm{Ad}\alpha(D) = \mathrm{Ad}D$ and therefore $\alpha(D) = D$.

Example: Let $n = 2$. Then $F \cong M_{2^\infty}$, the fermion algebra. For $(t_1,t_2) \in \mathbb{T}^2$, define $\lambda_{(t_1,t_2)}$ by

$$\lambda_{(t_1,t_2)}S_1 = t_1S_1 \quad \text{and} \quad \lambda_{(t_1,t_2)}S_2 = t_2S_2.$$

Then $\lambda_{(t_1,t_2)} \in \lambda_{\mathcal{D}^u}$ and $\lambda_{(t_1,t_2)}$ leaves F invariant. The restriction of $\lambda_{(t,1)}$ to F is exactly the group of gauge automorphisms for F and the fixed point algebra for $\lambda_{(t_1,t_2)}$ $((t_1,t_2) \in \mathbb{T}^2)$ is the current algebra. If (t_1,t_2) runs through a one-parameter subgroup of $\mathbb{T}^2$, then we get a one-parameter group of automorphisms which is a "mixture" of λ_t and gauge automorphisms for F.

The normalizer $N(\mathcal{D})$ of $\mathcal{D}$ is the group of unitaries X in $\mathcal{O}_n$ such that $XDX^{-1} \in \mathcal{D}$ for all $D \in \mathcal{D}$.

It can be shown that $\mathcal{D}$ is a regular maximal abelian subalgebra of $\mathcal{O}_n$, i.e. $N(\mathcal{D})$ generates $\mathcal{O}_n$.

1.4 Proposition: Let $X \in N(\mathcal{D})$. Then $\lambda_X(\mathcal{D}) \subset \mathcal{D}$.

Proof: This follows from Proposition 2(a).

We say that $X \in \mathcal{O}_n$ is an eigenvector for $\lambda_{\mathcal{D}^u}$ if there is a homomorphism χ from $\lambda_{\mathcal{D}^u}$ into the unitary group of $\mathcal{O}_n$ such that

$\alpha(X) = \chi(\alpha)X$ for all $\alpha \in \lambda_{\mathcal{D}^u}$.

<u>1.5 Proposition:</u> Let λ_X be an automorphism of $\mathcal{O}_n$. The following are equivalent

(a) $X \in N(\mathcal{D})$

(b) X is an eigenvector for $\lambda_{\mathcal{D}^u}$

(c) $\lambda_X(\mathcal{D}) = \mathcal{D}$

(d) λ_X is in the normalizer $N(\lambda_{\mathcal{D}^u})$ of $\lambda_{\mathcal{D}^u}$ in Aut $\mathcal{O}_n$.

<u>Proof:</u>

(a) ⇒ (c) by Prop. 4.

(c) ⇒ (d) For all $D \in \mathcal{D}$ we have $\lambda_X\lambda_D\lambda_X^{-1}\Big|_{\mathcal{D}} = \mathrm{id}$. Therefore, by Prop. 3(b) $\lambda_X\lambda_D\lambda_X^{-1} \in \lambda_{\mathcal{D}^u}$.

(d) ⇒ (c) If $\lambda_X \in N(\lambda_{\mathcal{D}^u})$, then the restriction of λ_D to $\lambda_X(\mathcal{D})$ is is trivial for all $D \in \mathcal{D}$, whence $\lambda_X(\mathcal{D}) = \mathcal{D}$.

(c) ⇒ (a) If $\lambda_X(\mathcal{D}) = \mathcal{D}$, then $\lambda_X(S_i\,\mathcal{D}\,S_i^*) = X^*S_i\,\mathcal{D}\,S_i^*X \subset \mathcal{D}$ so that $X \in N(\mathcal{D})$.

(a) ⇒ (b) follows from the relation $\mathrm{Ad}X\Big|_{\mathcal{D}} = \mathrm{Ad}\lambda_D(X)\Big|_{\mathcal{D}}$ $(D \in \mathcal{D}^u)$.

(b) ⇒ (a) If X is an eigenvector for the character χ, then $\chi(\lambda_{\mathcal{D}^u})$ and, since $\chi(E)\chi(D)X = \lambda_D\lambda_E(X) = \lambda_D(\chi(E)X) = \lambda_D(\chi(E))\chi(D)X$ we get $\lambda_D(\chi(E)) = \chi(E)$ for all $D, E \in \mathcal{D}^u$.

Thus $\chi(E) \in \mathcal{D}$ and $\lambda_E(X^*DX) = X^*DX$ $(E \in \mathcal{D})$ implies $X^*DX \in \mathcal{D}$ for each $D \in \mathcal{D}$.

The characterization of $N(\mathcal{D})$ as the set of eigenvectors for $\lambda_{\mathcal{D}^u}$ can be used to give a complete list of the elements in $N(\mathcal{D})$ (modulo elements in $\mathcal{D}$). In particular, every such element is a polynomial in the generators $S_1,\dots,S_n$, $S_1^*,\dots,S_n^*$ of $\mathcal{O}_n$.

The quotient group $N(\lambda_{\mathcal{D}^u}) / \lambda_{\mathcal{D}^u}$ is discrete and may be considered as the Weyl group of Aut $\mathcal{O}_n$ with respect to the maximal commutative subgroup $\lambda_{\mathcal{D}^u}$.

§2 One-Parameter Groups of Automorphisms

The spectrum of the algebra $\mathcal{D}$ described in §1 is the compact space $K = \{1,\dots,n\}^{\mathbb{N}}$. The canonical state ϕ on $\mathcal{O}_n$ induces on K the measure ϕ which is the infinite product of the probability measure p defined on $\{1,\dots,n\}$ by $p(\{i\}) = \frac{1}{n}$ $(i=1,\dots,n)$. There is another natural way to construct a state of $\mathcal{D}$ as the infinite product of the probability measure p on $\{1,\dots,n\}$ defined by $p(\{i\}) = p_i$ $(i=1,\dots,n)$ where $p_1 + \dots + p_n = 1$. If ψ_o is such a state for $\mathcal{D}$, then $\psi = \psi_o d$, where $d: \mathcal{O}_n \to \mathcal{D}$ is the conditional expectation, will be called a canonical state of $\mathcal{O}_n$.

Let Θ be the shift defined by $\Theta\xi = (\xi_2,\xi_3,\dots)$ for $\xi = (\xi_1,\xi_2,\xi_3,\dots) \in K$. Then $\Phi(H) = H \circ \Theta$ for $H \in \mathcal{D}$ considered as a continuous function on K. It is known and easy to see that with respect to a canonical measure ψ, the shift Θ is ergodic, i.e. if $M \subseteq K$ is a measurable subset such that $\Theta(M) = M$ then $\psi(M)$ is either 0 or 1.

If $\{\lambda_{X_t}\}$ is a continuous one-parameter group of automorphisms of $\mathcal{O}_n$, which leaves $\mathcal{D}$ invariant, then, since $N(\lambda_{\mathcal{D}^u}) / \lambda_{\mathcal{D}^u}$ is discrete, $\{\lambda_{X_t}\}$ must lie in the connected component $\lambda_{\mathcal{D}^u}$ of the identity of $N(\lambda_{\mathcal{D}^u})$. Therefore $X_t \in \mathcal{D}$ for all t and $\{X_t\}$ is a one-parameter group of unitaries in $\mathcal{D}$.

<u>2.1 Proposition:</u> For a one-parameter group $\{\lambda_{X_t}\} \subset \mathrm{Aut}\, \mathcal{O}_n$ the following conditions are equivalent:

(a) $\{\lambda_{X_t}\}$ admits a canonical state ψ as a 1-KMS state.

(b) $\lambda_{X_t}(S_k) = p_k^{it} S_k$ where $p_1,\dots,p_n$ define ψ.

Condition (b) in Proposition 2.1 could also be expressed by saying $X_t = H^{it}$ where $H = \sum_{k=1}^{n} p_k(S_k S_k^*) \in \mathcal{D}_o$.

<u>2.2 Proposition:</u> Let H_1, H_2 be positive elements of $\mathcal{D}_o$ as above. Then $\lambda_{H_1^{it}}$ is not exterior equivalent to $\lambda_{H_2^{it}}$, i.e. there are no unitaries $U_t \in \mathcal{O}_n$ such that $\lambda_{H_1^{it}} = \mathrm{Ad}U_t\, \lambda_{H_2^{it}}$.

In the opposite direction we have the following result.

<u>2.3 Proposition:</u> Let ψ be a canonical state and $\psi_1(\cdot) = \psi(h\cdot)$ (more formally $\psi_1(X) = \psi(hd(X))$ where h is a positive function in $L^1(K, \psi_o)$. Suppose that $\{\alpha_t\}$ is a one-parameter group of automorphisms of $\mathcal{O}_n$ which admits ψ_1 as a KMS-state and that h is continuous on some open subset of K. Then h is continuous, i.e. $h \in \mathcal{D}$ and $\alpha_t = \mathrm{Ad}\, h^{it}\, \lambda_{H^{it}}$ where H is associated to ψ as in 2.2.

This proposition follows from the fact that h must then be continuous on a set which is invariant under Θ, and the ergodicity of Θ.

§3 Automorphisms in $N(\lambda_{\mathcal{D}^u})$

A point $\xi = (\xi_1, \xi_2, \ldots) \in K$ is called rational if the sequence $\xi_1, \xi_2, \ldots$ becomes finally periodic, irrational otherwise. Every $\xi \in K$ can be considered as a pure state of $\mathcal{D}$ and this pure state can be extended to a pure state of $\mathcal{O}_n$.

<u>3.1 Proposition:</u> If $\xi \in K$ is irrational, then ξ has a unique extension to a pure state of $\mathcal{O}_n$. If ξ is rational, then for every $t \in \mathbb{T}$ (= {complex numbers of modulus one}), there is a unique extension ξ_t to a pure state of $\mathcal{O}_n$ such that $\xi_t(W_\xi) = t$, where for $\xi = \mu_1, \ldots, \mu_k, \eta_1, \ldots, \eta_p, \eta_1, \ldots, \eta_p, \eta_1, \ldots,$ W_ξ is defined as $W_\xi = S_{\mu_1} \ldots S_{\mu_k} S_{\eta_1} \ldots S_{\eta_p} S^*_{\mu_k}, \ldots, S^*_{\mu_1}$.

If α is an automorphism in $N(\lambda_{\mathcal{D}^u})$, let α_o be the homeomorphism of K induced by α.

<u>3.2 Corollary:</u> If $\alpha \in N(\lambda_{\mathcal{D}^u})$ then α_o preserves the set of rational points in K.

For $\alpha \in N(\lambda_{\mathcal{D}^u})$ and $\xi \in K$ rational, define $\xi(\alpha) = t$ if $\xi_1 \circ \alpha^{-1} = (\alpha_o \xi)_t$.

<u>3.3 Proposition:</u> Let $\alpha \in N(\lambda_{\mathcal{D}^u})$ inner. Then

(a) $\xi(\alpha) = 1$ for all rational $\xi \in K$

(b) For every $\xi = (\xi_1, \xi_2, \ldots) \in K$ the point $\alpha_o(\xi)$ is obtained by deleting finitely many $\xi_1, \ldots, \xi_r$ and replacing them by $\eta_1, \ldots, \eta_s$ with $\eta_i \in \{1, \ldots, n\}$. In other words $\alpha_o(\xi) \in \Theta^{-s}(\Theta^r \xi)$.

<u>3.4 Proposition:</u> Assume $\alpha \in \mathrm{Aut}\, \mathcal{O}_n$ leaves both $\mathcal{D}$ and F invariant. Then for every rational $\xi \in K$, the period of $\alpha_o(\xi)$ is the same as the period of ξ.

The classification of all automorphisms of $\mathcal{O}_n$ which leave $\mathcal{D}$ and F invariant, modulo $\lambda_{\mathcal{D}^u}$, is a combinatorical problem, and should be possible.

References

[1] J. Cuntz, Simple C*-algebras generated by isometries, Commun.Math.Phys. 57 (1977), 173-185.

[2] J. Cuntz, Murray-von Neumann equivalence of projections in infinite simple C*-algebras, Rev.Math. Pures et Appl. 23 (1978), 1011-1014.

[3] D. Olesen and G.K. Pedersen, Some C*-dynamical systems with a single KMS-state, Math.Scand. 42 (1978), 111-118.

[4] J. Roberts, Crossed products by group duals, Symposia Mathematica XX.

Non-Commutative Group Duality

and

The Kubo-Martin-Schwinger Condition

Masamichi Takesaki*)

Department of Mathematics

University of California

Los Angeles, California, USA

Abstract

Given a C*-algebra A equipped with an action α of a locally compact group G, a necessary and sufficient condition for an α-invariant state ω on A to satisfy the Kubo-Martin-Schwinger condition with respect to a suitable one parameter subgroup of G is formulated in terms of a w*-closed right translation invariant isomorphism of a w*-dense subalgebra $\mathcal{D}$ of the Fourier-Stieltjes algebra B(G) into B(G) itself.

---

*) The present work is supported in part by the National Science Foundation.

Introduction

The Kubo-Martin-Schwinger boundary condition, so-called the KMS-condition, involves always a continuous homomorphism $\{\beta(t)\}$ of the additive topological group $\mathbb{R}$ into $\hat{\mathbb{R}}$ itself. To be more precise, let A be a C*-algebra equipped with a one parameter automorphism group $\{\alpha_t\}$. Given a state ω on A, we set

$$f_{x,y}(t) = \omega(\alpha_t(x)y) \quad \text{and} \quad g_{x,y}(t) = \omega(y\alpha_t(x)), \; x,y \in A.$$

We say that ω is β-KMS for $\{\alpha_t\}$ if

$$\hat{f}_{x,y}(p) = e^{\beta(p)}\, \hat{g}_{x,y}(p),$$

where $\hat{f}$ and $\hat{g}$ mean the Fourier transform of f and g as tempered distributions. Since $\beta(p) = \beta p$, $p \in \mathbb{R}$, for some fixed number β, we identify $\beta(\cdot)$ with the number β and consider it as $1/kT$, the inverse temperature of the system described by the state ω.

In the search of the algebraic structure behind the presence of chemical potentials as a parameter of equilibrium states, one is forced to consider more general locally compact groups other than just $\mathbb{R}$, [1]. Here one considers a time evolution automorphism group $\{\alpha_t\}$ together with a gauge transformation group $\{\gamma_g: g \in G\}$, then looks for a one parameter subgroup of the product group $\{\alpha_t\gamma_g: t \in \mathbb{R}, g \in G\}$ for which a given state satisfies the KMS-condition.

Generally speaking, one can interpret the duality theorem for locally compact groups, abelian or not, as a method of identifying an object as an element of a preassigned group. Therefore, if one fixes a locally compact automorphism group $\{\alpha_g: g \in G\}$ of a C*-algebra A and a state ω, then the duality theorem provides a natural tool in the search of a one parameter subgroup of $\{\alpha_g: g \in G\}$ for which ω satisfies the KMS-condition. In this paper, I would like to exhibit how this approach works in general.

The material presented here is taken from a joint work of D. Kastler and myself, [6].

1. Duality for Locally Compact Groups

Let G be a locally compact group with a left Haar measure ds. If G is abelian, then the classical Kampen-Pontrjagin duality theorem takes care of the problem. For non-abelian groups, there are several formulations of the duality theorem which look superficially different. But, in my opinion, all of the proposed duality theorems are equivalent to the following:

Let W be the unitary on $L^2(GXG)$ defined by

$(W\xi)(s,t) = \xi(s,st), \ s,t \in G, \quad \xi \in L^2(GXG).$

Then the following statements for a non-zero bounded operator x on $L^2(G)$ are euqivalent:

$$x = \lambda(s) \quad \text{for some } s \in G; \tag{1}$$

$$W^*(x\otimes 1)W = x\otimes x, \tag{2}$$

where λ is of course the left regular representation of G on $L^2(G)$.

For the proof, we refer to [7] and [10]. This result is known as the Tannaka-Stinespring-Eymard-Tatsuuma duality theorem.

For the proof as well as for applications, it is convenient to reformulate the theorem in terms of operator algebras. Let $\mathcal{R}(G)$ be the von Neumann algebra on $L^2(G)$ generated by the left regular representation λ. Since $\mathcal{R}(G)$ is generated by $\lambda(s)$, $s \in G$, and since $W^*(\lambda(s)\otimes 1)W = \lambda(s)\otimes\lambda(s)$ trivially, one has the map π of $\mathcal{R}(G)$ into $\mathcal{R}(G)\bar{\otimes}\mathcal{R}(G)$ defined by

$$\pi(x) = W^*(x\otimes 1)W, \quad x \in \mathcal{R}(G).$$

Since

$$(\pi\otimes\iota)\cdot\pi(\lambda(s)) = \lambda(s)\otimes\lambda(s)\otimes\lambda(s) = (\iota\otimes\pi)\cdot\pi(\lambda(s)), \ s \in G,$$

one has

$$(\pi\otimes\iota)\cdot\pi = (\iota\otimes\pi)\cdot\pi,$$

which gives rise to a Banach algebra structure in the predual $\mathcal{R}_*(G)$ of $\mathcal{R}(G)$ by the following operation:

$$\langle x,\phi\cdot\psi\rangle = \langle\pi(x),\phi\otimes\psi\rangle,\ x\in\mathcal{R}(G),\ \phi,\psi\in\mathcal{R}_*(G).$$

We then see that

$$\langle\lambda(s),\phi\cdot\psi\rangle = \langle\pi(\lambda(s)),\phi\otimes\psi\rangle = \langle\lambda(s)\otimes\lambda(s),\phi\otimes\psi$$

$$= \langle\lambda(s),\phi\rangle\langle\lambda(s),\psi\rangle = \langle\lambda(s),\psi\cdot\phi\rangle.$$

Hence $\lambda(s)$ gives rise to a character of $\mathcal{R}_*(G)$ and $\mathcal{R}_*(G)$ is abelian since $\lambda(s)$, $s\in G$, generate $\mathcal{R}(G)$. Thus, $\mathcal{R}_*(G)$ is an abelian Banach algebra. Now, the Gelfand theory for commutative Banach algebras comes into play, and the above duality theorem is equivalent to the following:

$$\langle x,\phi\cdot\psi\rangle = \langle x,\phi\rangle\ \langle x,\psi\rangle,\ \phi,\ \psi\in\mathcal{R}_*(G). \qquad (3)$$

Therefore, the spectrum of $\mathcal{R}_*(G)$ is indeed identified with the given group G. The range of the Gelfand representation of $\mathcal{R}_*(G)$ is denoted by A(G) and called the Fourier algebra of G. If G is abelian, then A(G) is precisely the Fourier inverse transform of $L^1(\hat{G})$ with $\hat{G}$ the dual group of G.

If G is not compact, then A(G) is not unital. Here we have another important abelian Banach algebra, called the Fourier-Stieljes algebra B(G), which is the linear span of the cone of all continuous positive definite functions on G. Let C*(G) denote the C*-group algebra of G, i.e. the enveloping C*-algebra of $L^1(G)$. Then a positive linear functional on C*(G) is naturally identified with a continuous positive definite functions, and hence the conjugate space C*(G)* is identified with B(G). One can then prove that B(G) is an abelian Banach algebra with respect to the norm inherited from C*(G)* and the pointwise multiplication as a function algebra. Of course, A(G) is a subalgebra of B(G). Furthermore, the fact that $\lambda\otimes v\simeq\lambda\otimes\mathrm{id}$ for any unitary representation v of G reflects to the property that A(G) is an ideal of B(G).

For the non-compact case, the spectrum of $B(G)$ is far much bigger than G itself. However, G is characterized as the set of all characters of $B(G)$ which does not annihilate $A(G)$. It should be noted that if G is abelian, then $B(G)$ is exactly the Fourier-Stieltjes transform of the measure algebra $M^1(\hat{G})$ over the dual group $\hat{G}$.

We note here that $A(G)$ and $B(G)$ are both invariant under translations by an element of G, which means that each element $s \in G$ gives rise to the left (resp. right) translation automorphism λ_s (resp. ρ_s) of $A(G)$ and $B(G)$ by the following:

$$(\lambda_s\phi)(t) = \phi(s^{-1}t), \quad (\rho_s\phi)(t) = \phi(ts), \quad \phi \in B(G), \ s,t \in G.$$

It is then straightforward to check the following statement:

An automorphism α of $A(G)$ (resp. $B(G)$) is of the form $\alpha = \lambda_s$ for some $s \in G$ if and only if α commutes with $\rho(G)$.

With some more work, one can replace $A(G)$ or $B(G)$ by the abelian von Neumann algebra $L^\infty(G)$ in the above result, [9].

2. One Parameter Subgroups

Given a locally compact group G, a one parameter subgroup of G means a continuous homomorphism of the additive group $\mathbb{R}$ into G, i.e. a continuous map: $t \in \mathbb{R} \to g(t) \in G$ such that $g(s+t) = g(s)g(t)$, $s, t \in \mathbb{R}$. One should note here that the range of a one parameter subgroup is not necessarily closed. However, if the range $\{g(t): t \in \mathbb{R}\}$ of a one parameter subgroup is not closed, then its closure must contain a compact subgroup. Namely, an easy example of a non closed one parameter subgroup of the two dimensional torus group is more or less generic.
Suppose that a one parameter subgroup $\{g(t): t \in \mathbb{R}\}$ of a locally compact group G is given. Putting

$$\alpha_t = \lambda_{g(t)},$$

one has a one parameter autmorphism group $\{\alpha_t\}$ of $A(G)$ as well as $B(G)$. Here one should note that the map: $t \in \mathbb{R} \to \alpha_t\phi$, $\phi \in B(G)$, is not continuous in norm, but continuous in the weak* topology with respect to the duality $B(G) = C^*(G)^*$. Let $\mathfrak{A}$ be the set of those elements ϕ of $B(G)$ such that the associated $B(G)$-valued function: $t \in \mathbb{R} \to \alpha_t\phi \in B(G)$ extends to a $B(G)$ valued entire function on $\mathbb{C}$. By the standard smoothing operation using approximate identities in $L^1(\mathbb{R})$, say $\{\sqrt{\frac{n}{2\pi}}\, e^{-t^2/n}\}$, one can show that $\mathfrak{A}$ (resp. $\mathfrak{A} \cap A(G)$ is weak* (resp. norm) dense in $B(G)$ (resp. $A(G)$). Set

$$T_\phi = \alpha_{-i}\phi \qquad \phi \in \mathfrak{A}.$$

Then T is an automorphism of $\mathfrak{A}$ (resp. $\mathfrak{A} \cap A(G)$) commuting with the right translation $\rho(G)$. Furthermore, it is not hard to see that T is preclosed with respect to the weak* topology (resp. the norm topology) in $B(G)$ (resp. $A(G)$). We denote the closure of T by the same notation T again with domain $\mathfrak{D}(T)$. We have the following:

Proposition 2.1: In the above situation, $\mathfrak{D}(T)$ consists of those $\phi \in B(G)$ such that for each $s \in G$, the function: $t \in \mathbb{R} \to \phi(g(t)s) \in \mathbb{C}$ is extended to a bounded holomorphic function $F_{\phi,s}$ in the strip, $0 \le \operatorname{Im} z \le 1$ and continuous on the boundary, $\operatorname{Im} z = 1$, and ϕ and $T\phi$ give the boundary values of $F_{\phi,s}$ in the sense that

$$\phi(g(t)s) = F_{\phi,s}(t)\ , \quad (T\phi)(g(t)s) = F_{\phi,s}(t+i),\ t \in \mathbb{R}.$$

If we define actions of $C^*(G)$ on $B(G)$ (resp. $A(G)$ by $\langle y,\phi x\rangle = \langle xy,\phi\rangle = \langle x,y\phi\rangle$, $x,y \in C^*(G)$, $\phi \in B(G)$ or $\phi \in A(G)$, then one can find a sequence $\{a_n\}$ in $C^*(G)$ with the properties:

$\{a_n\}$ is an approximate identity of $C^*(G)$; i)

$a_n B(G) \subset \mathfrak{D}(T)$ (moreover $a_n B(G) \subset \mathfrak{A}$); ii)

$\phi \in B(G)$ (resp. $A(G)$ belongs to $\mathfrak{D}(T)$ if and only if $\{Ta_n\}$ has a w*-limit (resp. norm limit) in $B(G)$ (resp. $A(G)$). iii)

Theorem 2.2 Suppose $\mathfrak{D}$ is a w*-dense (resp. norm dense) right translation invariant subalgebra of $B(G)$ (resp. $A(G)$). Suppose T is an isomorphism of $\mathfrak{D}$ into $B(G)$ (resp. $A(G)$) commuting with right translation. If T is preclosed with respect to the w*-topology (resp. norm topology) and admits a sequence $\{a_n\}$ with above three properties, then there exist an elements $s_o \in G$ and a one parameter subgroup $\{g(t)\}$ of G such that $\lambda_{s_o} T$ corresponds to a subgroup $\{g(t)\}$ in the above fashion.

3. The Kubo-Martin-Schwinger Condition

We have seen above that the KMS-like condition enters naturally in the duality theory for one parameter subgroups. Now, we shall see a direct link between the genuine KMS-condition in the algebraic approach to equilibrium statistical mechanics and the above mechanism.

Suppose A is a C*-algebra equipped with an action $\{\alpha_s: s \in G\}$ of a locally compact group G. Suppose an α-invariant state ω on A satisfies the KMS-condition with respect to $\alpha_{g(t)}: t \in \mathbb{R}\}$, where $\{g(t)\}$ is a one parameter subgroup of G. Namely, for any $x, a \in A$, there exists a function F on $0 \leq \text{Im } z \leq 1$ holomorphic inside the strip with boundary values:

$$F(t) = \omega(\alpha_{g(t)}(x)y), \quad F(t+i) = \omega(y\alpha_{g(t)}(x)), \quad t \in \mathbb{R}.$$

We define functions $\phi_{x,y}$ and $\psi_{x,y}$ on G by

$$\phi_{x,y}(s) = \omega(\alpha_s(x)y), \quad \psi_{x,y}(s) = \omega(y\alpha_s(x)).$$

With the GNS-representation $\{\pi_\omega, \mathfrak{H}_\omega, \xi_\omega, V_\omega\}$, we rewrite $\phi_{x,y}$ and $\psi_{x,y}$ as follows:

$$\phi_{x,y}(s) = (\pi_\omega(y)\xi_\omega | V_\omega(s)\pi_\omega(x^*)\xi_\omega), \quad s \in G;$$

$$\psi_{x,y}(s) = (V_\omega(s)\pi_\omega(x)\xi_\omega | \pi_\omega(y^*)\xi_\omega).$$

This shows that $\phi_{x,y}$ and $\psi_{x,y}$ both belong to B(G). We then get

$$\phi_{x,y}(g(t)s) = \omega(\alpha_{g(t)}(\alpha_s(x))y), \quad \psi_{x,y}(g(t)s) = \omega(y\alpha_{g(t)}(\alpha_s(x))).$$

Therefore we see that the correspondence $\phi_{x,y} \to \psi_{x,y}$ is extended to the associated closed isomorphism T with domain $\mathcal{D}(T)$ by the procedure in the last section. The converse is the following:

<u>Theorem 3.1</u> An α-invariant state ω on A satisfies the KMS-condition with respect to a one parameter subgroup of G if and only if the correspondence $\phi_{x,y} \to \psi_{x,y}$ is extended to a w*-closed right translation invariant homomorphism T of a w*-dense right translation invariant subalgebra $\mathcal{D}$ of B(G).

The existence of a sequence $\{a_n\}$ with properties (2.i) - (2.iii) follows from the invariance of ω. The absence of the phase factor s_o in Theorem 2.2 follows from the fact that the positive definiteness of $\phi_{x,y}$ implies the same of $\psi_{x,y}$.

References

[1] H. Araki, R. Haag, D. Kastler and M. Takesaki, Extension of KMS-states and chemical potential, Comm.Math.Phys., 53 (1977), 97-134.

[2] P. Eymard, L'algèbre de Fourier d'un group localement compact, Bull.Soc.Math. France, 92 (1964), 181-236.

[3] R. Haag, N.M. Hugenholtz and M. Winnink, On the euqilibrium states in quantum statistical mechanics, Comm.Math.Phys., 5 (1967), 215-236.

[4] R. Haag, D. Kastler and E.B. Trych-Pohlmeyer, Stability and equilibrium states, Comm.Math.Phys., 38(1974), 173.

[5] D. Kastler, Equilibrium states of matter and operator algebras, Symposia Math., 20 (1976), 49-107.

[6] D. Kastler and M. Takesaki, Group duality and the Kubo-Martin-Schwinger Condition, to appear.

[7] K. Saito, On a duality for locally compact groups, Tôhoku Math.J., 20 (1968), 355-367.

[8] M. Takesaki, Tomita's theory of modular Hilbert algebras and its applications, Lecture Notes in Math., Springer-Verlag, 128 (1970).

[9] M. Takesaki and N. Tatsuuma, Duality and subgroups, Ann. Math., 93 (1971), 344-364.

[10] N. Tatsuuma, A duality theorem for locally compact groups, J.Math. Kyoto Univ., 6 (1967), 187-293.

[11] N. Tatsuuma, An extensions of AKTH-theory to locally compact groups, preprint, Kyoto University.

[12] M.E. Walter, Group duality and isomorphisms of Fourier and Fourier-Stieltjes algebras from a W*-algebra point of view, Bull.Amer.Math.Soc., 76 (1970), 1321-1325.

A Uniqueness Theorem for Central Extensions of Discrete Products of Cyclic Groups

Ph. Combe

R. Rodriguez

Faculté des Sciences de Luminy, and
Centre de Physique Théorique, CNRS Marseille, France

M. Sirugue

Centre de Physique Théorique, CNRS Marseille, France

M. Sirugue-Collin

Université de Provence, and
Centre de Physique Théorique, CNRS Marseille, France

Abstract

We prove that all central extensions of $Z_n^\Lambda \times Z_n^\Lambda$ where Λ is a a countable infinite set and Z_n the cyclic group of prime order n, which satisfy an homogeneity condition with respect to permutations of the lattice Λ are isomorphic.

1. Introduction

Central extensions of abelian groups appeared at the very beginning of quantum theory, in particular in connection with the Weyl group of systems with finite number of degrees of freedom and field theory of Bose particles. Also it has been noticed [1] that quantized Fermi field can be viewed as central extension of an abelian group and this was developed in [2].

Surprisingly enough, there is up to isomorphism only one non trivial central extension of the abelian group $\mathbb{R}^2$ and in [3] we gave a somewhat similar result for the groups attached to the Fermi fields, and to the quantum spins 1/2 on a lattice.

Again, and this is the aim of our paper, we can extend these results to more general cases which are of some importance in physics, namely to the groups $(Z_n \times Z_n)^{\Lambda}_R$ which are the direct sum over Λ of $Z_n \times Z_n$ where Z_n is the cyclic group of order n.

More precisely, let b be the bicharacter which is canonically attached to a central extension of an abelian group G, then for Λ infinite countable and n prime there is up to isomorphism only one central extension of $(Z_n \times Z_n)^{\Lambda}_R$ such that

i) b is non-degenerated

ii) b is invariant with respect to the automorphisms of $(Z_n \times Z_n)^{\Lambda}_R$ induced by permutations of finite number of points on Λ

This is Theorem (3.3).

On the other hand, let us notice that $(Z_n \times Z_n)^{\Lambda}_R$ is discrete, hence its dual $(Z_n \times Z_n)^{\Lambda}$ is compact. Then previous mentioned results give a description of some ergodic actions of the compact group $(Z_n \times Z_n)^{\Lambda}$ on some von Neumann algebras (see e.g. [4]).

For the sake of completeness, in Section 2 we quote some results about central extensions of abelian groups. Also we give another description of the "gauge groups" G_R^Λ where G is a finite abelian group and Λ a countable lattice.

Actually there is a one to one correspondance between functions from Λ to G and the partitions of Λ into n disjoints subsets of Λ where n is the order of G. In a similar way functions from Λ to G with finite support correspond to a subclass of the previous partitions. This description is particularly suited to the results of Section 3 where we prove the above mentioned result about uniqueness of central extensions of $(Z_n \times Z_n)_R^\Lambda$.

We are grateful to Prof. R. Høegh-Krohn for stimulating discussions and one of us (M.S) is indebted to Prof. S. Albeverio and L. Streit for their hospitality in Bielefeld where some parts of this work were finished.

2. Finite Group on a Lattice

Let Λ be an at most countable set. $\Lambda = \{x_i\}_{i\in Z}$. $\mathfrak{F}(\Lambda)$ is the family of finite subsets of Λ and $\Gamma(\Lambda)$ the family of all subsets of Λ. If $X \in \mathfrak{F}(\Lambda)$, $|X|$ denotes the cardinality of X. $\mathfrak{S}_\Lambda$ is the group of finite permutations of points in Λ viz $\sigma \in \mathfrak{S}_\Lambda$ iff:

- σ is a bijection of the points of Λ onto itself. (2.1)

- there exists a finite subset $S_\sigma \in \mathfrak{F}(\Lambda)$ such that: (2.2)

$$\sigma(x_i) = x_i \quad \forall\, x_i \notin S_\sigma .$$

In a natural way $\sigma \in \mathfrak{S}_\Lambda$ extends to a bijection of $\mathfrak{F}(\Lambda)$ onto itself, denoted by the same symbol.

Let G be a finite abelian group. G_R^Λ is the set of functions f from Λ to G, such that there exists a set $S_f \in \mathfrak{F}(\Lambda)$ with the property that

$$f(x_i) = e \qquad \forall\, x_i \in S_f \tag{2.3}$$

e is the identity of G .

G_R^Λ is an abelian group with respect to pointwise multiplication. We denote by E the identity of G_R^Λ:

$$E(x_i) = e \qquad \forall\, x_i \in \Lambda \tag{2.4}$$

$\mathfrak{S}_\Lambda$ acts in a natural way as a group of automorphisms of G_R^Λ. Indeed if $\sigma \in \mathfrak{S}_\Lambda$ and $f \in G_R^\Lambda$:

$$[\sigma]f = f \circ \sigma^{-1}. \tag{2.5}$$

In a similar way Aut(G), the group of automorphisms of G, acts as group of automorphisms of G_R^Λ. Indeed if $\alpha \in \mathrm{Aut}(G)$ and $f \in G_R^\Lambda$ then

$$\{(\alpha)f\}(x_i) = \alpha(f(x_i)) \quad \forall\, x_i \in \Lambda. \tag{2.6}$$

In what follows we shall need another description of G_R^Λ and it is the reason for which we introduce $P_f(G,\Lambda)$. $P_f(G,\Lambda)$ is the set of functions $\mathbb{X}$ from G to $P(\Lambda)$ such that

$$\mathbb{X}(g) \cap \mathbb{X}(g') = \emptyset \quad \text{if } g \neq g' \tag{2.7}$$

$$\bigcup_{g\in G} \mathbb{X}(g) = \Lambda \tag{2.8}$$

$$\bigcup_{g\neq e} \mathbb{X}(g) \in \mathfrak{P}(\Lambda) . \tag{2.9}$$

<u>Lemma:</u> $P_f(G,\Lambda)$ is a group with respect to the following operations

$$\begin{aligned} &\text{i)} \quad \mathbb{X} \blacktriangle \mathbb{Y}(g) = \bigcup_{g'\in G} (\mathbb{X}(g') \cap \mathbb{Y}(g'^{-1}g)) \\ &\text{ii)} \quad \mathbb{X}^{-1}(g) = \mathbb{X}(g^{-1}) \ , \mathbb{X}, \mathbb{Y} \in P_f(G,\Lambda) \end{aligned} \tag{2.10}$$

with the identity (which we denote also by E):

$$E(g) = \emptyset \quad \text{if } g \neq e$$

moreover G_R^Λ and $P_f(G,\Lambda)$ are isomorphic, and the isomorphism F is defined by

$$\{Ff\}(g) = \{x \in \Lambda \ ; \ f(x) = g\} \quad , \forall \ g \in G \quad f \in G_R^\Lambda .$$

It is important to notice that

$$F(f_1 \cdot f_2) = F(f_1) \mathbin{\triangle} F(f_2) \qquad \forall \ f_1, \ f_2 \in G_R^\Lambda \tag{2.11}$$

$$F(f^{-1}) = F(f)^{-1}. \tag{2.12}$$

Consequently actions on G_R^Λ of $\mathfrak{S}_\Lambda$ and $\mathrm{Aut}(G)$ can be transposed to actions of $\mathfrak{S}_\Lambda$ and $\mathrm{Aut}(G)$ on $P_f(G,\Lambda)$.

Actually G_R^Λ is embedded in the larger group G^Λ of all functions from Λ to G. This group is compact by Tichonov theorem. In a similar way $P_f(G,\Lambda)$ is embedded in $P(G,\Lambda)$ which is the set of functions $\mathbb{X}$ from G to $P(\Lambda)$ satisfying

$$\mathbb{X}(g) \cap \mathbb{X}(g') = \emptyset \quad \text{if } g \neq g' \tag{2.13}$$

$$\bigcup_{g \in G} \mathbb{X}(g) = \Lambda. \tag{2.14}$$

G as a finite abelian group is isomorphic to its dual. Consequently G^Λ (resp. $P(G,\Lambda)$ is the dual of G_R^Λ (resp. $P_f(G,\Lambda)$). Let us consider $\{g_1,\ldots g_p\}$ a system of generators of the finite group G i.e. any element g in G can be written in a unique way as:

$$g = g_1^{\alpha_1} \ g_2^{\alpha_2} \ldots . g_p^{\alpha_p} \ .$$

Then define the following elements $\{X\}_{\alpha,k}$ of $P_f(G,\Lambda)$ for any $X \in \mathcal{P}(\Lambda)$ and integers α and k, $1 \leq k \leq p$

$$\{X\}_{\alpha,k}(g) = X \qquad \text{if } g = g_k^\alpha$$

$$\{X\}_{\alpha,k}(g) = \emptyset \qquad \text{if } g \neq g_k^\alpha \quad \text{and } g \neq e \tag{2.15}$$

$$\{X\}_{\alpha,k}(e) = C_\Lambda(X).$$

It is clear that the elements $\{X_i\}_{1,k}$ generate $P_f(G,\Lambda)$.

In particular if $G = Z_n$ the cyclic group of order n generated by an element a:

$$Z_n = \{a, a^2, \ldots, a^n = e\}$$

then if $\mathbf{X} \in P_f(G,\Lambda)$ one has the following decomposition

$$\mathbf{X} = \mathop{\blacktriangle}_{p=1}^{n-1} \left(\mathop{\blacktriangle}_{x \in \mathbf{X}(a^p)} \{x\}_1 \right)^p \tag{2.16}$$

where the power is understood in the sense of the $\blacktriangle$ operation. In our case there is only one generator and $\{X\}_p \equiv \{X\}_{p1}$.

Let us come to the central extensions of an abelian group G by the torus T. It is given by a multiplier ζ on G, viz a function from GxG to T, such that

$$\zeta(g_1,g_2)\ \zeta(g_1g_2,g_3) = \zeta(g_1,g_2g_3)\ \zeta(g_2,g_3). \tag{2.17}$$

A multiplier ζ allows to define a group operation on the set GxT according to:

$$(g_1,\alpha_1)(g_2,\alpha_2) = (g_1g_2,\ \zeta(g_1,g_2)\ \alpha_1\alpha_2) \tag{2.18}$$

$$(g,\alpha)^{-1} = (g^{-1},\ \overline{\zeta(e,e)}\ \overline{\zeta(g,g^{-1})\alpha}),\ g_1,g_2,g \in G,\ \alpha_1,\alpha_2,\alpha \in T. \tag{2.19}$$

Let us denote by $G \boxtimes_\zeta T$ the corresponding group. It is known that given a multiplier ζ, b_ζ defined by:

$$b_\zeta(g,g') = \zeta(g,g')\overline{\zeta(g',g)}, \quad g,g' \in G, \tag{2.20}$$

is a bicharacter which satisfies:

$$b_\zeta(g,g') = \overline{b_\zeta(g',g)}, \quad g,g' \in G, \tag{2.21}$$

$$b_\zeta(g,g) = 1 \quad , \quad g \in G. \tag{2.22}$$

Moreover a multiplier ζ (resp. the associated bicharacter b_ζ) is non-degenerated iff:

$$G_\zeta = \{g \in G:\ b(g,g') = 1,\ \forall g' \in G\} = \{e\} \tag{2.23}$$

in such a case the center of $G \boxtimes_\zeta T$ is exactly the set $\{(e,\alpha): \alpha \in T\}$.

The automorphism group of G acts within the multipliers ζ (and correspondingly within the associated bicharacters) according to $\alpha \in \mathrm{Aut}(G)$: $\xi \to {}^{\alpha}\zeta$ where

$$^{\alpha}\zeta(g,g') = \zeta(\alpha(g), \alpha(g')) \quad g,g' \in G. \tag{2.24}$$

Such an action defines isomorphisms between the different central extensions of an abelian group G, more precisely ([3]),

<u>Proposition (2.25):</u>

If G is an abelian group, ζ and ξ two multipliers on G; α an automorphism of G such that there exists a function $\lambda : G \to T$ with the property:

$$^{\alpha}\xi\,(g,g') = \lambda(g)\,\lambda(g')\,\overline{\lambda(gg')}\,\zeta(g,g')$$

$\forall g,g' \in G$, then $G \boxtimes_{\xi} T$ and $G \boxtimes_{\zeta} T$ are isomorphic; the isomorphism is given by

$$\hat{\alpha}(g,\beta) = (\alpha(g), \overline{\lambda(g)}\beta) \quad \forall g \in G,\ \beta \in T.$$

Accordingly if a multiplier ζ is invariant with respect to some subgroup of the group $\mathrm{Aut}(G)$ then this subgroup is a subgroup of the group $\mathrm{Aut}(G \boxtimes_{\zeta} T)$.

Unitary projective representations of G with multiplier ζ are in bijection with *-representations of a C*-algebra $\overline{\Delta(G,\zeta)}$ which is unique if C_{ζ} is trivial (see e.g.[5]). This is the norm completion of the *-algebra generated by unitary elements δ_g , $g \in G$ which satisfy:

$$\delta_g \delta_{g'} = \zeta(g,g')\,\delta_{gg'} \quad g,g' \in G$$

$$\delta_g^* \delta_g = \delta_g \delta_g^* = 1.$$

It is important to realize that isomorphisms of G such that $^{\alpha}\zeta(g,g') = \lambda(g)\,\lambda(g')\,\overline{\lambda(gg')}\,\xi(g,g')$ extend to *-isomorphisms of $\overline{\Delta(G,\zeta)}$ to $\overline{\Delta(G,\xi)}$.

3. Central Extensions of $(Z_n \times Z_n)_R^\Lambda$

Let Z_n be the cyclic group with n elements namely $Z_n = \{a, a, \ldots, a^n = e\}$. Let Λ be an at most countable set, then any bicharacter of $P_f(Z_n,\Lambda)$ invariant with respect to $\mathfrak{S}_\Lambda$ is of the form

$$b(\mathbb{X}, \mathbb{Y}) = z^{\alpha \sum_{p,q=1}^{n-1} pq\,|\mathbb{X}(a^p)\cap\mathbb{Y}(a^q)| + \beta \sum_{p,q=1}^{n-1} pq|\mathbb{X}(a^p)|\,|\mathbb{Y}(a^q)|} \quad (3.1)$$

with

$\alpha = 1,2,\ldots,n \quad \beta = 1,2,\ldots n$

$\mathbb{X}, \mathbb{Y} \in P_f(Z_n,\Lambda)$, $z = \exp(\frac{2i\pi}{n})$.

Indeed, according to formula (2.16) the elements $\{x_i\}_1$, $x_i \in \Lambda$, are generators of $P_f(Z_n,\Lambda)$. Since b is a bicharacter, it is completely defined by its values $b(\{x_i\}_1, \{x_j\}_1)$, x_i, $x_j \in \Lambda$, on the pairs of elements of Λ. But $\mathfrak{S}_\Lambda$ divides the set of pairs of points of Λ into two orbits namely the pairs (x_i,x_i), $x_i \in \Lambda$ and the pairs (x_i,x_j), $i \neq j$. Consequently if X, Y $\in \mathfrak{P}(\Lambda)$:

$$b(\{X\}_1, \{Y\}_1) = \prod_{x_j\in X} \prod_{x_j\in Y} b(\{x_i\}_1, \{x_j\}_1)$$

$$= \prod_{x_i\in X\cap Y} b(\{x_i\}_1, \{x_i\}_1) \prod_{\substack{x_i\in X\\ x_j\in Y\\ x_i\neq x_j}} b(\{x_i\}_1, \{x_j\}_1).$$

Now by invariance $b(\{x_i\}_1,\{x_i\}_1)$ (resp. $b(\{x_i\}_1,\{x_j\}_1)$ for $i \neq j$) does not depend on x_i (resp. on (x_i,x_j), $i \neq j$); hence there exist two integers μ and ν such that:

$$b(\{x_i\}_1,\{x_i\}_1) = z^\mu$$

$$b(\{x_i\}_1.\{x_j\}_1) = z^\nu \qquad i \neq j .$$

Consequently:

$$b(\{X\}_1, \{Y\}_1) = z^{(\mu-\nu)|X\cap Y|+\nu|X||Y|} \quad (3.2)$$

and the general result (3.1) follows by direct application of (2.16).

This shows that the invariance with respect to $\mathfrak{S}_\Lambda$ is too stringent in general, since as a corollary, it implies there is no non trivial multiplier on $(Z_n)^\Lambda_R$ invariant with respect to $\mathfrak{S}_\Lambda$ for odd n, since none of the bicharacters (3.1) are antisymmetric. For $(Z_n \times Z_n)^\Lambda_R$ with n prime greater than 2, one has the following theorem:

Theorem (3.3):

Any bicharacter b of $P_f(Z_n \times Z_n, \Lambda)$ for n prime greater than 2, Λ strictly infinite countable, such that

i) $b(g,g') = \overline{b(g',g)}$, $g, g' \in P_f(Z_n \times Z_n, \Lambda)$

ii) invariant with respect to $\mathfrak{S}_\Lambda$

iii) non degenerated

is of the form:

$$b((\mathbb{X}, \mathbb{Y}); (\mathbb{X}', \mathbb{Y}'))$$

$$= z^{\alpha \sum_{p,q=1}^{n-1} pq(|\mathbb{X}(a^p)\cap\mathbb{Y}'(a^q)| - |\mathbb{X}'(a^p)\cap\mathbb{Y}(a^q)|)} \cdot z^{\beta \sum_{p,q=1}^{n-1} pq(|\mathbb{X}(a^p)|\,|\mathbb{Y}'(a^q)| - |\mathbb{X}'(a^p)|\,|\mathbb{Y}(a^q)|)}$$

for $\alpha = 1,2,\dots,n-1$, $\beta = 1,2,\dots n$ and $z = \exp(\frac{2i\pi}{n})$.

First notice that $P_f(Z_n \times Z_n, \Lambda)$ is isomorphic to $P_f(Z_n,\Lambda) \times P_f(Z_n,\Lambda)$ but any bicharacter b on $P_f(Z_n,\Lambda) \times P_f(Z_n,\Lambda)$ splits into four bicharacters on $P_f(Z_n,\Lambda)$:

$$b((\mathbb{X},\mathbb{Y}); (\mathbb{X}',\mathbb{Y}')) = b^1(\mathbb{X},\mathbb{X}')b^2(\mathbb{X},\mathbb{Y}')b^3(\mathbb{Y},\mathbb{X}')b^4(\mathbb{Y},\mathbb{Y}')$$

where:

$b^1(\mathbb{X},\mathbb{X}') = b((\mathbb{X},E); (\mathbb{X}',E))$, $b^4(\mathbb{Y},\mathbb{Y}') = b((E,\mathbb{Y}); (E,\mathbb{Y}'))$

$b^2(\mathbb{X},\mathbb{Y}') = b((\mathbb{X},E); (E,\mathbb{Y}'))$, $b^3(\mathbb{Y},\mathbb{X}') = b((E,\mathbb{Y}); (\mathbb{X}',E)$

for $\mathbb{X}, \mathbb{X}', \mathbb{Y}, \mathbb{Y}' \in \mathcal{P}_f(Z_n,\Lambda)$.

Antisymmetry of b implies that

$$b^1(\mathbb{X},\mathbb{X}') = \overline{b^1(\mathbb{X}',\mathbb{X})} \quad , \quad b^4(\mathbb{Y},\mathbb{Y}') = \overline{b^4(\mathbb{Y}',\mathbb{Y})}$$

$$b^2(\mathbb{X},\mathbb{Y}') = \overline{b^3(\mathbb{Y}',\mathbb{X})} .$$

Moreover the b^i's are $\mathfrak{S}_\Lambda$ invariant if b is $\mathfrak{S}_\Lambda$ invariant; for odd n this implies that $b^1 = b^4 = 1$, the trivial bicharacter, and

$$b((\mathbb{X},\mathbb{Y});\ (\mathbb{X}',\mathbb{Y}')) = b^2(\mathbb{X},\mathbb{Y}')\ \overline{b^2(\mathbb{X}',\mathbb{Y})} \ .$$

Again for odd n this ensures that

$$b((\mathbb{X},\mathbb{Y});\ (\mathbb{X},\mathbb{Y})) = 1 \quad , \quad \mathbb{X},\mathbb{Y} \in \mathcal{P}_f(Z_n,\Lambda) .$$

Now we can apply the result (3.1) to get the general form (3.3). Finally to prove non degeneracy we assume that:

$$b((\mathbb{X},\mathbb{Y});\ (\mathbb{X}',\mathbb{Y}')) = 1 \qquad \text{for all } \ \mathbb{X}',\ \mathbb{Y}' \in \mathcal{P}_f(Z_n,\Lambda)$$

and in particular for $\mathbb{X}' = E$; consequently

$$z^{\alpha \sum\limits_{p,q}^{n-1} pq|\mathbb{X}(a^p)\cap \mathbb{Y}'(a^q)| + \beta \sum\limits_{pq}^{n-1} pq|\mathbb{X}(a^p)|\,|\mathbb{Y}'(a^q)|} = 1 .$$

This is also true for all $\mathbb{Y}'$ such that $|\mathbb{Y}'(a^q)| = 0 \bmod n$, $q \neq n$. Hence

$$z^{\alpha \sum\limits_{p,q}^{n-1} pq|\mathbb{X}(a^p)\cap \mathbb{Y}'(a^q)|} = 1$$

for those $\mathbb{Y}'$, and in particular for $\mathbb{Y}' = \{Y_o\}_r$, $Y_o \in \mathcal{P}_f(\Lambda)$, $|Y_o| = 0 \bmod n$

$$z^{\alpha r \sum\limits_{p=1}^{n-1} p\,|\mathbb{X}(a^p)\cap Y_o|} = 1 \qquad 1 \le r \le n-1 .$$

If $\mathbb{X}(a^p) \neq \emptyset$ for some p, under the assumption that $|\Lambda| = \infty$ we can choose Y'_o such that $|Y'_o| = 0 \bmod n$ and

$$|\mathbb{X}(a^p)\cap Y'_o| = |\mathbb{X}(a^p)\cap Y_o| - 1, \quad |\mathbb{X}(a^{p'})\cap Y'_o| = |\mathbb{X}(a^{p'})\cap Y_o|, p' \neq p, n.$$

Consequently $z^{\alpha rp} = 1$ which turns out to be impossible if n is prime.

Previous theorem does not apply to the simplest case n = 2, since then b^1 and b^4 are not necessarily trivial. Nevertheless one has the

Theorem (3.4):

Any bicharacter b of $P_f(Z_2 \times Z_2, \Lambda)$, Λ strictly infinite, which is such that

i) $b((\mathbb{X},\mathbb{Y}), (\mathbb{X}',\mathbb{Y}')) = \overline{b((\mathbb{X}',\mathbb{Y}'), (\mathbb{X},\mathbb{Y}))}$

II) $b((\mathbb{X},\mathbb{Y}), (\mathbb{X},\mathbb{Y})) = 1$

iii) invariant with respect to $\mathfrak{S}_\Lambda$

iv) non degenerate

v) $b((\mathbb{X},\mathbb{Y}); (\mathbb{X}', \mathbb{Y}')) = b((\mathbb{Y},\mathbb{X}), (\mathbb{Y}',\mathbb{X}'))$, $\mathbb{X},\mathbb{X}',\mathbb{Y},\mathbb{Y}' \in P_f(Z_2,\Lambda)$

is one of the four following ones:

$$b^S((\mathbb{X},\mathbb{Y}); (\mathbb{X}',\mathbb{Y}')) = (-1)^{|X\cap Y'|+|X'\cap Y|}$$

$$b^\sigma((\mathbb{X},\mathbb{Y}); (\mathbb{X}',\mathbb{Y}')) = (-1)^{|X\cap Y'|+|X'\cap Y|+|X||Y'|+|X'||Y|}$$

$$b^\varphi((\mathbb{X},\mathbb{Y}); (\mathbb{X}',\mathbb{Y}')) = (-1)^{|X\cap X'|+|Y\cap Y'|+|X||X'|+|Y||Y'|}$$

$$b^F((\mathbb{X},\mathbb{Y}); (\mathbb{X}',\mathbb{Y}')) = (-1)^{|X\cap X'|+|Y\cap Y'|+(|X|+|Y|)(|X'|+|Y'|)}$$

where $X \equiv \mathbb{X}(a)$, $Y \equiv \mathbb{Y}(a)$, $X' \equiv \mathbb{X}'(a)$ and $Y' \equiv \mathbb{Y}'(a)$.

Indeed, if n = 2 any $\mathbb{X} \in P_f(Z_2,\Lambda)$ is completely defined by its value on a ($a^2 = e$). Still one can apply formula (3.1) to get:

$$b((\mathbb{X},\mathbb{Y}), (\mathbb{X}',\mathbb{Y}')) = (-1)^{\alpha_1|X\cap X'|+\beta_1|X||X'|+\alpha_4|Y\cap Y'|+\beta_4|Y||Y'|}$$
$$(-1)^{\alpha_2|X\cap Y'|+\beta_2|X||Y'|+\alpha_3|X'\cap Y|+\alpha_4|X'||Y|}.$$

Antisymmetry of b (in our case symmetry) implies that $\alpha_2 = \alpha_3$, $\beta_2 = \beta_3$. The second condition b(g,g) = 1 implies that $\alpha_1 = \beta_1$, $\alpha_4 = \beta_4$. Moreover if $\alpha_1 = \alpha_2 = \alpha_3 = \alpha_4$ then b is degenerated indeed

$$b((\mathbb{X},\mathbb{Y});\ (\mathbb{X}',\mathbb{Y}')) = (-1)^{\alpha|X\Delta Y\cap X'\Delta Y'|+\alpha(|X||X'|+|Y||Y'|)}$$

$$(-1)^{\beta_2|X||Y'| + \beta_3|X'||Y|}$$

is identically 1 for X = Y, $|X| = 0 \bmod 2$.

Finally invariance with respect to the exchange (XY) $\to$ (YX) implies $\alpha_1 = \alpha_4 = \beta_1 = \beta_4$.

Consequently we are left with the four possibilities:

i) $\alpha_1 = \alpha_4 = \beta_1 = \beta_4 = 0 \quad \alpha_2 = \alpha_3 = 1 \quad \beta_2 = \beta_3 = 0$

ii) $\alpha_1 = \alpha_4 = \beta_1 = \beta_4 = 0 \quad \alpha_2 = \alpha_3 = 1 \quad \beta_2 = \beta_3 = 1$

iii) $\alpha_1 = \alpha_4 = \beta_1 = \beta_4 = 1 \quad \alpha_2 = \alpha_3 = 0 \quad \beta_2 = \beta_3 = 0$

iv) $\alpha_1 = \alpha_4 = \beta_1 = \beta_4 = 1 \quad \alpha_2 = \alpha_3 = 0 \quad \beta_2 = \beta_3 = 1$.

Non degeneracy of the corresponding bicharacters is essentially proved along the same lines as previously.

Let us make some remarks:

i) The general case for arbitrary cardinality of Λ and arbitrary n can be treated as well with the same techniques. Non degeneracy depends strongly on both $|\Lambda|$ and n.

ii) Similarly one can drop for n = 2 the invariance through the exchange of X and Y.

iii) In every case, these bicharacters correspond to actual central extensions of $(Z_n \times Z_n)^{\Lambda}_{R}$ for n = 2; this has been done already in [3]. For n prime greater than 2 one has the following construction.

Let $H = \ell_2(P_f(Z_n,\Lambda))$ be the Hilbert space of square integrable functions with respect to the discrete Haar measure. H is countable. Then

$$\{U(\mathbb{X},\mathbb{Y})\}\ (\mathbb{Z}) = z^{\alpha \sum_{p,q}^{n-1} pq|\mathbb{Y}(a^p)\cap\mathbb{Z}(a^q)| + \beta \sum_{p,q}^{n-1} pq|\mathbb{Y}(a^p)||\mathbb{Z}(a^q)|} \tag{3.5}$$

$$f(\mathbb{X}\Delta\mathbb{Z}) \qquad \mathbb{X},\ \mathbb{Y},\ \mathbb{Z} \in P_f(Z_n,\Lambda).$$

is an unitary projective representation of $P_f(Z_n \times Z_n,\Lambda)$ whose associated bicharacter is precisely $b^{\alpha\beta}$.

If $\zeta^{\alpha\beta}$ is the associated multiplier on $P_f(Z_n \times Z_n,\Lambda)$

$$\zeta^{\alpha\beta}((\mathbb{X},\mathbb{Y}),\ (\mathbb{X}',\mathbb{Y}')) = z^{\alpha \sum_{p,q}^{n-1} pq|\mathbb{X}(a^p)\mathbb{Y}'(a^q)| + \beta \sum_{p,q}^{n-1} pq|\mathbb{X}(a^p)||\mathbb{Y}'(a^q)|} \tag{3.6}$$

then the C*-algebra $\overline{\Delta(P_f(Z_n \times Z_n,\Lambda),\zeta^{\alpha\beta})}$ is isomorphic to the U.H.F. algebra constructed from the nxn matrices.

Proof is relatively easy for ζ^{1n} and for general $\zeta^{\alpha\beta}$ it follows from the theorem

<u>Theorem (3.7):</u>

Let n be prime greater than 2. Let $\zeta^{\alpha\beta}$ and $\zeta^{\alpha'\beta'}$ be two multipliers of $P_f(Z_n \times Z_n,\Lambda)$ defined in (3.6), then there exists an isomorphism τ of $P_f(Z_n \times Z_n,\Lambda)$ such that:

$$\zeta^{\alpha\beta}(\tau(\mathbb{X},\mathbb{Y}),\ \tau(\mathbb{X}',\mathbb{Y}')) = \zeta^{\alpha'\beta'}((\mathbb{X},\mathbb{Y}),\ (\mathbb{X}',\mathbb{Y}')).$$

$$\alpha,\ \alpha' = 1,2,\dots,\ n-1.$$

The proof is divided in two steps: first of all we describe the automorphism σ_{rs} which connects $\zeta^{\alpha n}$ to $\zeta^{\alpha' n}$. On the generating elements $(\{X\}_1,\{Y\}_1)$ $X,Y \in \mathfrak{F}(\Lambda)$:

$$\sigma_{rs}(\{X\}_1,\{Y\}_1) = (\{X\}_r,\{Y\}_s) \qquad 1 \le r < n,\ 1 \le s < n,$$

where r,s is any pair of integers less or equal to n such that:

$$\alpha rs = \alpha' \mod n.$$

The solution is not unique, but in general one cannot choose <u>r=s</u>.

The second step shows that there is an isomorphism to connect $\zeta^{\alpha n}$ to $\zeta^{\alpha\beta}$. We give explicitly the action of this isomorphism on the generators $(\{x_i\}_1, E)$ and $(E, \{x_i\}_1)$: since Λ is countable let us choose an order $i \in Z \rightarrow x_i \in \Lambda$ then:

$$\tau_t(E, \{x_{np+r}\}_1) = (E, \{X_{np}\}_t \Delta \{x_{np+r}\}_1), \quad 1 \leq r \leq n, \; p \geq 0,$$

where $X_{np} = \{x_i \in \Lambda;\; i \leq np\}$, $X_o = \emptyset$,

$$\tau_t(\{x_{nq+s}\}_1, E) = (\{X_{n(q+1)}\}_t \Delta \{x_{nq+s}\}_1, E), \quad 1 \leq s \leq n, \; q \geq 0,$$

and t is chosen such that $\alpha t = \beta \bmod n$.

It can be shown that σ and τ have the following properties

$$\sigma_{11} = 1, \quad \sigma_{rs}\sigma_{r's'} = \sigma_{rr' \bmod n,\; ss' \bmod n}$$

$$\tau_n = 1, \quad \tau_t \tau_{t'} = \tau_{t+t' \bmod n}$$

$$\sigma_{rs}\tau_t = \tau_t \sigma_{rs}.$$

The same theorem holds for n = 2 (see [1]) and shows that, up to isomorphisms, there exists only one non trivial central extension of $(Z_n \times Z_n)^{\Lambda}_{R}$ n prime, whose associated bicharacter is permutation invariant.

U.H.F. algebra built over nxn matrices appears in this way as an algebra of canonical commutation relations and one can derive some interesting properties in this way (see [2] for the case n = 2).

References

[1] C. Chevalley, Theory of Lie Groups, Princeton University Press (1946).

[2] Ph. Combe, R. Rodriguez, M. Sirugue, M. Sirugue-Collin, On the Quantization of Quantum Spins and Fermi Systems, Journ. of Math. Phys. 20, 611 (1979).

[3] Ph. Combe, R. Rodriguez, M. Sirugue, M. Sirugue-Collin, Comm.Math.Phys. 63, 219 (1978).

[4] S. Albeverio, R. Høegh-Krohn, Ergodic Action by Compact Group on von Neumann Algebras, Preprint Marseille.

[5] J. Manuceau, M. Sirugue, D. Testard, D. Verbeure, Comm. math. Phys. 38, 29 (1973).

Introduction to W*-Categories

R. Lima

Centre de Physique Théorique

CNRS Marseille, France

Abstract

A W*-category is the natural generalization of a von Neumann algebra where, instead of taking the bounded linear mappings of a fixed Hilbert space as a model, we take the bounded linear mappings between a collection of Hilbert spaces.

Together with the elementary theory of W*-categories we present a list of examples which motivates and illustrates the general notion developed in this lecture.

1. Introduction

The aim of this lecture is to introduce you to W^*-categories. Roughly speaking a W^*-category is the natural generalization of a von Neumann algebra but taking bounded linear mappings between a collection of Hilbert spaces as model.

As we shall see, this structure is implicit when we deal with the theory of unitary representations of groups or when we want to compare objects such as weights or cocycles in the von Neumann algebra context.

Also this structure emerges from the analysis of the superselection structure of elementary particle physics ([1], [2], [3], [4]) and it is therefore not surprising that it appears also in Roberts' lecture.

Most of the constructions described here should be contained in a joint paper with P. Ghez and J.E. Roberts.

Section 2 gives a list of examples which motivates the exact definitions and further developments supplied in the subsequent sections 3, 4 and 5

Additional remarks are given in section 6 for the examples in section 2 and section 7 is a conclusion.

2. Examples of W*-Categories

Example 1: Let $\mathcal{H}$ be the category of Hilbert spaces. The objects of H are (separable) Hilbert spaces belonging to some universe; if H_1 and H_2 are two such Hilbert spaces the set of arrows from H_1 to H_2 is denoted by (H_1, H_2) and $t \in (H_1, H_2)$ if t is a bounded linear operator from H_1 to H_2. The linear structure and the composition, adjoint and norm of linear mappings are defined as usual. Note that (H_1, H_2) is a Banach space which is a right -

$-(H_1, H_1) = \mathcal{B}(H_1)$ module and a left-$(H_2, H_2) = \mathcal{B}(H_2)$ module.

Example 2: Let M be a von Neumann algebra. M can be viewed as a W*-category with a unique object whose arrows are the elements of M. The remaining structure is derived from the corresponding structure in M.

Example 3: If M is a von Neumann algebra then we may consider the category W*(M) whose objects are the projections in M and if e,f are two such objects the arrows $t \in (e,f)$ are the elements of M such that $ft = t = te$. Again the remaining structure is inherited from M.

Example 4: Let G be a locally compact group and Rep(G) be the category whose objects are continuous unitary representations of G and whose arrows are bounded linear intertwining operators. The remaining structure is just derived from the corresponding structure in $\mathcal{H}$.

Example 5: If A is A C*-algebra, let Rep(A) be the category of non degenerate representations of A and bounded linear intertwining operators. Again, the remaining structure is derived from the corresponding structure in $\mathcal{H}$. If A is a von Neumann algebra we can consider the category of normal representations of A. These are the W*-categories studied by M. Rieffel in [5].

Example 6: The category End(M) whose objects are endomorphisms of a von Neumann algebra M (i.e. normal *-homomorphisms ρ of M such that $\rho(1) = 1$). An arrow from ρ_1 to ρ_2 is a $t \in M$ that verifies

$$t\rho_1(x) = \rho_2(x)t \quad \forall x \in M.$$

The algebraic and topological structure of the arrows is that inherited from M. This is one of the W*-categories introduced by J. Roberts in [4].

Example 7: Let M be a von Neumann algebra and let $\mathcal{H}(M)$ denote the category of Hilbert spaces in M (a Hilbert space in M is a norm closed linear subspace h of M such that $a^*a \in \mathbb{C}\,1$ for all $a \in h$ and if $x \in M$ and $xa = 0$ for all $a \in h$ then $x = 0$). If h_1 and h_2 are two such objects an arrow $t \in (h_1, h_2)$ is a $t \in M$ such that $th_1 \subset h_2$.

The algebraic and topological structure of the arrows is that inherited from M.

This W*-category has been studied in [4]. If M is acted upon by a locally compact group, any (globally) invariant Hilbert space in M carries a representation of the group, induced by the action.

Example 8: $\omega(M)$ is the category of weights on a type III factor M. The objects of $\omega(M)$ are normal semi-finite weights on M and an arrow from from ϕ_1 to ϕ_2 in $\omega(M)$ is an element $x \in M$ that verifies the following conditions:

(i) $x\, s(\phi_1) = x = s(\phi_2)x$ if $s(\phi_i)$, i=1,2

is the support of ϕ_i, i.e. $s(\phi_i)$ is a projection in M and $\phi_i(1-s(\phi_i)) = 0$, ϕ_i is faithful on the reduced von Neumann algebra $M_{s(\phi_i)}$.

(ii) $x\,(D\phi_1 : D\psi)_t = (D\phi_2 : D\psi)_t \quad \tau_t^{\psi}(x) \qquad \forall t \in R$

for one - and therefore all - faithful normal semi-finite weight ψ on M. Here $(D\phi_i : D\psi)$ denote the Radon-Nicodym derivative (see [6]). The algebraic and topological structure is induced by M.

$\omega(M)$ has been studied by A. Connes and M. Takesaki in [6].

Example 9: Let α be a continuous action af a locally compact group G by automorphisms of a σ-finite von Neumann algebra M. Let $Z_{\alpha}^{1}(M)$ be the category whose objects are 1-cocycles, i.e. σ-strong* continuous function $a: g \in G \to a(g) \in M$ such that

(i) $a(gh) = a(g)\alpha_g(a(h)) \qquad \forall g,h \in G$

(ii) $a(g^{-1}) = \alpha_g^{-1}(a(g)^*) \qquad \forall g \in G$.

As arrow $t \in (a,b)$ in $Z_{\alpha}^{1}(M)$ is an element of M such that:

(1) $t\, a(1) = t = b(1)t$, where 1 is the unit element in G
and
(2) $t\, a(g) = b(g)\, \alpha_g(t)$ for all $g \in G$.

The algebraic and topological structure is derived from the corresponding structure of M.

This W*-category is studied in [6].

Example 10: Let $\mathcal{G}$ be a measurable groupoid and Λ be a transverse measure on G (see [7] for exact definitions). Let $\tau_\Lambda(g)$ bc the category whose objects are the square integrable representations of $\mathcal{G}$ and each arrow is a class mod Λ of essentially bounded intertwiniers between such representations.

This W*-category appears in [7] as the representation theory of "non commutative measure spaces."

Example 11: Let X be a topological space and M be a von Newmann algebra. Let Δ^n denote the standard n-simplex, i.e.
$\Delta^n \equiv \{(t^o, t^1,\dots, t^n) \in R^{m+1}: t^i \geq 0$ and $\sum_{i=o}^{n} t^i = 1\}$ and
$\Sigma_n(X)$ the set of n-simplexes of X, i.e. the continuous mappings from Δ^n to X. Let $Z^1_X(M)$ be the category whose objects are the (unitary) 1-cocycles i.e. the mappings form $\Sigma_1(X)$ to the group of unitary operators in M, such that for all $c \in \Sigma_2(X)$

$$z(\partial_o c)\ z(\partial_2 c) = z(\partial_1 c)$$

where ∂_i: $\Sigma_2(X) \to \Sigma_1(X)$ are the boundary operations, i.e.

$$\partial_o c(t^o, t^1) = c(o, t^o, t^1)$$

$$\partial_1 c(t^o, t^1) = c(t^o, o, t^1)$$

$$\partial_2 c(t^o, t^1) = c(t^o, t^1, o) \qquad (t^o, t^1) \text{ in } \Delta^1.$$

The arrows from the 1-cocycle z to the 1-cocycle z' are the mappings y: $\Sigma_o \to M$ (recall that $\Sigma_o \sim X$) such that

$$z'(b)\ y(\partial_1 b) = y(\partial_o b)\ z(b)\ ,\ b \in \Sigma_1.$$

This W*-category has been presented, for instance, in [8].

3. W*-Categories

In order to stress the connection with the theory of von Neumann algebras we introduce first the notion of C*-category. After this the definition of W*-category follows naturally.

A C*-category is a category whose objects we denote by A, B, ... such that the set of arrows from A to B, denoted by (A,B), is endowed with a structure of Banach space from all A and B. The composition law of arrows is bilinear; there is an antilinear isometric involution $x \in (A,B) \to x^* \in (B,A)$ such that x^*x is a positive element in the Banach algebra (A,A) and finally

(1) $||x\ y|| \leq ||x||\ \ ||y||$

whenever the left hand side is defined, and

(2) $||x||^2 = ||x^*x||$ for all arrows x.

Any set of bounded linear operators between a collection of Hilbert spaces, containing the identities on each space, closed under the usual composition laws and involution which is also closed in norm topology provides an example for C*-category. In fact any C*-category can be represented in this way. A representation of a C*-category $\mathcal{A}$ is a field F of Hilbert spaces on the set of objects of $\mathcal{A}$, together with bounded linear mappings F(x) from F(A) to F(B) for each $x \in (A,B)$ so that $x \to F(x)$ preserves composition and involution (such a map is automatically continuous).

Using a generalization of the GNS construction we can see that every C*-category can be realized, via a faithful representation, as a C*-subcategory of $\mathcal{H}$ (ex. 1) so that the arrows are represented by bounded linear mappings.

For convenience, we may introduce the following notation: if $A_1, A_2, \dots A_n$ are objects of a C*-category $\mathcal{A}$, then $M(A_1,\dots A_n)$ denote the *-algebra of $n \times n$ matrices $x = (x_{ij})_{i,j=1,2\dots n}$ where the x_{ij} are arrows in $\mathcal{A}$, with $x_{ij} \in (A_j, A_i)$. $M(A_1,\dots A_n)$ is endowed with the usual product rule and involution of matrices. Then $M(A_1,\dots A_n)$ can be made into a C*-algebra.

Now a W*-category is a C*-category such that each Banach space (A,B) is the dual of a Banach space, denoted by $(A,B)_*$. This clearly generalizes the corresponding abstract definition of a von Neumann Algebra (see ex. [10]). Hence (A,A) is, for each A, a von Neumann algebra and therefore the dual of a unique Banach space $(A,A)_*$. The following characterization of W*-categories implies, in particular, that $(A,B)_*$ is also unique, for all A and B.

<u>Theorem:</u> Let $\mathcal{A}$ be a C*-category. The following are equivalent:

(1) $\mathcal{A}$ is a W*-category

(2) M(A,B) is a von Neumann algebra for all A,B

(3) There is a faithful representation of $\mathcal{A}$ (into $\mathcal{H}$) with a weakly closed image ($\mathcal{A}$ is isomorphic to a weakly closed sub *-category of $\mathcal{H}$)

(4) (A,A) is a von Neumann algebra for all A and (A,B) is a right self-dual (A,A) module, i.e. for all bounded module maps R: $(A,B) \to (A,A)$ there is an $x \in (A,B)$ such that $R(a) = x^*a$ for all $a \in (A,B)$.

Here the module structure is just the right multiplication of the elements of (A,B) by the elements of the von Neumann algebra (A,A).

Condition (2) makes it clear that there will be a close connection between the properties of von Neumann algebras and those of W*-categories. Condition (3) is just the precise sense of the definition implicit in the introduction and condition (4) gives the connection with categories studied by Rieffel in [5], and modules studied by Paschke [9].

The existence of polar decomposition of arrows and normal forms and the validity of a version of Kaplansky's density theorem can be easily understood on the basis of this Theorem.

We turn now to the problem of defining the commutant and bicommutant of a representation F of a, say, C*-category $\mathcal{A}$.

The commutant $F(\mathcal{A})'$ is just the von Neumann algebra of the bounded intertwiners between F and F, namely elements of the commutant are fields on the set of objects of $\mathcal{A}$:

$$\eta : A \to \eta_A$$

each η_A being a bounded linear operator on the Hilbert space F(A) and such that

$$F(t)\eta_A = \eta_B F(t) \qquad \text{if } t \in (A,B)$$

and

$$||\eta|| = \sup_A ||\eta_A|| < +\infty .$$

Multiplication and involution of these fields are defined pointwise:

$$(\eta\cdot\xi)_A = \eta_A \cdot \xi_A$$

and

$$(\eta^*)_A = (\eta_A)^* .$$

In the same way we can define the bicommutant of F which is, then, a W*-category with the same objects as $\mathcal{A}$ and arrows in $\mathcal{H}$, namely

$$(A,B)_{F(\mathcal{A})''} = \{x \in (F(A),\ F(B)) \quad x\eta_A = \eta_B x \quad \forall \eta \in F(\mathcal{A})'\}.$$

The von Neumann double commutant theorem can then be proved in this context.

4. Center and Comparison of Objects in a W*-Category

One of the most interesting problems common to the above examples is the comparison of objects in a W*-category. We shall return to these examples later but it is not hardly surprising that we should try to generalize one of the most powerful techniques in the theory of von Neumann, the comparison of projections, these are just the objects of the W*-category W*(M) described in section 2.

With this example in mind we say that A is a subobject of B, denoted A ≤ B, if there is a partial isometry u ∈ (A,B) with $u^*u = 1_A$. If A ≤ B and B ≤ A we say that A is equivalent to B.

It is clear from the examples that a weaker partial ordering is also needed, "subobject up to multiplicity": we say that A is quasi-contained in B if for each non null subobject A' of A there is a non nul subobject A" of A' such that A" ≤ B. The concept of center of a W*-category helps to clarify these definitions. In particular the notion of central support is very useful because it takes into account the relation between different objects.

The center of a W*-category $\mathcal{A}$ is the von Neumann algebra of the bounded natural transformations from the identity functor of $\mathcal{A}$ to itself, so that an element η in the center of $\mathcal{A}$ is just a field on the set of objects of $\mathcal{A}$, $\eta\colon A \to \eta_A \in (A,A)$ such that

$$t\eta_A = \eta_B t \quad \text{if } t \in (A,B)$$

and

$$||\eta|| = \sup_A ||\eta_A|| < +\infty .$$

Notice that η, in fact, takes values in the center of each von Neumann algebra (A,A). Again multiplication and involution are defined pointwise.

Since we have notion of center we can introduce the central support of an arrow in $\mathcal{A}$.

Lemma: For each arrow t of a W*-category there is a smallest projection c(t) in the center of $\mathcal{A}$, called the central support of t, such that

$$t\, c(t)_A = t \quad \text{if} \quad t \in (A,B).$$

The central support of an object is the central support of the identity arrow on it.

There is a well known way of building new objects from a given object without changing its central support, and this is to take direct sums. We write $A \sim \oplus_j A_j$ if there are isometries $\omega_j \in (A_j,A)$ with $\Sigma\omega_j\omega_j^* = 1_A$ and in fact we have, then

$$c(A) = \bigvee_j c(A_j).$$

In a W*-category closed under countable direct sums and with "enough" sub-objects, i.e. if or each projection e there is an object B and an arrow u such that $u^*u = 1_B$ and $uu^* = e$, the relation between quasi-equivalence and central support is very easily stated because the following conditions are equivalent

(1) A is quasi-contained in B

(2) $A \leq \bigoplus_{\mathbb{N}} B$

(3) $c(A) \leq c(B)$.

Here $\bigoplus_{\mathbb{N}} B$ denotes a direct sum of countable many copies of B and the last condition means that $c(B) - c(A)$ is a positive element in the center of the W*-category or, equivalently that for each object X, we have

$$c(A)_X \leq c(B)_X$$

and because $c(A)_A = 1_A$ we can see that a quasi-maximal object ought to have a central support which is the identity. If such an object A exists, the center of the W*-category is isomorphic to the center of the von Neumann algebra (A,A) and for each normal representation F of the W*-category the commutant of F is isomorphic to the commutant of the image of (A,A) under F.

5. Modular Theory

Let us just notice in what sense the Modular theory of von Neumann algebras (Tomita-Takesaki theory) can be extended to the case of a W*-category $\mathcal{A}$.

A weight on a W*-category is, as it is to be expected, a linear map from the positive part of the W*-category to $[0,+\infty]$. The attributes normal, semi-finite, or faithful have their obvious extensions to weights on a W*-category.

Associated to each faithful, normal, semi-finite weight ϕ on $\mathcal{A}$ there is a representation of the additive group R by automorphisms of $\mathcal{A}$

$$t \in R \to \sigma_t^{\phi}.$$

The identities in $\mathcal{A}$ are pointwise invariant by the σ_t^{ϕ}. The relation between ϕ and σ^{ϕ} is the KMS-boundary condition.

Given two faithful, normal, semi-finite weights ϕ and ψ there is a unique bounded unitary natural transformation from σ^{ψ} to σ^{ϕ} and, as for von Neumann algebras, a converse of this property holds.

Finally, associated to each faithful, normal, semi-finite weight there is a normal standard representation of $\mathcal{A}$ where we can define Δ and J in analogy with Tomita-Takesaki theory.

6. Further Comments

We return now to the examples, hoping that this will help the reader to understand how simple and natural the structures described in the previous sections really are.

Example 1: The W*-category $\mathcal{H}$ of (separable) Hilbert spaces is, in some sense, the model for the theory. $\mathcal{H}$ has sufficient subobjects in the same sense as section 4 and also is closed under (countable) direct sums. Non-zero objects are quasi-equivalent, therefore generators. All infinite dimensional objects are equivalent and the comparison of objects is just dimension theory.

Example 2: If we think of a von Neumann algebra as a W*-category with a single object, everything reduces to the usual theory of von Neumann algebras.

Example 3: The W*-category W*(M) of projections of a von Neumann algebra M has a maximal object and has sufficient subobjects. Conversely any W*-category with these two properties is equivalent to a W*-category of projections of a (unique) von Neumann algebra. W*(M) is closed for (countable) direct sums of objects if and only if M is properly infinite. For W*(M) the comparison of objects is the Murray and von Neumann theory of comparison of projections.

Indeed, for e and f projections in M, $e < f$ is the sense of Murray and von Neumann if there is a partial isometry ω in M such that $\omega^*\omega = e$ and $\omega\omega^* f = \omega\omega^*$, and because ω is then an arrow in W*(M) we can see that this notion coincides with the comparison of objects described in section 4. The same is true for the notion of an object quasi-contained in another object.

The center of W*(M) is isomorphic to the center of M.

<u>Example 4:</u> The comparison of objects in the W*-category Rep(G) of the unitary representations of a locally compact group G is the ususal comparison of representations. Notice that there is a natural embedding of Rep(G) into $\mathcal{H}$ and the Tannaka-Tasuuma duality theorem characterizes a homeomorphic injection of G into the commutant of this embedding.

<u>Example 5:</u> Again the comparison of objects in the W*-category Rep(A) of non degenerated (resp. normal) representations of a C*-algebra (resp. a von Neumann algebra) is the usual comparison of representations.

Rep(A) as sufficient subobjects and is complete for countable direct sum. The universal representation of a C*-algebra and any normal faithful representation of a von Neumann algebra are generators. Hence there is, again, a natural embedding of Rep(A) into $\mathcal{H}$ and the commutant of this embedding is the enveloping von Neumann algebra of A in case of C*-algebras and isomorphic to A in case of von Neumann algebras. If A is a von Neumann algebra any faithful representation with properly infinite commutant is maximal in Rep(A), so that it is equivalent to a W*-category of projections as in example 3. It is then clear that two von Neumann algebras have equivalent categories of normal representations (i.e. they are Morita equivalent) if and only if they have faithful representations with isomorphic commutant [5].

<u>Examples 6 and 7:</u> If we consider M to be a W*-category with a single object, as in example 2, then End(M) is the category whose objects are normal *-endofunctors of M and the arrows are the bounded natural transformations between these endofunctors. End(M) is closed for direct sums if and only if M is properly infinite.

If M is a factor, the characterization property of section 3 says that for each endomorphism ρ in the class of quasi-equivalence of the identity endomorphism i, the set of arrows (i, ρ) is a Hilbert space in M, in the sense of example 7. Conversely [4], for each Hilbert space h in M, there is an unique endomorphism ρ such that

$$\rho(x)a = ax \qquad \forall x \in M \ , \ \forall a \in h \ .$$

Further, an element of M is an arrow between two Hilbert spaces in M if and only if it is an arrow between the corresponding endomorphisms.

Notice that an automorphism of M is a special endomorphism and it corresponds to an Hilbert space in M if and only if it is an inner automorphism.

Example 8: Let first to indicate how the W*-category $\omega(M)$ of the weights of a properly infinite von Neumann algebra M can be obtained in a canonical way from a W*-category equivalent to W*(M) (the projections of M).

To this end, consider the W*-category with objects labelled by the normal, semi-finite weights of M, the arrows between two such objects ϕ_1 and ϕ_2, are the elements x of M such that

$$xs(\phi_1) = x = s(\phi_2)x \qquad (*)$$

where $s(\phi_i)$, i=1,2 is the support of the weight ϕ_i.

Now this W*-category is equivalent to W*(M) and has a canonical weight ψ, in the sense of section 5, namely the weight ψ defined by:

$$\psi(x) = \phi(x) \quad \text{ if } x \in (\phi,\phi)^+ .$$

The centralizer of ψ is just the W*-category $\omega(M)$, namely an element x of M verifying (*) is in $\omega(M)$ if and only if $\sigma_t^\psi(x) = x$, $\forall t \in R$, where σ^ψ is the modular action of section 5.

The comparison of objects in $\omega(M)$ in the sense of the previous section is the theory of comparison of weights developed by Connes-

Takesaki [6]. The full subcategory $\omega_i(M)$ of integrable weights as a maximal object, namely any dominant weight, and it has sufficient subobjects, so that it is equivalent to the W*-category of projections of the von Neumann algebra N appearing in any continuous decomposition of M.

Accordingly, the center of $\omega_i(M)$ is isomorphic to the center of N. The central support is a surjection of the set of integrable weights (objects of $\omega_i(M)$) onto the σ-finite projections of the center of $\omega_i(M)$. The action τ of the multiplicative group $\hat{R}$ of positive real numbers on the center of $\omega_i(M)$ defined by

$$\tau_\lambda \; c(\phi) = c(\lambda\phi) \quad \lambda \in R^*_+$$

is a realisation of the (smooth) flow of weights, [6].

<u>Example 9:</u> The category $Z^1_\alpha(M)$ of a 1-cocycles for an action α of a locally compact group G on a von Neumann algebra M can also be defined from a W*-category equivalent to W*(M) in a similar way to that of example 8.

Again the comparison of objects in this case is the comparison of 1-cocycles developed by Connes-Takesaki in [6].

The full subcategory of the square integrable 1-cocycles is then equivalent to the W*-category of the projections of the cross-product $W^*(M,G,\alpha)$.

<u>Example 11:</u> The equivalence classes of 1-cocycles in $Z^1_X(M)$ are related with homotopy classes of paths in the topological space X. In the case of a space X with trivial cohomology (ex. Minkowski space) all 1-cocycles are equivalent (to the identity 1-cocycle). This can be easily seen using a contracting homotopy (see [8] for the proof).

In connection with Quantum Field Theory, where cohomology plays an important role, the non triviality of the cohomology comes from additional conditions imposed on the coefficients (ex: locality, see [8]).

7. Conclusion

This concludes our discussion about general structure and known examples of W*-categories. This is probably the best place to stress that these examples usually have additional structure and this could prove to be a good starting point for further research into this subject.

Acknowledgments

The work presented here has been done in collaboration with P. Ghez and J.E. Roberts. I thank both colleagues for the joy of collaboration and for the possibility of the announcement of the results prior to publication.

There results were obtained while the author stayed at the Fakultät für Physik - Universität Bielefeld. The author would like to express his gratitude to Prof. L. Streit and his colleagues for their hospitality extended to him and, finally to the A. von Humboldt-Stiftung for a research grant.

References

[1] S. Doplicher, R. Haag, J.E. Roberts, Local Observables and Particle Statistics
I, Commun.Math.Phys., 23 (1971) 199-230
II, Commun.Math.Phys., 35 (1974) 49-85.

[2] S. Doplicher, R. Haag, J.E. Roberts, Fields, Observables and Gauge Transformations
I, Commun.Math.Phys., 13 (1969) 1-23
II, Commun.Math.Phys., 15 (1969) 173-200.

[3] S. Doplicher, J.E. Roberts, Fields, Statistics and Non-Abelian Gauge Group, Commun.Math.Phys., 28 (1972), 331-348.

[4] J.E. Roberts, Cross Products of von Neumann Algebras by Group Duals, Symposia Math. XX (1976).

[5] M. Rieffel, Morita Equivalence for C*-Algebras and W*-Algebras, Journal of Pure and Applied Algebra, 5 (1974), p. 51-96.

[6] A. Connes, M. Takesaki, The Flow of Weights on Factors of Type III, Tôhoku Math. Journal, Vol. 29, no. 4 (1977).

[7] A. Connes, Sur la Théorie non Commutative de l'Intégration, Lecture Notes in Mathematics 725, Springer, Berlin-Heidelberg-New York 1979.

[8] J.E. Roberts, Local Cohomology and its Structural Implications for Field Theory, R.C.P. no. 25, Strasbourg, Novembre 1976.

[9] W.L. Paschke, Inner Product Modules over B*-Algebras, Transactions of the American Mathematical Society, vol. 182, (1973), p. 443.

[10] S. Sakai, C*-Algebras and W*-Algebras, Springer, Berlin-Heidelberg-New York 1971.

Net Cohomology and Its Applications to Field Theory

John E. Roberts

Fachbereich 5, Universität Osnabrück

Osnabrück, Federal Republic of Germany

Abstract

Nets form a convenient basic mathematical language for field theory and their cohomological invariants can give qualitative information on the field theory under investigation. Emphasis is placed here on computing their low dimensional cohomologies. The results are applied to the 0- and 1-cohomologies of the field nets and observable nets of quantum field theory and the associated superselection structure. The place of soliton sectors in two space-time dimensions is pointed out but no detailed analysis of this exceptional case is presented.

§1 Nets and Field Theory

The cohomology theory presented here was devised as a means of analysing certain specific structural phenomena in quantum field theory. Whilst these phenomena, which include the theory of superselection structure developed in [1,2] and the fact that the electromagnetic field tensor cannot be derived from a local electromagnetic potential, serve to motivate this cohomology, the time has come to put it into a wider context so as to stress its intimate relation with the basic mathematical structure of field theory.

Quantum fields are so singular that they cannot be considered as operator-valued functions on space-time. Thus some new mathematical language is needed for field theory. Let me begin by considering singular classical fields; such a field ϕ will be a distribution rather than a function.

$$\phi(f) = \int \phi(x) f(x) dx.$$

The distribution ϕ specifies the configuration of the system and corresponds quantum mechanically to a state. The test function f corresponds to an observable and can be thought of as the response function of some apparatus measuring the field strength.

There is one element of structure here which seems to provide a convenient basic language for field theory. If $\mathcal{D}(\mathcal{O})$ denotes the set of real C^∞-functions on $\mathbb{R}^d$ with support in the compact set $\mathcal{O}$, then

$$\mathcal{O}_1 \subset \mathcal{O}_2 \quad \text{implies} \quad \mathcal{D}(\mathcal{O}_1) \subset \mathcal{D}(\mathcal{O}_2). \tag{1.1}$$

Such a strict inductive system will be referred to as a <u>net</u> over the set $\mathcal{C}$ of compact subsets of $\mathbb{R}^d$.

A field theory could reasonably be defined as a dynamical theory whose basic dynamical variables form a net. This definition stresses the spatial dependence of the dynamical variables without making any a priori statement about their degree of singularity or otherwise specifying the mathematical framework. Thus we meet nets defined over subsets of a spacelike hyperplane in dealing with the Cauchy problem (Hamiltonian formalism), nets defined over subsets of Minkowski space in dealing with the solutions of the equations of motion and nets defined over Euclidean space-time in dealing with Euclidean field theory (Lagrangian formalism). In classical field theory one might take the net of dynamical variables to be $\mathcal{D}$ itself or some algebra generated by $\mathcal{D}$ such as the polynomial algebra over $\mathcal{D}$. In Wightman field theory we have the net unbounded field operators $\mathcal{O} \to \{\phi(f): f \in \mathcal{D}(\mathcal{O})\}$ or the net of polynomial algebras in these fields, in algebraic field theory, the net $\mathcal{O} \to \widetilde{\mathfrak{F}}(\mathcal{O})$, where $\widetilde{\mathfrak{F}}(\mathcal{O})$ is the von Neumann algebra generated by $\{\phi(f): f \in \mathcal{D}(\mathcal{O})\}$. In fact, R. Haag [3] was led to develop algebraic field theory after grasping the basic importance of net structure in quantum field theory.

The terminology adopted for dealing with nets is largely self-explanatory. We could talk about $\mathcal{D}$ as being a net of Abelian groups under addition or as a net of real vector spaces because the inclusion mappings in (1.1) respect the additive and linear structures. The inductive limit of a net will be denoted by A_∞ and a mapping $f: A_\infty \to B_\infty$ will be said to be a net morphism from A to B if

$$f(A(\mathcal{O})) \subset B(\mathcal{O}), \qquad \mathcal{O} \in C. \tag{1.2}$$

We shall have occasion to consider nets defined not on all compact sets but only on some set K of suitably regular subsets of $\mathbb{R}^d$. In dealing with Euclidean space $\mathbb{R}^d$, we take K to be the set of closed balls and in dealing with Minkowski space $\mathbb{R}^{s+1}$, the set of closed double cones. We let K_o denote the subset of K whose elements are centred on the origin. A mapping $f: A_\infty \to B_\infty$ is said to be a local mapping from A to B if there exists $\mathcal{O} \in K$ with

$$F(A(\mathcal{O}_1)) \subset B(\mathcal{O}+\mathcal{O}_1), \qquad \mathcal{O}_1 \in C. \tag{1.3}$$

We write $\mathcal{O}_1 \perp \mathcal{O}_2$ to denote either disjoint sets of Euclidean space or spacelike separated sets of Minkowski space.

Net theory can be regarded as a development running parallel to sheaf theory. A sheaf is a projective system over open sets, a net an inductive system over compact sets. In the motivating examples in sheaf theory, the open set is a domain of definition, in net theory, the closed set is a support region.

§2 The Cohomology of Nets of Abelian Groups

Let me begin by recalling the standard notions of singular cohomology. If X is a topological space then a p-simplex of X is a continuous map of the standard p-simplex

$$\Delta^p = \{(t^o,t^1,t^2,\dots,t^p) \in \mathbb{R}^{p+1}:\ t^i \geq 0,\ \sum_{i=0}^{p} t^i = 1\} \quad (2.1)$$

into X. Let $\Sigma_p(X)$ denote the set of p-simplexes of X. There are face maps $\partial_i\colon \Sigma_p(X) \to \Sigma_{p-1}(X)$, $i = 0,1,\dots,p$ defined by

$$(\partial_i c)(t^o,t^1,\dots,t^{p-1}) = c(t^o,t^1,\dots,t^{i-1},0,t^i,\dots,t^{p-1}),$$

$$c \in \Sigma_p(X). \quad (2.2)$$

A p-cochain of X with values in an Abelian group A is a mapping $f\colon \Sigma_p(X) \to A$. The set of p-cochains forms an Abelian group under addition denoted by $C^p(X,A)$. Using the face maps and the group structure of A one defines coboundary operators $d\colon C^p(X,A) \to C^{p+1}(X,A)$ by

$$(df)(c) = \sum_{i=0}^{p+1} (-1)^i f(\partial_i c). \quad (2.3)$$

One checks that $d^2 = 0$ and this gives rise to a cochain complex

$$C^o(X,A) \xrightarrow{d} C^1(X,A) \xrightarrow{d} C^2(X,A) \xrightarrow{d} \dots. \quad (2.4)$$

The p-cocycles $Z^p(X,A)$ and the p-coboundaries $B^p(X,A)$ are the subgroups of $C^p(X,A)$ defined by $Z^p(X,A) = \mathrm{Ker}d$, $B^p(X,A) = \mathrm{Im}d$, with the convention that $B^o(X,A) = 0$. The cohomology groups $H^p(X,A)$ are the quotient groups $Z^p(X,A)/B^p(X,A)$.

Here we are interested in $X = \mathbb{R}^d$ in which case these cohomology groups are trivial for $p \geq 1$. To see this, we define a contracting homotopy h based on x_o. If c is a p-simplex, then h(c) is the p+1-simplex which is a cone with vertex x_o and base c

$$h(c)(t^o,t^1,\ldots,t^{p+1}) = t^o x_o + (1-t^o)c(\tau t^1,\tau t^2,\ldots,\tau t^{p+1}), \quad t^o \neq 1$$

$$= x_o, \qquad t^o = 1. \tag{2.5}$$

Here we have written $\tau = (1-t^o)^{-1}$. We have

$$\partial_o h(c) = c, \quad \partial_i h(c) = h(\partial_{i-1} c), \quad i > 0 \tag{2.6}$$

except that if $c \in \Sigma_o(X)$, then $\partial_1 h(c) = x_o$.

Now given $f \in Z^p(X,A)$ define $g \in C^{p-1}(X,A)$ by $g(b) = f(h(b))$, $b \in \Sigma_{p-1}(X)$ and we get

$$0 = df(h(c)) = \sum_{i=0}^{p+1} (-1)^i f(\partial_i h(c)) = f(c) - \sum_{i=0}^{p} (-1)^i f(h(\partial_i c)), \; c \in \Sigma_p(X)$$

Hence $f = dg \in B^p(X,A)$ so $H^p(X,A) = 0$, $p \geq 1$.

For $p = 0$, the existence of a contracting homotopy based on x_o shows that $f \to f(x_o)$ is an isomorphism of $H^o(X,A) = Z^o(X,A)$ onto A.

One gets much more information about a topological space X if one studies the cohomology of sheaves of Abelian groups over X. The cohomology groups $H^p(X,A)$ defined above coincide with the cohomology groups of the constant sheaf defined by the Abelian group A. When $X = \mathbb{R}^d$, there are still plenty of sheaves whose cohomology is non-trivial. It is thus not surprising that, if one defines a cohomology theory for nets over the set of compact sets in $\mathbb{R}^d$, there are again nets with a non-trivial cohomology. This is the source of the interesting cohomological invariants in field theory. The definitions of net cohomology have been designed to incorporate the examples one meets in field theory.

These definitions are given below for the case of a net of Abelian groups and the version needed for quantum field theory when one has nets of von Neumann algebras is presented in the next section. This has the advantage that the novel features of net cohomology can be seen in the familiar context of Abelian cohomology without being obscured by structural details peculiar to quantum field theory.

If A is a net of Abelian groups over C, a p-cochain of A is a map $f: \Sigma_p \to A_\infty$ such that there exists $\mathcal{O} \in K_o$ with

$$f(c) \in A(\mathcal{O}+|c|), \quad c \in \Sigma_p \tag{2.7}$$

where $|c| = c(\Delta^p)$ is the support of c. The set $C^p(A)$ of p-cochains of A is a subgroup of $C^p(\mathbb{R}^d, A_\infty)$. Since $|\partial_i c| \subset |c|$, $f \in C^p(A)$ implies $df \in C^{p+1}(A)$ and we have a cochain complex

$$C^o(A) \xrightarrow{d} C^1(A) \xrightarrow{d} C^2(A) \xrightarrow{d} \ldots\ldots \tag{2.8}$$

The cohomology of A is the cohomology of this cochain complex.

Equivalently, we may regard Σ_p as a net, setting

$$\Sigma_p(\mathcal{O}) = \{c \in \Sigma_p : \; |c| \subset \mathcal{O}\} \tag{2.9}$$

and define $C^p(A)$ as the set of local mappings of Σ_p into A. The face mappings $\partial_i: \Sigma_{p+1} \to \Sigma_p$ are net morphisms so that $df = \sum_{i=0}^{p+1} (-1)^i f \circ \partial_i$ is local whenever f is local.

Much of the original motivation for net cohomology was drawn from the cohomology of differential forms considered in the context of quantum field theory. However, these ideas also apply to classical field theory. A singular totally antisymmetric covariant tensor field of rank p, an even p-courant in the language of deRham [4], may be considered as a linear functional on the net C_p of smooth totally antisymmetric functions with compact support:

$$\phi(f) = \sum_{\mu_1 \ldots \mu_p} \int \phi^{\mu_1 \ldots \mu_p}(x) \; f_{\mu_1 \ldots \mu_p}(x)\, dx \tag{2.10}$$

$\mathcal{C}_o$ is just the net we denoted by $\mathcal{D}$ in section 1.

If the field satisfies $d\phi = 0$ where d is the exterior derivation

$$(d\phi)^{\mu_1 \ldots \mu_{p+1}} = \partial^{[\mu_1} \phi^{\mu_2 \ldots \mu_{p+1}]} \tag{2.11}$$

then $(d\phi)(f) = \phi(\delta f) = 0$, where δ denotes the coderivative*)

$$(\delta f)_{\mu_1 \ldots \mu_p} = - \sum_{\mu} \partial^{\mu} f_{\mu\mu_1 \ldots \mu_p} \quad . \tag{2.12}$$

We have a chain complex of nets

$$\ldots \to \mathcal{C}_p \xrightarrow{\delta} \mathcal{C}_{p-1} \xrightarrow{\delta} \ldots \xrightarrow{\delta} \mathcal{C}_o \longrightarrow 0 \tag{2.13}$$

and denote the subnets ker δ and im δ of $\mathcal{C}_p$ by $\mathcal{Z}_p$ and $\mathcal{B}_p$ respectively. $\mathcal{Z}_p = \mathcal{B}_p$ for $p \neq 0$ as a consequence of Poincaré's Lemma and $\mathcal{Z}_o = \mathcal{C}_o$. We can consider our singular classical field ϕ satisfying $d\phi = 0$ as a linear functional on the net $\mathcal{B}_{p-1}$ since if $f \in \mathcal{B}_{p-1}$, we may write $f = \delta g$ and $\phi(g)$ is independent of the choice of g.

I would like to make it plausible that the net $\mathcal{B}_{p-1}$ has a non-trivial p-cohomology for $p \leq d$ and will argue in terms of smooth, or at least differentiable simplexes. Actually the classical arguments of algebraic topology are sufficient to show that net cohomology can equally well be defined in terms of the nets Σ_p^{∞} of smooth p-simplexes.

Let h be a smooth function with support in $\mathcal{O} \in K_o$ and $c \in \Sigma_{p-1}^{\infty}$ then there is an element $h*c$ of $\mathcal{C}_{p-1}$ such that

$$\sum_{\mu_1 \ldots \mu_{p-1}} \int (h*c)_{\mu_1 \ldots \mu_{p-1}}(x) f^{\mu_1 \ldots \mu_{p-1}}(x)\,dx = \int_c h*f \tag{2.14}$$

for each (p-1)-form f where the expression on the right is the integral of the (p-1)-form $h*f$, the convolution of h and f, over the (p-1)-simplex c. Now $h*c$ has its support in $\mathcal{O}+|c|$ so $c \to h*c$

*) The appearance of the Riemannian structure at this point is gratuitous and can be avoided by replacing $f \in \mathcal{C}_p$ by the dual tensor $*f$ which is an odd (d-p)-form.

is a local map of Σ^{∞}_{p-1} into C_{p-1}, i.e. a smooth (p-1)-cochain in C_{p-1} *). Its coboundary is the smooth p-coboundary in C_{p-1}

$$c \to \sum_{i=0}^{p} (-1)^i \, h * \partial_i c \equiv h * \partial c, \quad c \in \Sigma_p . \tag{2.15}$$

A simple computation using Stokes Theorem shows that

$$h * \partial c = \delta(h * c)$$

so $c \to h * \partial c$ actually takes values in the subnet B_{p-1}. As such it is a smooth p-cocycle of B_{p-1}, but not, in general, a p-coboundary.

This argument breaks down if $p-1 = d$ because $h * c = o$ for $c \in \Sigma^{\infty}_{d+1}$. However in this case $\Sigma_d \ni c \to h * c$ is a smooth d-cocycle in C_d which, in general, cannot be a smooth d-coboundary because it has the wrong support properties.

The above discussion illustrates two mechanisms for generating non-trivial cohomologies: they may either arise from particular constraints such as $d\phi = o$ or, as in the case of the d-cohomology, from a support obstacle. To actually compute cohomologies, a more systematic approach is needed and I refer to [5] for the details. The strategy is the same as for other cohomologies. The first step is to compute $H^o(A) = Z^o(A)$; this is easily done, because since $\mathbb{R}^d$ is path-connected

$$H^o(A) = \bigcup_{\mathcal{O} \in K_o} \; \bigcap_{a \in \Sigma_o} A(\mathcal{O}+a) . \tag{2.16}$$

The second step is to find sufficiently large classes of nets with $H^p(A) = 0$ for $p > 0$ to allow one to compute non-trivial cohomologies effectively. One such class consists of trivial nets; otherwise only partial results are known which suffice to compute H^1 and H^2.

The triviality results in [5] were derived for nets constructed from sheaves. If S is a sheaf of Abelian groups on $\mathbb{R}^d$, then $\Gamma_b(S)$ denotes the net constructed from the global sections of

---

*) This example explains the appearance of $\mathcal{O}$ in (2.7); it is the support of some regularizing function.

$\mathcal{S}$ by setting

$$\Gamma_b(\mathcal{S})(\mathcal{O}) = \{f \in \mathcal{S}(\mathbb{R}^d) : f = 0 \text{ on } \mathcal{O}^c\} \tag{2.17}$$

for $\mathcal{O} \in \mathcal{C}$. One should try to express the triviality results directly in terms of properties of the nets. In particular, a net A will be said to have the cosheaf property if

$$\bigcap_i A(\mathcal{O}_i) = A(\bigcap_i \mathcal{O}_i). \tag{2.18}$$

<u>2.1 Theorem</u> If A is a net of Abelian groups over $\mathcal{C}$ satisfying (2.18), $A(\emptyset) = 0$ and

$$A(\mathcal{O}_1 \cup \mathcal{O}_2) = A(\mathcal{O}_1) \oplus A(\mathcal{O}_2), \quad \mathcal{O}_1 \perp \mathcal{O}_2 \tag{2.19}$$

then $H^0(A) = 0$ and, if $d > 1$, $H^1(A) = 0$.

The nets $\Gamma_b(\mathcal{S})$ can be easily seen to satisfy the hypotheses of this Theorem, which can be proved by adapting the proof of [5; Thm. 2.2].

For a sheaf $\mathcal{S}$ of Abelian groups over Minkowski space $\mathbb{R}^{s+1}$ we define nets $\Gamma_c(\mathcal{S})$ by setting

$$\Gamma_c(\mathcal{S})(\mathcal{O}) = \{f \in \mathcal{S}(\mathbb{R}^{s+1}) : f = 0 \text{ on } \mathcal{O}'\}. \tag{2.20}$$

In place of (2.18) we now have

$$\bigcap_i A(\mathcal{O}_i) = A(\mathcal{O}), \quad \text{if } \bigcup_i \mathcal{O}_i' = \mathcal{O}' \tag{2.21}$$

and instead of Theorem 2.1

<u>2.2 Theorem</u> If A is a net of Abelian groups over $\mathcal{C}$ satisfying (2.21), $A(\emptyset) = 0$ and

$$A(\mathcal{O}_1 \cup \mathcal{O}_2) = A(\mathcal{O}_1) \oplus A(\mathcal{O}_2), \quad \mathcal{O}_1 \perp \mathcal{O}_2, \quad \mathcal{O}_1, \mathcal{O}_2 \in \mathcal{O} \tag{2.22}$$

then $H^0(A) = 0$ and, if $s > 1$, $H^1(A) = 0$.

Although this can also be proved by adapting the proof of [5;

Thm. 2.2], it does not contain this result because $\Gamma_c(S)$ does not satisfy (2.22) in general. However (2.22) is satisfied whenever it is possible to argue in terms of Cauchy data on a spacelike hyperplane as in the case of smooth solutions of the wave equation or the Klein-Gordon equation. Note that the support obstacle now appears for $s = 1$.

These theorems can be applied to the nets of closed real subspaces of the 1-particle spaces which can be used to generate the nets of von Neumann algebras of the free fields. Theorem 2.1 can be applied to nets defined on a spacelike hyperplane and Theorem 2.2 to a manifestly covariant formulation.

It would be nice to have triviality results for the second cohomology on the lines of [5; Thm. 2.10] expressed directly in terms of properties of the nets.

The last step in computing cohomologies is to find a notion of exact sequence of nets which leads to a long exact sequence in cohomology. This is achieved in Section 3 of [5]. The most important result here is that the cohomology of a net is entirely determined by its restriction to K and we get a long exact sequence in cohomology if

$$0 \to A(\mathcal{O}) \to B(\mathcal{O}) \to C(\mathcal{O}) \to 0 \tag{2.23}$$

is an exact sequence of Abelian groups for each $\mathcal{O} \in K$ [5; Thm. 3.2].

Cohomologies can now be computed in the standard way using exact sequence. To return to our example of test function nets,

$$\begin{array}{ll} 0 \to B_p \to C_p \xrightarrow{\delta} B_{p-1} \to 0 & p \neq 0 \\ 0 \to B_o \to C_o \xrightarrow{\int} \mathbb{R} \to 0 & \end{array} \tag{2.24}$$

where $C_o \xrightarrow{\int} \mathbb{R}$ denotes the integration of a function over $\mathbb{R}^d$, give rise to long exact sequences in cohomology. This is equivalent to [5; Eq. (4.4)] if one passes to dual tensors. Combined with the triviality results $H^o(C_p) = 0$, $H^1(C_p) = 0$ $d > 1$, and $H^2(C_p) = 0$ $d > 2$, one concludes that $\mathbb{R} \cong H^o(\mathbb{R}) \cong H^1(B_o) \cong H^2(B_1)$ provided $d > 1$ or $d > 2$ respectively. Under this isomorphism the cohomology class of $c \to \delta(h * c)$, $c \in \Sigma_p^\infty$ in $H^p(\mathcal{B}_{p-1})$ corresponds to $\int h\,dx \in \mathbb{R}$ for $p = 1,2$.

Similar techniques allow one to compute more interesting cohomologies such as those of the wave equation Klein-Gordon equation, vector wave equation with Lorentz condition and Maxwell's equations. The details may be found in section 4 of [5]. These results provide a useful guide in trying to compute the cohomologies of the corresponding free fields. We shall see in the next section how this may be done for the 0-cohomology and the 1-cohomology.

§3 The Cohomology of Nets of von Neumann Algebras

We now discuss the cohomology of a net $\mathfrak{A}$ of von Neumann algebras*). This is the typical type of net arising in all forms of quantum field theory. Here we are dealing with non-Abelian cohomology and use the basic scheme of [6] with refinements to cater for the specific algebraic structure of $\mathfrak{A}$.

3.1 Definition A o-cocycle in $\mathfrak{A}$ is a map $z: \Sigma_o \to \mathfrak{A}_\infty$ satisfying the cocycle identity

$$z(\partial_o b) = z(\partial_1 b) , \quad b \in \Sigma_1 \tag{3.1}$$

and the locality condition, there exists $\mathcal{O} \in K_o$ with

$$z(a) \in \mathfrak{A}(\mathcal{O}+a) , \quad a \in \Sigma_o . \tag{3.2}$$

Since any two points can be joined by a path, $z(a)$ is actually independent of a and we set

$$Z^o(\mathfrak{A})(\mathcal{O}) = \bigcap_{a \in \Sigma_o} \mathfrak{A}(\mathcal{O}+a). \tag{3.3}$$

Thus $Z^o(\mathfrak{A})$ is a net of von Neumann algebras over K_o. In practice $Z^o(\mathfrak{A})$ usually has a trivial net structure. Indeed in the examples arising in quantum field theory one usually finds that $Z^o(\mathfrak{A})$ reduces to $\mathbb{C}$, the complex numbers, e.g. by showing that $Z^o(\mathfrak{A})$

*) We always suppose that the inclusion mappings $\mathfrak{A}(\mathcal{O}_1) \subset \mathfrak{A}(\mathcal{O}_2)$ preserve the identities.

commutes with some irreducible algebra.

3.2 Definition A 1-cocycle in $\mathfrak{A}$ is a map $z\colon \Sigma_1 \to U(\mathfrak{A}_\infty)$, the unitary group of $\mathfrak{A}_\infty$, satisfying the cocycle identity

$$z(\partial_0 c)\, z(\partial_2 c) = z(\partial_1 c)\ , \qquad c \in \Sigma_2 \tag{3.4}$$

and the locality condition, there exists $\mathcal{O} \in K_o$ with

$$z(b) \in \mathfrak{A}(\mathcal{O} + |b|), \qquad b \in \Sigma_1 \ . \tag{3.5}$$

We consider the 1-cocycles as the objects of a net $Z^1(\mathfrak{A})$ of W*-categories,*) over K_o. An object of $Z^1(\mathfrak{A})(\mathcal{O})$ is a 1-cocycle satisfying (3.5) and an arrow in $Z^1(\mathfrak{A})(\mathcal{O})$ from z' to z is a triple $(z|w|z')$ where $w\colon \Sigma_o \to \mathfrak{A}_\infty$ satisfies

$$z(b)w(\partial_1 b) = w(\partial_0 b)z'(b), \qquad b \in \Sigma_1 \tag{3.6}$$

and

$$w(a) \in \mathfrak{A}(\mathcal{O} + a) \quad , \qquad a \in \Sigma_o \ . \tag{3.7}$$

The composition law in $Z^1(\mathfrak{A})$ is defined by

$$(z|w|z') \circ (z'|w'|z'') = (z|ww'|z'') \tag{3.8}$$

where $(ww')(a) = w(a)w'(a)$. The involution on $Z^1(\mathfrak{A})$ is defined by

$$(z|w|z')^* = (z'|w^*|z) \tag{3.9}$$

where $w^*(a) = w(a)^*$. The complex linear structure is also defined similarly by pointwise operations. The norm is defined by $||(z|w|z')|| = ||w(a)||$ and this is independent of $a \in \Sigma_o$ by (3

3.3 Definition Two 1-cocycles of $\mathfrak{A}$ are cohomologous if they are unitarily equivalent as objects of $Z^1(\mathfrak{A})$. z is a 1-coboundary if it is cohomologous to the trivial 1-cocycle 1 defined by $1(b) = I$, $b \in \Sigma_1$. $Z^0(\mathfrak{A})$ can be identified with the arrow from 1 to 1 in $Z^1(\mathfrak{A})$ using the map $w \to (1|w|1)$.

) A survey of the basic definitions and results in the theory of W-categories may be found in these proceedings [7].

Some of the main results of a structural nature for $Z^1(\mathfrak{A})$, where $\mathfrak{A}$ is the net of observables in quantum field theory, have been announced in [8]. Here, by way of contrast, we present results enabling us to compute the 1-cohomology in concrete cases. The basic ideas involved are just those used to compute the cohomology of Abelian nets except that the details have now to be done by hand.

We first need a triviality result, i.e. conditions which ensure that $Z^1(\mathfrak{A})$ is, in some sense, trivial. The basic lemma concerns the cohomology of certain nets M of von Neumann algebras defined on the set of subsimplexes of the standard 2-simplex Δ^2. With an obvious notation, denote the 2-simplex by c, the 1-simplexes by $\partial_o c$, $\partial_1 c$, $\partial_2 c$ and the o-simplexes by $\partial_o\partial_1 c$, $\partial_o\partial_2 c$ and $\partial_1\partial_2 c$. The algebras associated with each subsimplex will be supposed to be canonically isomorphic to the tensor product of the algebras associated with its vertices. Thus

$$M(c) = M(\partial_o\partial_1 c) \otimes M(\partial_o\partial_2 c) \otimes M(\partial_1\partial_2 c).$$

To save writing we write M_i in place of $M(\partial_i c)$ and M_{ij} in place of $M(\partial_i\partial_j c)$. A 1-cocycle of M is a set of unitary operators $z_i \in M_i$ satisfying

$$z_o z_2 = z_1.$$

Now there is a conditional expectation $\Phi: M \to M_o$ with $\Phi(z_2) \neq 0$ and $\Phi(M_2) \subset M_{o2}$, $\Phi(M_1) \subset M_{o1}$. We may take Φ to be the Tomiyama slice map associated with a normal state on M_{12} [9; §2.6].

Thus $\quad z_o\, \Phi(z_2) = \Phi(z_1)$

so $\quad \Phi(z_2)^*\Phi(z_2) = \Phi(z_1)^*\Phi(z_1) \in M_{o2} \cap M_{o1} = \mathbb{C}$.

If M is an Abelian algebra, we deduce that there are unitary operators $u_{o2} \in M_{o2}$ and $u_{o1} \in M_{o1}$ with

$$z_o\, u_{o2} = u_{o1}$$

and it is not difficult to see that there exists a $u_{12} \in M_{12}$

with

$$z_2\, u_{12} = u_{02}\ ;\ z_1\, u_{12} = u_{01}.$$

In other words, if M is Abelian, every 1-cocycle in M is actually a 1-coboundary, a result which must be well known in various contexts.

When M is not Abelian, the result is a little more complicated.

<u>3.4 Lemma</u> Given any 1-cocycle in M, there are unique Hilbert spaces $H_{ij} \subset M_{ij}$ with

$$z_2\, H_{12} = H_{02}\ ,\ z_1\, H_{12} = H_{01} \quad \text{and} \quad z_o\, H_{02} = H_{01}.$$

A Hilbert space H in a von Neumann algebra [10; §2] is a norm-closed linear subspace with left support I such that $x,y \in H$ implies $y^*x \in \mathbb{C}$. The lemma implies that there are isometries $\psi^\alpha_{ij} \in M_{ij}$ with $\sum_\alpha \psi^\alpha_{ij}\, \psi^{\alpha*}_{ij} = I$ and

$$z_2\, \psi^\alpha_{12} = \psi^\alpha_{02}\ ,\ z_1\ \psi^\alpha_{12} = \psi^\alpha_{01}\ ,\ z_o\, \psi^\alpha_{02} = \psi^\alpha_{01}.$$

It means that any object of $Z^1(M)$, although not necessarily unitarily equivalent to the trivial 1-cocycle is a direct sum of trivial 1-cocycles. We shall say that the 1-cohomology of M is <u>quasitrivial</u>. If any of the algebras M_{ij} is not properly infinite, the Hilbert spaces are necessarily 1-dimensional and each 1-cocycle is a 1- coboundary as in the Abelian case.

To prove the Lemma, we first just exploit the fact that the net M is a lattice homomorphism of the lattice of subsimplexes of Δ^2 into the lattice of von Neumann subalgebras of M(c). We set

$$\begin{aligned} H_{12} &= \{x \in M_{12} : z_2 x \in M_{02}\} \\ H_{02} &= \{x \in M_{02} : z_o x \in M_{01}\} \\ H_{01} &= \{x \in M_{01} : z_1^* x \in M_{12}\} \quad . \end{aligned} \tag{3.10}$$

We first show that these norm-closed linear subspaces are mapped in the right way by the unitary operators z_i. If $x \in H_{12}$, then $z_2 x \in M_{02}$ and $z_o(z_2 x) = z_1 x \in M_o \cap M_1 = M_{01}$ thus $z_2 x \in H_{02}$. Conversely if $y \in H_{02}$, then $z_2^* y = z_1^* z_o y \in M_2 \cap M_1 = M_{12}$ so $z_2^* y \in H_{12}$. Hence $z_2 H_{12} = H_{02}$ and similarly $z_1 H_{12} = H_{01}$, $z_o H_{02} = H_{01}$. Now if $x, y \in H_{02}$, then $y^* x = y^* z_o^* z_o x \in M_{02} \cap M_{01} = \mathbb{C}$. It only remains to show that the left support of one of these spaces, say H_{02} is I. Here we make use of the existence of sufficient conditional expectations Φ as above. Let $I-e_{02}$ denote the left support of H_{02}, then $e_{02} \in M_{02}$ and $z_o e_{02} = e_{01} z_o$ where $I-e_{01}$ is the left support of H_{01} so that $e_{01} \in M_{01}$. $z_o e_{02}\Phi(z_2) = e_{01} z_o \Phi(z_2) = e_{01} \Phi(z_1) \in M_{01}$ hence $e_{02} \Phi(z_2) \in H_{02}$. Thus $0 = e_{02}\Phi(z_2) = \Phi(e_{02}z_2)$ for all such conditional expectations. Hence $e_{02}z_2 = 0$, and $e_{02} = 0$ as required. The uniqueness of the Hilbert spaces H_{ij} is a trivial consequence of their explicit construction.

We now present conditions which ensure that a net $\mathfrak{F}$ of von Neumann algebras over $\mathcal{C}$ has a quasitrivial 1-cohomology. The net is denoted by $\mathfrak{F}$ rather than $\mathfrak{A}$ because in quantum field theory it is field nets rather than observable nets which have quasitrivial 1-cohomology.

<u>3.5 Theorem</u> Let $\mathfrak{F}$ be a net of von Neumann algebras over $\mathcal{C}$ satisfying for any $\mathcal{O} \in K_o$ the following two conditions:

a) $\bigcap_{\partial b' = \partial b} \tilde{\mathfrak{F}}(\mathcal{O}+|b'|) = \tilde{\mathfrak{F}}(\mathcal{O}+\partial_o b) \vee \tilde{\mathfrak{F}}(\mathcal{O}+\partial_1 b)$, $b \in \Sigma_1$

b) If $(\mathcal{O}+\partial_o b) \perp (\mathcal{O}+\partial_1 b)$, then $\tilde{\mathfrak{F}}(\mathcal{O}+\partial_o b) \vee \tilde{\mathfrak{F}}(\mathcal{O}+\partial_1 b)$ is canonically isomorphic to $\tilde{\mathfrak{F}}(\mathcal{O}+\partial_o b) \otimes \tilde{\mathfrak{F}}(\mathcal{O}+\partial_1 b)$, $b \in \Sigma_1$.

Then $Z^o(\tilde{\mathfrak{F}}) = \mathbb{C}$ and given any object z of $Z^1(\tilde{\mathfrak{F}})(\mathcal{O})$ there are unique Hilbert spaces $H(a) \subset \tilde{\mathfrak{F}}(\mathcal{O}+a)$ such that $z(b)\, H(\partial_1 b) = H(\partial_o b)$, $b \in \Sigma_1$. In particular any object of $Z^1(\tilde{\mathfrak{F}})$ is a direct sum of trivial 1-cocycles. If the algebras $\tilde{\mathfrak{F}}(\mathcal{O})$ are not properly infinite then any 1-cocycle in $\mathfrak{F}$ is a 1-coboundary.

<u>Proof</u> If $w \in Z^o(\tilde{\mathfrak{F}})(\mathcal{O})$ then pick $b \in \Sigma_1$ with $\mathcal{O}+\partial_o b \perp \mathcal{O}+\partial_1 b$, so that $w(\partial_o b) = w(\partial_1 b) \in \tilde{\mathfrak{F}}(\mathcal{O}+\partial_o b) \cap \tilde{\mathfrak{F}}(\mathcal{O}+\partial_1 b) = \mathbb{C}$ by b). Hence $Z^o(\tilde{\mathfrak{F}}) = \mathbb{C}$. Now let z be an object of $Z^1(\tilde{\mathfrak{F}})(\mathcal{O})$. The 1-cocycle identity implies that $z(b) = z(b')$ whenever $\partial b = \partial b'$. Thus $z(b) \in \tilde{\mathfrak{F}}(\mathcal{O}+\partial_o b) \vee \tilde{\mathfrak{F}}(\mathcal{O}+\partial_1 b)$ by a). Given $a \in \Sigma_o$, pick $b \in \Sigma_1$ with $\partial_o b = a$ and $\mathcal{O}+\partial_o b \perp \mathcal{O}+\partial_1 b$ and set

$$H(a) = \{F \in \mathfrak{F}(\mathcal{O}+a) : z(b)^* F \in \mathfrak{F}(\mathcal{O}+\partial_1 b)\}.$$

We claim that H(a) is a Hilbert space in $\mathfrak{F}(\mathcal{O}+a)$ independent of the choice of b. To see this, pick $c \in \Sigma_2$ with $\partial_o c = b$ and $\partial_1\partial_2 c$ sufficiently far away so that Lemma 3.4 can be applied with $M_{ij} = \mathfrak{F}(\mathcal{O}+\partial_i\partial_j c)$. For example, we can arrange that there is an $\mathcal{O}_1 \in K_o$ such that $(\mathcal{O}_1+\partial_1\partial_2 c) \perp (\mathcal{O}_1+\partial_o\partial_1 c)$ and $(\mathcal{O}+\partial_o c) \subset (\mathcal{O}_1+\partial_o\partial_1 c)$. Lemma 3.4 shows that H(a) is the Hilbert space H_{01} and if $\partial_1 c = b'$

$$H(a) = \{F \in \mathfrak{F}(\mathcal{O}+a) : z(b')^* F \in \mathfrak{F}(\mathcal{O}+\partial_1 b')\}$$

so that H(a) is actually independent of the choice of b. We further deduce that if we define H(a) in this way for all $a \in \Sigma_o$, then whenever $b \in \Sigma_1$ with $(\mathcal{O}+\partial_o b) \perp (\mathcal{O}+\partial_1 b)$, $z(b)\, H(\partial_1 b) = H(\partial_o b)$. This now holds for arbitrary $b \in \Sigma_1$ since picking $c \in \Sigma_2$ with $\partial_o c = b$ and $(\mathcal{O}+\partial_o c) \perp (\mathcal{O}+\partial_1\partial_2 c)$, we have

$$z(b)\, H(\partial_1 b) = z(\partial_o c)\, H(\partial_o\partial_2 c) = z(\partial_o c)\, z(\partial_2 c)\, H(\partial_1\partial_2 c) = z(\partial_1 c)\, H(\partial_1\partial_2 c) = H(\partial_o b).$$

The remaining statements are obvious.

The full force of the hypotheses is not used in the proof. It suffices, for example, if a) holds whenever $(\mathcal{O}+\partial_o b) \perp (\mathcal{O}+\partial_1 b)$ and other minor modifications are possible. More interesting is the question of when these hypotheses can be expected to be fulfilled for the nets arising in quantum field theory. It seems reasonable to suppose that b) has a wide range of validity: if we consider the vacuum state restricted to $\mathfrak{F}(\mathcal{O}_1) \vee \mathfrak{F}(\mathcal{O}_2)$, it is not a product state because there are correlations in the relativistic vacuum. What b) means is that these correlations are not so strong that the product of the local vacua is no longer a normal state of $\mathfrak{F}(\mathcal{O}_1) \vee \mathfrak{F}(\mathcal{O}_2)$. For the free field, b) has been proved by Buchholz [11] even for unbounded regions that are properly spacelike separated. Recently, Driessler [12] has made the interesting observation that for free fields b) can be reduced to the question of the essential self-adjointness of certain current operators in the tensor product net $\mathfrak{F} \otimes \mathfrak{F}$. In any case b) is a technical assumption because it holds equally well for subnets such as the observable nets whose cohomology we shall examine in a moment.

The crucial assumption is a); the idea here is that $F \subset \tilde{\mathfrak{F}}(\mathcal{O})$ means that F is supported in $\mathcal{O}$ so that we would expect $\partial b' \overset{\cap}{=} \partial b \quad \tilde{\mathfrak{F}}(\mathcal{O}+|b'|) = \tilde{\mathfrak{F}}((\mathcal{O}+\partial_0 b) \cup (\mathcal{O}+\partial_1 b))$. Here there are two provisos: we need $d > 1$, or in the case of casual support as in quantum field theory $s > 1$, otherwise we have the support obstacle discussed in the case of Abelian nets. The second proviso is that for disconnected regions such as $(\mathcal{O}+\partial_0 b) \cup (\mathcal{O}+\partial_1 b)$ when $(\mathcal{O}+\partial_0 b) \perp (\mathcal{O}+\partial_1 b)$, there are, in principle, different ways of defining $\tilde{\mathfrak{F}}((\mathcal{O}+\partial_0 b) \cup (\mathcal{O}+\partial_1 b))$ and we need to have chosen the right one. In fact, the content of assumption a) is that $\tilde{\mathfrak{F}}((\mathcal{O}+\partial_0 b) \cup (\mathcal{O}+\partial_1 b))$ is unambiguously defined because the definitions suggested by support considerations coincides with the minimal definition $\tilde{\mathfrak{F}}(\mathcal{O}+\partial_0 b) \vee \tilde{\mathfrak{F}}(\mathcal{O}+\partial_1 b)$ suggested by additivity. a) is violated by the observable algebra when there are localized charges present. For the algebra of the free field, a) is easily checked using Araki's results [13].

Under the hypotheses of Theorem 3.5, every 1-cocycle z of $\tilde{\mathfrak{F}}$ is characterized up to a unitary equivalence by an integer d(z) which is just the dimension of the associated field $a \to H_z(a)$ of Hilbert spaces. We show that if $\mathfrak{A}$ is the fixed-point net of $\tilde{\mathfrak{F}}$ under a group G of gauge automorphisms then the 1-cocycles of $\mathfrak{A}$ can be characterized up to an equivalence by a unitary equivalence class of representations of G.

If $\mathfrak{A}$ is a subnet of $\tilde{\mathfrak{F}}$, the group G of gauge automorphisms of $\tilde{\mathfrak{F}}$ relative to $\mathfrak{A}$ is defined to be the subgroup of automorphisms β of $\tilde{\mathfrak{F}}_\infty$ such that

$$\beta(\tilde{\mathfrak{F}}(\mathcal{O})) \subset \tilde{\mathfrak{F}}(\mathcal{O}), \qquad \mathcal{O} \in C \tag{3.11}$$

and $\beta(A) = A$ if $A \in \mathfrak{A}_\infty$. We suppose that $\mathfrak{A}$ is the fixed-point net of $\tilde{\mathfrak{F}}$ under G.

$$\mathfrak{A}(\mathcal{O}) = \{F \in \tilde{\mathfrak{F}}(\mathcal{O}) \, : \, \beta(F) = F \, , \quad \beta \in G\} \quad \mathcal{O} \in C. \tag{3.12}$$

We suppose that the 0-cohomology of $\mathfrak{F}$ is trivial, $Z^0(\mathfrak{F}) = \mathbb{C}$ and that the 1-cohomology of $\mathfrak{F}$ is quasitrivial, i.e. that every object z of $Z^1(\mathfrak{F})(\mathcal{O})$ is a direct sum of trivial 1-cocycles. This is equivalent to saying that there is a field $a \to H_z(a)$ of Hilbert spaces in $\mathfrak{F}(\mathcal{O}+a)$ such that

$$z(b)\ H_z(\partial_1 b) = H_z(\partial_0 b),\quad b \in \Sigma_1. \tag{3.13}$$

Under these circumstances we have

<u>3.6 Lemma</u>

$$H(a) = \{F \in \mathfrak{F}(\mathcal{O}+a) : z(b)F \in \mathfrak{F}(\mathcal{O}+\partial_0 b),\ b \in \Sigma_1,\ \partial_1 b = a\}.$$

<u>Proof</u> If $F, F' \in H(a)$, $z(b)F$, $z(b)F' \in \mathfrak{F}(\mathcal{O}+\partial_0 b)$ so

$$F^*F' \in \bigcap_{a' \in \Sigma_0} \mathfrak{F}(\mathcal{O}+a') = Z^0(\mathfrak{F})(\mathcal{O}) = \mathbb{C}\ .$$

This is easily seen to imply the Lemma (cf. [10; Eq. (2.2)]). This Lemma shows that when $Z^0(\mathfrak{F}) = \mathbb{C}$, the field $a \to H_z(a)$ associated with a 1-cocycle z is necessarily unique. Furthermore, if $(z'|t|z)$ is an arrow of $Z^1(\mathfrak{F})$, then

$$t(a)\ H_z(a) \subset H_z'(a),\ a \in \Sigma_0. \tag{3.14}$$

Now let z be an object of $Z^1(\mathfrak{A})(\mathcal{O}) \subset Z^1(\mathfrak{F})(\mathcal{O})$. Since $\beta(z(b)) = z(b)$, $b \in \Sigma_1$, $\beta \in G$, we conclude that

$$\beta(H_z(a)) = H_z(a),\quad \beta \in G. \tag{3.15}$$

Thus each H(a) carries a unitary representation of $\mathcal{G}$ (cf. [10; Ex. 3.5]). Since z(b) intertwines the representations of G on $H_z(\partial_1 b)$ and $H_z(\partial_0 b)$, the unitary equivalence class of this representation is independent of $a \in \Sigma_0$. If $(z'|t|z)$ is an arrow of $Z^1(\mathfrak{A})(\mathcal{O})$, then t(a) intertwines the representations of G on $H_z(a)$ and $H_{z'}(a)$, $a \in \Sigma_0$. Thus the unitary equivalence classes of 1-cocycles are in 1-1 correspondence with a certain set of unitary equivalence classes of representations of G.

Let $\hat{G}(\mathcal{O})$ denote the set of unitary equivalence classes of G realized on Hilbert spaces in $\mathfrak{F}(\mathcal{O})$. $\mathcal{O} \to \hat{G}(\mathcal{O})$ is a net over C and

<u>3.7 Theorem</u> Let $\mathfrak{F}$ be a net of von Neumann algebras over C with

trivial 0-cohomology and quasitrivial 1-cohomology. Let $\mathfrak{A}$ be the fixed-point net of $\mathfrak{F}$ under a group G of gauge automorphisms, then the unitary equivalence classes of 1-cocycles are in 1-1 correspondence with $Z^0(\hat{G})$.

Proof: If z is a 1-cocycle of $\mathfrak{A}$: let $\hat{z}(a)$ be the unitary equivalence class of the representation of G on $H_z(a)$. Then $\hat{z} \in Z^0(\hat{G})$ and we need only show that this map is onto. Given $w \in Z^0(\hat{G})(\mathcal{O})$, pick a Hilbert space $H(a)$ in $\mathfrak{F}(\mathcal{O}+a)$ carrying a representation of class w for each $a \in \Sigma_o$. There is a unitary $u(a)$ in $\mathfrak{A}_\infty$ such that $u(a)\ H(o) = H(a)$. Setting $z(b) = u(\partial_o b)u(\partial_1 b)^*$, $z(b)\ H(\partial_1 b) = H(\partial_o b)$. This implies that $z(b)$ is a gauge-invariant element of $\mathfrak{F}(\mathcal{O}+\partial_o b) \vee \mathfrak{F}(\mathcal{O}+\partial_1 b)$. Hence $z(b) \in \mathfrak{A}(\mathcal{O}+|b|)$ and $z \in Z^1(\mathfrak{A})$ with $\hat{z} = w$.

Note that we have only assumed that $\mathfrak{A}$ is the fixed-point net under G, not that G consists of all gauge automorphisms of $\mathfrak{F}$ relative to $\mathfrak{A}$. The problem of computing the 1-cohomology of $\mathfrak{A}$ has therefore been reduced to the much simpler problem of computing the 0-cohomology of $\hat{G}$.

We shall see in the next section that in quantum field theory G is often a compact group and $Z^0(\hat{G})$ contains all equivalence classes of continuous irreducible representations*) of G. Under these circumstances we may conclude that G consists of <u>all</u> gauge automorphisms of $\mathfrak{F}$. To see this consider the representation theory of G realized on the Hilbert spaces of the form $H_z(a)$ where z is a 1-cocycle of $\mathfrak{A}$ and $a \in \Sigma_o$. Any gauge automorphism β leaves $H_z(a)$ invariant as a set and thus induces a unitary operator on $H_z(a)$. If we consider these unitary operators $\beta(\rho)$, say, as functions of the corresponding unitary representations ρ of G on the same Hilbert space, they satsify

$$t\beta(\rho) = \beta(\rho')t \qquad (3.16)$$

for each intertwining operator t between ρ and ρ', since β leaves

*) Unless G is Abelian, $Z^0(\hat{G})$ is automatically closed under countable direct sums. However, when G is Abelian, the Pontryagin duality Theorem suffices for the argument given in the text.

$\mathfrak{A}$ pointwise invariant. Furthermore

$$\beta(\rho \boxtimes \rho') = \beta(\rho) \boxtimes \beta(\rho') \tag{3.17}$$

since β is an automorphism. Thus β coincides with some $\beta' \in G$ on these Hilbert spaces by the Tannaka duality Theorem. However, these Hilbert spaces generate the net $\widetilde{\mathfrak{F}}$ so β and β' coincide on the whole net. Details of the analogous argument in the case of a single von Neumann algebra, including a proof of the Tannaka duality Theorem can be found in [15; Appendix C].

Before going on to apply these results to quantum field theory, one further question must be discussed in the context of nets of von Neumann algebras. Up to now, we have assumed that the nets in question have been defined on the set C of all compact sets. In practice, there are often different ways of defining $\mathfrak{A}(\mathcal{O})$ for a general compact set $\mathcal{O}$. This has already been touched on in the discussion following Theorem 3.5. Fortunately such ambiguities can be ignored for the purpose of computing cohomologies because, as we shall see, the cohomology of a net over C is essentially determined by its restriction to K. As far as $Z^0(\mathfrak{A})$ goes, the definition has been phrased so that only sets of K appear. $Z^1(\mathfrak{A})$ is, however, in the first instance not defined for nets over K. To define it, we first extend the net from K to C; this may be done in many different ways. In particular, we have the minimal extension defined by

$$\mathfrak{A}_0(\mathcal{O}) = \vee\{\mathfrak{A}(\mathcal{O}_1) : \mathcal{O}_1 \subset \mathcal{O},\ \mathcal{O}_1 \in K\} \tag{3.18}$$

and the maximal extension defined by

$$\mathfrak{A}_1(\mathcal{O}) = \cap\{\mathfrak{A}(\mathcal{O}_1) : \mathcal{O} \subset \mathcal{O}_1,\ \mathcal{O}_1 \in K\}. \tag{3.19}$$

We prove

<u>3.8 Lemma</u> Let $\mathfrak{A}$ and $\mathfrak{B}$ be nets of von Neumann algebras over C whose restrictions to K coincide so that, in particular, $\mathfrak{A}_\infty = \mathfrak{B}_\infty$. Then if $\mathcal{O}_1, \mathcal{O}_2 \in K_0$ with $\mathcal{O}_1 \subset \operatorname{int} \mathcal{O}_2$

$$Z^1(\mathfrak{A})(\mathcal{O}_1) \subset Z^1(\mathfrak{B})(\mathcal{O}_2).$$

<u>Proof</u> If $\mathfrak{A} \subset \mathfrak{B}$, we even have $Z^1(\mathfrak{A})(\mathcal{O}_1) \subset Z^1(\mathfrak{B})(\mathcal{O}_1)$, so it

suffices to consider the case $\mathcal{B} \subset \mathcal{A}$, and furthermore to show that the inclusion holds for 1-cocycles since the extension to arrows is then trivial. So suppose z is an object of $Z^1(\mathcal{A})(\mathcal{O}_1)$ and $b \in \Sigma_1$. Then we may pick a 'dissection' of b, i.e. $b_o, b_1, \ldots, b_n \in \Sigma_1$ with $\partial_o b_o = \partial_o b$, $\partial_o b_i = \partial_1 b_{i-1}$, $i=1,2,\ldots n$ and $\partial_1 b_n = \partial_1 b$, with $|b_i| + \mathcal{O}_1 \subset \partial_o b_i + \mathcal{O}_2$, for example by iterating the barycentric decomposition as in [5, §3]. Thus $z(b_i) \in \mathcal{A}(\mathcal{O}_1 + |b_i|) \subset \mathcal{A}(\mathcal{O}_2 + \partial_o b_i)$ $= \mathcal{B}(\mathcal{O}_2 + \partial_o b_i) \subset \mathcal{B}(\mathcal{O}_2 + |b|)$. Since $z(b) = z(b_o) z(b_1) \ldots z(b_n)$ by the cocycle identity, $z(b) \in \mathcal{B}(\mathcal{O}_2 + |b|)$ and z is an object of $Z^1(\mathcal{B})(\mathcal{O}_2)$ as required.

We conclude that, however a net $\mathcal{A}$ is extended from K to C, $Z^1(\mathcal{A})$ is well defined up to an infinitesimal change in localization. Inner or outer regularization would then lead to a definition of $Z^1(\mathcal{A})$ independent of the chosen extension. In what follows, we shall work with nets over K without bothering to specify a precise localization for Z^1.

§4 Applications to Quantum Field Theory

Many concepts have been introduced into field theory without it being clear from the outset that they describe intrinsic features of the theory under examination rather than just expressions of the way we choose to describe the theory. Concepts such as gauge group, parastatistics, spontaneously broken gauge symmetry, phase transitions, Higgs mechanism and soliton sector spring to mind. The principal source of trouble here is that the fields themselves are not intrinsic quantities and the same physical theory is capable of being described in terms of different fields. The most spectacular example is the equivalence of the Sine-Gordon model and the Thirring model [16] but there are many other examples, e.g. the use of fields with anomalous instead of Bose-Fermi commutation relations or the possibility of describing theories with parastatistics either in terms of parafields or in terms of fields with Bose-Fermi statistics and an unobservable colour index. The method which has, in the past, proved successful in resolving problems of this nature is to adopt the point of view of Haag (see for example [17; p. 12]) that a physical theory is intrinsically determined by its observable net.*) In this case, the cohomology of the observable net becomes an intrinsic feature of the theory in question.

The 0-cohomology of the observable net is $\mathbb{C}$ under very general assumptions; this is an important structural feature of $\mathfrak{A}$ and a universal, hence uninteresting invariant. As we shall see the 1-cohomology is an interesting invariant which may be used to distinguish one theory from another. The first step to compute it in a concrete case is to look at the field net $\mathfrak{F}$ and try to show that its 1-cohomology is quasitrivial for example by exploiting Theorem 3.5. If the 1-cohomology of $\mathfrak{F}$ is not quasitrivial, $\mathfrak{F}$ is probably not a maximal field net and one should try to enlarge it. In practice, $\mathfrak{A}$ is usually defined by applying a principle of gauge invariance so that $\mathfrak{A}$ is the fixed-point net of $\mathfrak{F}$ under a group of gauge automorphisms G and we are in a position to apply Theorem 3.7 to conclude that $Z^1(\mathfrak{A})$ is equivalent to the category of unitary representations of G determined by $Z^0(\hat{G})$. The present dearth of explicit models in the continuum means that the problem of computing $Z^0(\hat{G})$ really only arises when G is a gauge group of the first kind. In this case, the translation covariance of $\mathfrak{F}$ means that there are automorphisms α_x with

$$\alpha_x(\mathfrak{F}(\mathcal{O})) = \mathfrak{F}(\mathcal{O}+x), \qquad x \in \mathbb{R}^{s+1}, \quad \mathcal{O} \in K.$$

commuting with the elements of G. Thus $Z^0(\hat{G}) = \hat{G}_\infty$. If G is a compact group leaving the vacuum state of $\mathfrak{F}$ invariant, $\omega_0 \circ \beta = \omega_0$, $\beta \in G$, then $\hat{G}(\mathcal{O})$ consists of all unitary equivalence classes of representations of G on separable Hilbert spaces for all $\mathcal{O} \in K$. This information may be extracted from [14; §3]. In fact this result may be extended to compact groups of spontaneously broken gauge automorphisms; the proof, which uses the Reeh-Schlieder property, positivity of the energy and W*-category techniques will not be given here. We conclude that when G is a compact gauge group of the first kind, $Z^1(\mathfrak{A})$ is equivalent to the W*-category of separable continuous unitary representations of G.

---

*) In view of these successes, it is regrettable how rarely, in the discussion of models, is attention given to the question of what the observable net of the model is. Even the word 'observable' is occasionally used in quite different senses.

In particular, we see that if G is non-trivial, the field net and the observable net are not isomorphic and condition a) of Theorem 3.5 even tells us how these nets fail to be isomorphic. We have also come a long way towards answering the question of whether the gauge group of the first kind is an intrinsic feature of a theory. At least the abstract representation theory of G has an intrinsic meaning. I do not know whether this suffices to determine a compact group uniquely although one does have enough information to separate the two non-Abelian groups of order 8 with isomorphic multiplication table because the 6-j symbols can be computed and are different for these two groups.

There are good reasons for believing that a non-trivial 2-cohomology will turn out to be an intrinsic way of defining gauge theories, i.e., theories with gauge transformations of the second kind. If we consider a G-gauge theory, e.g. G = U(1), SU(n), then it appears that the 2-cohomology will turn out to be the representation theory of G. The basic ideas are outlined in [8]. Although there are a number of abstract results on the 2-cohomology going beyond [8] and treating tensor products and conjugation of 2-cocycles, they will not be discussed here as the framework of the 2-cohomology is not yet in a definitive form.

Now there is an intimate relationship between gauge groups of the first kind and superselection sectors and I turn to the applications of net cohomology to superselection structure. If π_o is the vacuum representation of the observable net $\mathfrak{A}$, I consider representations π of $\mathfrak{A}$ satisfying the following selection criterion:

$$\pi_o \upharpoonright \mathfrak{A}(\mathcal{O}'+a) \simeq \pi \upharpoonright \mathfrak{A}(\mathcal{O}'+a) , \quad a \in \Sigma_o . \tag{S}$$

Here $\mathfrak{A}(\mathcal{O}')$ denotes the C*-algebra generated by the algebras $\mathfrak{A}(\mathcal{O}_1)$ with $\mathcal{O}_1 \in \mathcal{K}$ and $\mathcal{O}_1 \subset \mathcal{O}'$. This is the selection criterion proposed and motivated physically in [1], where it is also pointed out that it is too stringent for quantum electrodynamics because by Gauss' law, it excludes states with non-vanishing electric charge. The sectors of gauge theories will instead be generically associated with certain 2-cohomology classes.

To simplify notation, we suppose that $\mathfrak{A}$ has been defined as a concrete net on the vacuum Hilbert space $\mathcal{H}_o$ so that we can omit the symbol π_o. We shall further suppose that our basic nets are defined over K since we know from Lemma 3.8 that the 1-cohomology is essentially independent of the extension to C. We define a dual net

$$\mathfrak{A}^d(\mathcal{O}) = \mathfrak{A}(\mathcal{O}')', \qquad \mathcal{O} \in K . \tag{4.1}$$

The operation $\mathfrak{A} \to \mathfrak{A}^d$ may be regarded as the net-theoretic version of a commutant: $\mathfrak{A}_1 \subset \mathfrak{A}_2$ implies $\mathfrak{A}_1^d \supset \mathfrak{A}_2^d$ and $\mathfrak{A} \subset \mathfrak{A}^{dd}$. If $\mathfrak{A}$ is a local net then $\mathfrak{A} \subset \mathfrak{A}^d$.

Now suppose π is a representation of $\mathfrak{A}$ satisfying (S), then there are unitary operators $\psi(a)$: $H_\pi \to H_o$ with

$$\psi(a)\pi(A) = A\psi(a), \quad A \in \mathfrak{A}(\mathcal{O}'+a), \quad a \in \Sigma_o.$$

We can define an object z of $Z^1(\mathfrak{A}^d)(\mathcal{O})$ by setting

$$z(b) = \psi(\partial_o b)\, \psi(\partial_1 b)^*, \quad b \in \Sigma_1 \tag{4.2}$$

Changing $\psi(a)$ merely changes z to a unitary equivalent 1-cocycle. If π' also satisfies (S), t intertwines π and π', and z' is a 1-cocycle corresponding to π' then $(z'|t|z)$ is an arrow of $Z^1(\mathfrak{A}^d)(\mathcal{O})$ where $t(a) = \psi'(a)t\psi(a)^*$. Conversely if $(z'|t|z)$ is an arrow of $Z^1(\mathfrak{A}^d)$ then $\psi'(a)^*t(a)\psi(a)$ is independent of a and intertwines π and π'.

<u>4.1 Theorem</u> The W*-category $S(\pi_o)$ of representations of $\mathfrak{A}$ satisfying (S) and their intertwining operators is equivalent to a full subcategory of $Z^1(\mathfrak{A}^d)$. If $s > 1$, $S(\pi_o)$ is equivalent to $Z^1(\mathfrak{A}^d)$ and every representation satisfying (S) has a unique extension to a representation $\tilde{\pi}$ of $\mathfrak{A}^{dd}$ on the same Hilbert space which also satisfies (S).

<u>Proof</u> The first statement has been proved above. To show that $S(\pi_o)$ is equivalent to $Z^1(\mathfrak{A}^d)$ for $s > 1$, we must show that any 1-cocycle z of $Z^1(\mathfrak{A}^d)(\mathcal{O})$ is associated with some representation π_z of $\mathfrak{A}$ satisfying (S). We set

$$\pi_z(A) = z(b)Az(b)^*, \quad A \in \mathfrak{A}(\mathcal{O}'+\partial_1 b) \tag{4.3}$$

where $b \in \Sigma_1$ with $\partial_o b = 0$. The argument of [18; Thm 2.2] shows that π is well defined and hence a locally normal representation of $\mathfrak{A}$ provided $s > 1$. If h is a contracting homotopy based on 0, then (4.3) gives

$$z(ha)\ \pi_z(A) = A\ z(ha), \quad A \in \mathfrak{A}(\mathcal{O}'+a)$$

and since $z(b) = z(h\partial_o b)\ z(h\partial_1 b)^*$ we see that the cocycle z is attached to the representation π_z satisfying (S). If $z(b) = \psi(\partial_o b)\psi(\partial_1 b)^*$ where $a \to \psi(a)$ is associated with some given representation π as above, then (4.3) gives $\pi_z(A) = \psi(0)\ \pi(A)\psi(0)^*$. Since $(\mathfrak{A}^{dd})^d = \mathfrak{A}^d$, $\pi_z(A)$ is well defined for $A \in \mathfrak{A}^{dd}$ so

$$\tilde{\pi}(A) = \psi(0)^*\pi_z(A)\psi(0), \quad A \in \mathfrak{A}^{dd} \tag{4.4}$$

is an extension of π to a representation $\tilde{\pi}$ of $\mathfrak{A}^{dd}$ satisfying (S). Now by (4.3), $\tilde{\pi}(A)$ reduces to $\psi(a)^*A\psi(a)$ if $A \in \mathfrak{A}^{dd}(\mathcal{O}'+a)$. However $\psi(a)^*A\psi(a)$ is independent of the choice of $a \to \psi(a)$ when $A \in \mathfrak{A}^{dd}(\mathcal{O}'+a$, since $\psi'(a)\psi(a)^* \in \mathfrak{A}^d(\mathcal{O}+a)$. Thus $\tilde{\pi}$ is the only extension of π to $\mathfrak{A}^{dd}$ which satisfies (S).

This argument applies to a quite general net $\mathfrak{A}$. For it to be useful as a means of computing $S(\pi_o)$ we need to be able to identify $\mathfrak{A}^d$ and compute its 1-cohomology. A particularly simple case is when $\mathfrak{A}$ itself satisfies duality in the vacuum representation, i.e. when $\mathfrak{A} = \mathfrak{A}^d$ as nets over K.

<u>4.2 Corollary</u> If the observable net $\mathfrak{A}$ satisfies duality in the vacuum sector and $s > 1$ then $S(\pi_o)$ is equivalent to $Z^1(\mathfrak{A})$.

To prove duality in the vacuum sector in a concrete situation one uses the following result.

<u>4.3 Theorem</u> Let $\mathfrak{F}$ be a field net satisfying twisted duality in its vacuum sector and $\mathfrak{A}$ the fixed point net under a compact group G of gauge automorphisms of the first kind leaving the vacuum state of $\mathfrak{F}$ invariant, then $\mathfrak{A}$ satisfies duality in its vacuum sector provided $s > 1$.

<u>Proof</u> This result is proved in Section IV of [19] except that the equality $m(\mathfrak{F}(\mathcal{O}')^-) = \mathfrak{A}(\mathcal{O}')^-$ is not obvious in the present context because the set of double cones in $\mathcal{O}'$, used to define

$\mathfrak{A}(\mathcal{O}')$ and $\mathfrak{F}(\mathcal{O}')$ here, do not form a directed set under inclusion. To complete the proof, we draw on the results of [14, §3]. We first remark that in each $\mathfrak{F}(\mathcal{O})$, $\mathcal{O} \in K$ we may find a Hilbert space $H(\sigma)$ carrying any given irreducible representation σ of G and that as σ varies these Hilbert spaces generate $\mathfrak{F}(\mathcal{O})$ as a von Neumann algebra. Secondly if $\mathcal{O}_i \in K$, $\mathcal{O}_i \subset \mathcal{O}'$ and $H_i(\sigma)$ is such a Hilbert space in $\mathfrak{F}(\mathcal{O}_i)$, $i = 1,2$, $(H_1(\sigma), H_2(\sigma)) \cap \mathfrak{A} = \mathbb{C}\, U$ for some unitary U. Now $U \in \mathfrak{A}(\mathcal{O}')^-$; this in obvious if there is an $\mathcal{O}_3 \in K$ with $\mathcal{O}_1 \cup \mathcal{O}_2 \subset \mathcal{O}_3 \subset \mathcal{O}'$. But since $\mathcal{O}'$ is path-connected for $s > 1$, we may introduce intermediate double cones and factorize U as a finite product $U = U_1 U_2 \ldots U_n$ of unitaries of this same type. Since $H_2(\sigma) = UH_1(\sigma)$ we may conclude that $\mathfrak{F}(\mathcal{O}')^- = \mathfrak{A}(\mathcal{O}')^- \vee \mathfrak{F}(\mathcal{O}_1)$. Now let $A \in \mathfrak{A}(\mathcal{O}_2)$ and $\psi \in H_1(\sigma)$ then $\psi A = \rho(A)\psi$ where ρ is the morphism associated with $H_1(\sigma)$. We claim that $\rho(A) \in \mathfrak{A}(\mathcal{O}')^-$, because if $\mathcal{O}_4 \subset \mathcal{O}' \cap \mathcal{O}_2'$, $\mathcal{O}_4 \in K$ and U' is unitary in $(H_4(\sigma), H_1(\sigma)) \cap \mathfrak{A}$ then $\rho(A) = U'AU'^*$. It follows that $\rho(\mathfrak{A}(\mathcal{O}')^-) \subset \mathfrak{A}(\mathcal{O}')^-$. Hence using the decomposition theory of representations of a compact group we deduce that finite linear combinations of operators of the form $A\psi$ with $A \in \mathfrak{A}(\mathcal{O}')^-$ and $\psi \in H_1(\sigma)$ as σ varies form a weakly dense *-subalgebra of $\mathfrak{A}(\mathcal{O}')^- \vee \mathfrak{F}(\mathcal{O}_1) = \mathfrak{F}(\mathcal{O}')^-$. Since $m(A\psi) = Am(\psi)$ we conclude that $m(\mathfrak{F}(\mathcal{O}')^-) = \mathfrak{A}(\mathcal{O}')^-$ as required.

Now duality is too restrictive an assumption and we say that $\mathfrak{A}$ satisfies <u>essential duality</u> if $\mathfrak{A}^d = \mathfrak{A}^{dd}$.

<u>4.4 Theorem</u> Suppose that the gauge group G of Theorem 4.3 is spontaneously broken. Let $G_o = \{\beta \in G : \omega_o \circ \beta = \omega_o\}$ be the stability subgroup of the vacuum state ω_o of $\mathfrak{F}$ and $\mathfrak{B}$ the fixed-point subnet of $\mathfrak{F}$ under G_o. Then $\pi_o(\mathfrak{A})^d = \pi_o(\mathfrak{B})$ and $\mathfrak{A}$ satisfies essential duality in the vacuum sector if $s > 1$.

<u>Proof</u> We have $\mathfrak{A}(\mathcal{O}')^- = \mathfrak{B}(\mathcal{O}')^-$ [20; Prop. 9]. Thus $\pi_o(\mathfrak{A})^d = \pi_o(\mathfrak{B})^d$. But $\mathfrak{B}$ satisfies duality in the vacuum sector by Theorem 4.3 so $\mathfrak{A}$ satisfies essential duality in the vacuum sector.

Theorems 3.7, 4.1 and 4.4 give us reasonable hypotheses leading to the conclusion that $S(\pi_o)$ for an observable algebra is equivalent to Rep G_o, the W*-category of separable continuous unitary representations of G_o and their intertwining operators. They apply in particular to the usual range of massive free field models for $s > 1$. I do not anticipate difficulties for the mass zero case but I have not checked the details. In particular for the free field net itself a representation satisfying (S) must be an at most countable direct sum of copies of π_o. Independently Driessler [12] has recently given different (although related) criteria allowing one to conclude that the free field has no non-trivial sectors.

Essential duality is a very important property for an observable net $\mathfrak{A}$ because even if we cannot compute $S(\pi_o)$ explicitly, Theorem 4.1 shows that the results of [1,2] on superselection structure still apply. Fortunately the results of Bisognano and Wichmann [21] (see also [22]) allow one to conclude under rather general circumstances that $\mathfrak{A}$ satisfies essential duality.

There is some interest in computing the superselection structure for the exceptional dimension $s = 1$. In the first place, $s = 1$ is a proving ground for field-theoretical models so that there is a chance of applying the computations to interesting interacting models. Secondly, representations satisfying (S) include the soliton sectors for $s = 1$. Unfortunately my own computations are not sufficiently far advanced to be presented here although I believe they can probably be brought to a successful conclusion. Nevertheless there is one point to be made here; the examples computed in [5] in the context of Abelian cohomology suggest that there will be a very large 1-cohomology when $s = 1$. This does not mean that for $s = 1$ there will necessarily be very many irreducible representations satisfying (S) because in Theorem 4.1, $S(\pi_o)$ is no longer equivalent to the whole of $Z^1(\mathfrak{A}^d)$. The condition in (4.5) below is rather restrictive; it allows one to sharpen the results of [18; §3] and seems to bring the problem of computing $S(\pi_o)$ within reach.

<u>4.5 Theorem</u> If $s = 1$, the W*-category $S(\pi_o)$ of $\mathfrak{A}$ is equivalent to the full subcategory of $Z^1(\mathfrak{A}^d)$ whose objects satisfy

$$z(b) \in (\mathfrak{A}(\mathcal{O}'+\partial_0 b) \cap \mathfrak{A}(\mathcal{O}'+\partial_1 b))', \quad b \in \Sigma_1 \qquad (4.5)$$

for some $\mathcal{O} \in K_o$.

Proof We first remark that if z is defined by (4.2), then (4.5) is a consequence of the intertwining properties of $a \to \psi(a)$. Conversely, if z is an object of $Z^1(\mathfrak{A}^d)(\mathcal{O})$ satisfying (4.5) then Lemma 2.1 of [18] is valid for the inner automorphism generated by z(b). This is the only additional point needed to verify that π_z can be defined by (4.3).

References

[1] S. Doplicher, R. Haag, J.E. Roberts: Local Observables and Particle Statistics. I. Commun.Math.Phys. 23, 199-230 (1971).

[2] S. Doplicher, R. Haag, J.E. Roberts: Local Observables and Particle Statistics. II. Commun.Math.Phys. 35, 49-85 (1974).

[3] R. Haag: Discussion des "axiomes" et des propriétés asymptotiques d'une théorie des champs locale avec particules composées. 75. Colloques Internationaux du CNRS, Lille 1957. Paris: CNRS 1959.

[4] G. de Rham: Variétés différentiables. Hermann, Paris 1960.

[5] P. Leyland, J.E. Roberts: The Cohomology of Nets over Minkowski Space. Commun. Math. Phys. 62, 173-189 (1978).

[6] J.E. Roberts: Mathematical Aspects of Local Cohomology. Proceedings of the Colloquium on Operator Algebras and their Applications to Mathematical Physics, Marseille 1977. CNRS to appear.

[7] R. Lima: Introduction to W*-categories. These Proceedings.

[8] J.E. Roberts: A survey of Local Cohomology, Proceedings of the Conference on Mathematical Problems in Theoretical Physics, Rome 1977. Lecture Notes in Physics 80: Springer 1978.

[9] S. Sakai: C*-Algebras and W*-Algebras. Springer 1971.

[10] J.E. Roberts: Cross Products of von Neumann Algebras by Group duals, Symposia Mathematica 22, 335-363, Academic Press 1976.

[11] D. Buchholz: Product States for Local Algebras, Commun. Math. Phys. 36, 287-304 (1974).

[12] W. Driessler: Duality and Absence of Locally Generated Superselection Sectors for CCR-Type Algebras.

[13] H. Araki: Von Neumann Algebras of Local Observables for the free Scalar Field. J. Math. Phys. 5, 1-13 (1964).

[14] S. Doplicher, J.E. Roberts: Fields, Statistics and Non-Abelian Gauge Groups, Commun. Math. Phys. 28, 331-348 (1972).

[15] H. Araki, R. Haag, D. Kastler, M. Takesaki: Extensions of KMS-States and Chemical Potential, Commun. Math. Phys. 53, 97-134 (1977).

[16] S. Coleman: Quantum Sine-Gordon equation as the massive Thirring model. Phys. Rev. D11, 2088-2097 (1975).

[17] R. Haag: Quantum Field Theory. In "Mathematics of Contemporary Physics", Ed. R.F. Streater, Academic Press 1972.

[18] J.E. Roberts: Local Cohomology and Superselection Structure, Commun.Math.Phys. 51, 107-119 (1976).

[19] S. Doplicher, R. Haag, J.E. Roberts: Fields, Observables and Gauge Transformations. I. Commun. Math. Phys. 13, 1-23 (1969).

[20] J.E. Roberts: Spontaneously Broken Gauge Symmetries and Superselection Rules. Proceedings of the International School of Mathematical Physics, Camerino 1974. Università di Camerino 1976.

[21] J.J. Bisognano, E.H. Wichmann, On the Duality Condition for Quantum Fields. J. Math. Phys. 17, 303-321 (1976).

[22] C. Rigotti: Remarks on the Modular Operator and Local Observables. Commun. Math. Phys. 61, 267-273 (1978).

Construction of Specifications

Chris Preston

Fakultät für Mathematik

Universität Bielefeld

Bielefeld, Federal Republic of Germany

1. Introduction

Equilibrium behaviour in classical statistical mechanics can be described in terms of probability measures (called Gibbs ensembles) on an appropriate phase space. In their simplest form the Gibbs ensembles correspond to systems lying in a bounded region of space. However, in order to study phenomena such as phase transitions it is necessary to consider infinite volume limits of these ensembles. The work of Dobrushin (1968 a), (1968 b), (1969), Lanford and Ruelle (1969) and Ruelle (1970) shows that these infinite volume limits, called Gibbs states, can be defined as the probability measures which are compatible with a given collection of conditional probabilities. These conditional probabilities are defined in terms of the potential energy for the system being considered.

This approach using conditional probabilities can be formalized in the notion of a specification (due to Föllmer (1975 a), (1975 b)). It is the purpose of this paper to give a general method for constructing specifications. Let $(X,\mathcal{F})$ be a measurable space and I an index set equipped with a partial order $\leq$ that is directed from above and countably generated. Let $\mathbb{F} = \{\mathcal{F}_\Lambda\}_{\Lambda\in I}$ be a decreasing collection of sub-σ-algebras of $\mathcal{F}$ (i.e. $\mathcal{F}_{\Lambda'} \subset \mathcal{F}_\Lambda$ whenever $\Lambda \leq \Lambda'$), and for each $\Lambda \in I$ let $\pi_\Lambda : X \times \mathcal{F} \to \mathbb{R}^+$ be a kernel. $\mathcal{V} = \{\pi_\Lambda\}_{\Lambda\in I}$ is called an <u>$\mathbb{F}$-specification</u> if:

$$\text{for each } \Lambda \in I\ ,\ x \in X \quad \pi_\Lambda(x,X) \ \text{ is either 0 or 1;} \tag{1.1}$$

$$\text{for each } \Lambda \in I\ ,\ F \in \mathcal{F} \quad \pi_\Lambda(\cdot,F) \ \text{ is } \mathcal{F}\text{-measurable;} \tag{1.2}$$

$$\text{for each } \Lambda \in I\ ,\ F \in \mathcal{F},\ F' \in \mathcal{F}_\Lambda \quad \pi_\Lambda(\cdot,F'\,F) = \chi_{F'}\pi_\Lambda(\cdot,F); \tag{1.3}$$

$$\text{if } \Lambda \leq \Lambda \quad \text{then} \quad \pi_{\Lambda'} = \pi_{\Lambda'}\pi_\Lambda \ , \tag{1.4}$$

(where $(\pi_{\Lambda'}\pi_\Lambda)(x,F) = \int \pi_\Lambda(y,F)\pi_{\Lambda'}(x,dy)$). $\mathcal{G}(\mathcal{V})$, the set of <u>Gibbs states with specification $\mathcal{V}$</u>, is defined to be those probability measures μ on $(X,\mathcal{F})$ which are compatible with the kernels $\{\pi_\Lambda\}_{\Lambda\in I}$, i.e. those μ for which

$$E_\mu(\chi_F|\mathcal{F}_\Lambda) = \pi_\Lambda(\cdot,F) \quad \mu\text{-a.e. for each } F \in \mathcal{F},\ \Lambda \in I. \tag{1.5}$$

Let $P(X,\mathcal{F})$ denote the probability measures on $(X,\mathcal{F})$; because of (1.3) it is easily seen that we in fact have

$$\mathcal{G}(\mathcal{V}) = \{\mu \in P(X,\mathcal{F}) : \mu = \mu\pi_\Lambda \quad \text{for all} \quad \Lambda \in I\} ,$$

where the measure $\mu\pi_\Lambda$ is given by $(\mu\pi_\Lambda)(F) = \int \pi_\Lambda(x,F)\, d\mu(x)$.

Let $\mathcal{V} = \{\pi_\Lambda\}_{\Lambda\in I}$ be an $\mathbb{F}$-specification; we will give a method of constructing other $\mathbb{F}$-specifications from $\mathcal{V}$. When our results are applied to models from statistical physics $\mathcal{V}$ will usually correspond to a model with no interaction. The specifications we construct will then correspond to adding an interaction to the model defined by $\mathcal{V}$.

We will define specifications in terms of functions $G : X \times X \to \mathbb{R}^+$ satisfying

$$G(x_1, x_2) > 0 \quad \text{whenever } G(x_2, x_1) > 0 , \tag{1.6}$$

$$G(x_1,x_3) = G(x_1,x_2)\, G(x_2,x_3) \quad \text{whenever } G(x_1,x_2)\, G(x_2,x_3) > 0. \tag{1.7}$$

The simplest version of our construction (Theorem 3.2) gives: let G be $\mathcal{F} \times \mathcal{F}$-measurable and satisfy (1.6) and (1.7), suppose that

$\{(x,y) : G(x,y) > 0\} \in \mathcal{F}_\Lambda \times \mathcal{F}_\Lambda$ for each $\Lambda \in I$, and put

$B = \{x : G(x,x) > 0\}$.

Then $\int_B \{\int G(z,y)\, \pi_\Lambda(y,dz)\}^{-1}\, \pi_\Lambda(x,dy)$ is either 0 or 1

for each $\Lambda \in I$, $x \in X$, and if we define $\pi_\Lambda^G : X \times \mathcal{F} \to \mathbb{R}^+$ by

$$\pi_\Lambda^G(x,F) = \int_{F\cap B} \{\int G(z,y)\, \pi_\Lambda(y,dz)\}^{-1}\, \pi_\Lambda(x,dy) \tag{1.8}$$

then $\mathcal{V}_G = \{\pi_\Lambda^G\}_{\Lambda\in I}$ is an $\mathbb{F}$-specification.

Functions G satisfying (1.6) and (1.7) occur naturally as follows: For each $\Lambda \in I$ let $f_\Lambda : X \to \mathbb{R}^+$ be F-measurable and define $\xi_\Lambda : X \times X \to \mathbb{R}^+$ by

$$\xi_\Lambda(x_1,x_2) = \begin{cases} \dfrac{f_\Lambda(x_1)}{f_\Lambda(x_2)} & \text{if } f_\Lambda(x_2) > 0 , \\ 0 & \text{otherwise.} \end{cases}$$

Now choose an increasing, order-generating sequence $\{\Lambda_n\}_{n\geq 1}$ and define $G : X \times X \to \mathbb{R}^+$ by

$$G(x_1,x_2) = \begin{cases} \lim\limits_{n\to\infty} \xi_{\Lambda_n}(x_1,x_2) & \text{if the limit exists,} \\ 0 & \text{otherwise.} \end{cases}$$

Clearly G is $F \times F$-measurable and satisfies (1.6) and (1.7). With models from statistical physics we will usually take $f_\Lambda = \exp(-U_\Lambda)$, where U_Λ is the potential energy corresponding to the region Λ. In Section 4 we show that this construction gives the same result as the "standard" method of constructing specifications in terms of the conditional energy for the model.

As well as the examples from statistical physics there is a purely measure-theoretic way in which a collection of functions $\{f_\Lambda\}_{\Lambda\in I}$ arise and define a function G as above. Let $\mu \in P(X,F)$ and suppose $\mu << \mu\pi_\Lambda$ for all $\Lambda \in I$; take $f_\Lambda : X \to \mathbb{R}^+$ to be a version of the corresponding Radon-Nikodym derivative, and define G as before in terms of $\{f_\Lambda\}_{\Lambda\in I}$. In Section 2 we prove that

$$\mu(F) = \int \int_{F\cap B} \{ \int G(z,y)\ \pi_\Lambda(y,dz)\}^{-1}\ \pi_\Lambda(x,dy)\ d\mu(x) \qquad (1.9)$$

for all $\Lambda \in I$, $F \in F$. Thus we have

$$\mu(F) = \int \pi_\Lambda^G\,(x,F)\ d\mu(x) \quad \text{for all } \Lambda \in I,\ F \in F , \qquad (1.10)$$

where π_Λ^G is defined by (1.8). If $V_G = \{\pi_\Lambda^G\}_{\Lambda\in I}$ is an $\mathbb{F}$-specification (which will be true, for example, if $\{(x,y) : G(x,y) > 0\} \in F_\Lambda \times F_\Lambda$ for all $\Lambda \in I$) then we have constructed an $\mathbb{F}$-specification V_G with $\mu \in G(V_G)$.

In Section 5 we check that our construction can be used to give invariant specifications (i.e. specifications which are invariant under a group of transformations) when the situation is such that an invariant specification should be constructed.

2. Construction of Specifications

Let $V = \{\pi_\Lambda\}_{\Lambda\in I}$ be a given $\mathbb{F}$-specification. We are going to give a method of defining other $\mathbb{F}$-specifications from it. Starting with V there is really only one reasonable way to define specifications in terms of it: take a collection $w = \{w_\Lambda\}_{\Lambda\in I}$, where for each $\Lambda\in I$ $w_\Lambda : X \to \mathbb{R}^+$ is an F-measurable function, put

$$Z_\Lambda^{(w)}(x) = \int w_\Lambda(y)\, \pi_\Lambda(x,dy) \text{ , and define } \pi_\Lambda^{(w)} : X \times F \to \mathbb{R}^+ \text{ by} \tag{2.1}$$

$$\pi_\Lambda^{(w)}(x,F) = \begin{cases} \{Z_\Lambda^{(w)}(x)\}^{-1} \int_F w_\Lambda(y)\, \pi_\Lambda(x,dy) & \text{if } 0 < Z_\Lambda^{(w)}(x) < +\infty, \\ 0 & \text{otherwise.} \end{cases}$$

It is easily checked that (1.1), (1.2) and (1.3) hold for $\{\pi_\Lambda^{(w)}\}_{\Lambda\in I}$; thus $V_w = \{\pi_\Lambda^{(w)}\}_{\Lambda\in I}$ will be an $\mathbb{F}$-specification provided $\pi_{\Lambda'}^{(w)} = \pi_{\Lambda'}^{(w)}\pi_\Lambda^{(w)}$ whenever $\Lambda \le \Lambda'$. Proposition 2.1 gives sufficient conditions for this to happen.

Proposition 2.1 Suppose that if $\Lambda \le \Lambda'$

$$\int\int |w_{\Lambda'}(y_1)w_\Lambda(y_2) - w_{\Lambda'}(y_2)w_\Lambda(y_1)|\, \pi_\Lambda(x,dy_1)\pi_\Lambda(x,dy_2) = 0 \text{ for all } x\in X, \tag{2.2}$$

$$\int_{A_\Lambda} w_{\Lambda'}(y)\pi_{\Lambda'}(x,dy) = \int w_{\Lambda'}(y)\pi_{\Lambda'}(x,dy) \quad \text{for all } x \in A_{\Lambda'} \text{ ,} \tag{2.3}$$

where $A_\Lambda = \{x : 0 < Z_\Lambda^{(w)}(x) < +\infty\}$. Then $\pi_{\Lambda'}^{(w)} = \pi_{\Lambda'}^{(w)}\pi_\Lambda^{(w)}$.

Proof: At the end of the section. □

Remark:

The converse of Proposition 2.1 is almost true. It is not difficult to show that if $\pi_{\Lambda'}^{(w)} = \pi_{\Lambda'}^{(w)} \pi_{\Lambda}^{(w)}$ then (2.3) holds and (2.2) holds provided "for all $x \in X$" is replaced by "for $\pi_{\Lambda'}(z,\cdot)$-a.e. $x \in X$ for each $z \in X$".

Our method for constructing specifications will consist of a natural way of defining collections $w = \{w_\Lambda\}_{\Lambda \in I}$ which satisfy (2.2) and (2.3). In order to justify the construction we will start with $w = \{w_\Lambda\}_{\Lambda \in I}$ satisfying (2.2) and (2.3) and examine some of its properties. We first note that it is possible (and in what follows, convenient) to normalize the w_Λ. Define $v_\Lambda : X \to \mathbb{R}^+$ by

$$v_\Lambda(x) = \begin{cases} \{z_\Lambda^{(w)}(x)\}^{-1} \, w_\Lambda(x) & \text{if } 0 < z_\Lambda^{(w)}(x) < +\infty, \\ 0 & \text{otherwise;} \end{cases} \tag{2.4}$$

then it is easily checked that $\{v_\Lambda\}_{\Lambda \in I}$ satisfies (2.2) and (2.3), $\pi_\Lambda^{(v)} = \pi_\Lambda^{(w)}$ for all $\Lambda \in I$, and $z_\Lambda^{(v)}(x)$ is either 0 or 1. Therefore we can, and will, assume that

$$z_\Lambda^{(w)}(x) \quad \text{is either 0 or 1 for each } x \in X,\ \Lambda \in I. \tag{2.5}$$

Because of (2.5) we now have

$$\pi_\Lambda^{(w)}(x,F) = \int_F w_\Lambda(y) \pi_\Lambda(x,dy) \quad \text{for all } F \in \mathcal{F},\ x \in X,\ \Lambda \in I. \tag{2.6}$$

From Proposition 2.1 $\mathcal{V}_w = \{\pi_\Lambda^{(w)}\}_{\Lambda \in I}$ is an $\mathbb{F}$-specification; let us assume $\mathcal{G}(\mathcal{V}_w)$ is non-empty and take $\nu \in \mathcal{G}(\mathcal{V}_w)$. Using (2.6) we have

$$\int \pi_\Lambda^{(w)}(x,F) d\nu(x) = \int\int_F w_\Lambda(y) \pi_\Lambda(x,dy) d\nu(x) = \int_F w_\Lambda(y) d(\nu\pi_\Lambda)(y) \quad ,$$

thus $\nu \in \mathcal{G}(\mathcal{V}_w)$ means exactly that, for each $\Lambda \in I$, $\nu \ll \nu\pi_\Lambda$ and w_Λ is a version of the Radon-Nikodym derivative of ν with respect to $\nu\pi_\Lambda$. Therefore $\{w_\Lambda\}_{\Lambda \in I}$ must possess (up to the appropriate sets of measure zero) the properties of such Radon-Nikodym derivatives.

We now no longer assume that we have $\{w_\Lambda\}_{\Lambda \in I}$, but instead take μ to be any probability measure on $(X,\mathcal{F})$ with $\mu \ll \mu\pi_\Lambda$ for each

$\Lambda \in I$.

For each $\Lambda \in I$ let $f_\Lambda : X \to \mathbb{R}^+$ be a version of the Radon-Nikodym derivative of μ with respect to $\mu\pi_\Lambda$; we will look at the properties of $\{f_\Lambda\}_{\Lambda \in I}$.

For each $\Lambda \in I$ put $E_\Lambda = \{x : f_\Lambda(x) > 0\}$ and define $\xi_\Lambda : X \times X \to \mathbb{R}^+$ by

$$\xi_\Lambda(y_1,y_2) = \begin{cases} \dfrac{f_\Lambda(y_1)}{f_\Lambda(y_2)} & \text{if } y_2 \in E_\Lambda , \\ 0 & \text{otherwise.} \end{cases}$$

Choose an increasing, order-generating sequence $\{\Lambda_n\}_{n\geq 1}$ and define $G : X \times X \to \mathbb{R}^+$ by

$$G(y_1,y_2) = \begin{cases} \lim\limits_{n\to\infty} \xi_{\Lambda_n}(y_1,y_2) & \text{if the limit exists,} \\ 0 & \text{otherwise;} \end{cases}$$

let $E = \bigcup\limits_{m\geq 1} \bigcap\limits_{n\geq m} E_{\Lambda_n}$. Finally let us define $h : X \to [0,+\infty]$ and $\Delta_\Lambda : X \times X \to \mathbb{R}^+$ by

$$h_\Lambda(x,y) = \left[\int G(z,y)\ \pi_\Lambda(x,dz) \right]^{-1}, \tag{2.7}$$

$$\Delta_\Lambda(x,y) = \begin{cases} 1 & \text{if } \pi_\Lambda(x,E) > 0 \text{ and } \{z:G(z,y) > 0\} = E \ \pi_\Lambda(x,\cdot)\text{-a.e.,} \\ 0 & \text{otherwise.} \end{cases} \tag{2.8}$$

Note that if $\Delta_\Lambda(x,y) = 1$ then $h_\Lambda(x,y) < +\infty$.

<u>Proposition 2.2:</u>

For each $\Lambda \in I$ we have $f_\Lambda = w_\Lambda\ \mu\pi_\Lambda$-a.e., where $w_\Lambda : X \to \mathbb{R}^+$ is defined by

$$w_\Lambda(x) = \Delta_\Lambda(x,x)\ h_\Lambda(x,x) \quad \text{for all } x \in X. \tag{2.9}$$

<u>Proof:</u> At the end of the section. □

Now by its definition it is clear that G has the properties:

$$G(x_1,x_2) > 0 \quad \text{whenever} \quad G(x_2,x_1) > 0. \tag{2.10}$$

$$G(x_1,x_3) = G(x_1,x_2)\, G(x_2,x_3) \quad \text{whenever} \quad G(x_1,x_2)\, G(x_2,x_3) > 0. \tag{2.11}$$

These properties and the above representation of the Radon-Nikodym derivatives $\{f_\Lambda\}_{\Lambda\in I}$ form the basis for our method of constructing specifications. In the next section we will prove the following result: Let $G : X \times X \to \mathbb{R}^+$ be any $\mathcal{F} \times \mathcal{F}$-mesurable function satisfying (2.10) and (2.11) and let $E \in \mathcal{F}$; define h_Λ and Δ_Λ by (2.7) and (2.8) and let w_Λ be given by (2.9). Then $\{w_\Lambda\}_{\Lambda\in I}$ satisfy (2.2) and (2.3) (and in fact (2.5) as well) provided

$$\text{if } \Lambda \le \Lambda' \text{ and } \Delta_{\Lambda'}(x,y) = 1 \text{ then either } \Delta_\Lambda(x,y) = 1 \text{ or } \Delta_\Lambda(x,z) = 0 \text{ for all } z \in X. \tag{2.12}$$

This result, together with Proposition 2.1, enables us to construct an $\mathbb{F}$- specification $\mathcal{V}_w$ whenever we have G satisfying (2.10) and (2.11) and such that (2.12) holds. (Of course, (2.12) also involves the set $E \in \mathcal{F}$, but in most examples we will take $E = X$.) In models from statistical physics the function G will be defined in a similar way to that already used, that is, it is defined in terms of a collection of functions $\{f_\Lambda\}_{\Lambda\in I}$. However, instead of being Radon-Nikodym derivatives, they will now take the form $f_\Lambda = \exp(-U_\Lambda)$, where $U_\Lambda : X \to \mathbb{R}$ is the energy for the model corresponding to the region Λ.

<u>Proof of Proposition 2.1:</u>

Let $x \in X$, $F \in \mathcal{F}$; if $x \notin A_\Lambda$, then $\pi^{(w)}_{\Lambda'}(x,F)$ and $(\pi^{(w)}_{\Lambda'}\,\pi^{(w)}_{\Lambda})(x,F)$ are both zero, so for the rest of the proof we assume that $x \in A_{\Lambda'}$. If $Z^{(w)}_\Lambda(y)$ and $\int w_{\Lambda'}(z)\,\pi_\Lambda(y,dz)$ are both finite then (2.2) gives us

$$\int w_{\Lambda'}(z)\,\pi_\Lambda(y,dz) \int_F w_\Lambda(z)\,\pi_\Lambda(y,dz) = \int_F w_{\Lambda'}(z)\,\pi_\Lambda(y,dz)\, Z^{(w)}_\Lambda(y).$$

We thus have

$$Z_{\Lambda'}^{(w)}(x)\ (\pi_{\Lambda'}^{(w)}\pi_{\Lambda}^{(w)})(x,F) = \int \pi_{\Lambda}^{(w)}(y,F)\ w_{\Lambda'}(y)\ \pi_{\Lambda'}(x,dy)$$

$$= \iint \pi_{\Lambda}^{(w)}(z,F)\ w_{\Lambda'}(z)\ \pi_{\Lambda}(y,dz)\ \pi_{\Lambda'}(x,dy)$$

$$= \int\{\int w_{\Lambda'}(z)\ \pi_{\Lambda}(y,dz)\}\ \pi_{\Lambda}^{(w)}(y,F)\ \pi_{\Lambda'}(x,dy)$$

$$= \int_{A_\Lambda} \{\int w_{\Lambda'}(z)\ \pi_{\Lambda}(y,dz)\}\ [Z^{(w)}(y)]^{-1} \int_F w_{\Lambda}(z)\ \pi_{\Lambda}(y,dz)\ \pi_{\Lambda'}(x,dy)$$

(since $\int w_{\Lambda'}(z)\ \pi_{\Lambda}(\cdot,dz) < +\infty\ \pi_{\Lambda'}(x,\cdot)$ - a.e.)

$$= \int_{A_\Lambda} \int_F w_{\Lambda'}(z)\ \pi_{\Lambda}(y,dz)\ \pi_{\Lambda'}(x,dy) = \int_{A_\Lambda \cap F} \int w_{\Lambda'}(z)\ \pi_{\Lambda}(y,dz)\ \pi_{\Lambda'}(x,dy)$$

$$= \int_{A_\Lambda \cap F} w_{\Lambda'}(y)\ \pi_{\Lambda'}(x,dy) = \int_F w_{\Lambda'}(y)\ \pi_{\Lambda'}(x,dy) = Z_{\Lambda'}^{(w)}(x)\ \pi_{\Lambda'}^{(w)}(x,\ ),$$

and thus $\pi_{\Lambda'}^{(w)} = \pi_{\Lambda'}^{(w)}\ \pi_{\Lambda}^{(w)}$. □

Proof of Proposition 2.2:

If $\Lambda \le \Lambda'$ and $F \in \mathcal{F}$ then, since $\pi_{\Lambda'} = \pi_{\Lambda'}\ \pi_{\Lambda}$ and

$\int_F f_\Lambda\ d(\mu\pi_\Lambda) = \int_F f_{\Lambda'}\ d(\mu\pi_{\Lambda'})$, we get

$$\iint f_\Lambda(y)\ \pi_\Lambda(x,dy)\ d\mu(x) = \iint_F f_{\Lambda'}(y)\ \pi_\Lambda(x,dy)\ d(\mu\pi_{\Lambda'})(x). \tag{2.13}$$

From this it is not difficult to show (and the details can be found in Preston (1978), Lemma 4.3) that

$$\iiint |f_{\Lambda'}(y_1)\ f_\Lambda(y_2) - f_{\Lambda'}(y_2)\ f_\Lambda(y_1)|\ \pi_\Lambda(x,dy_1)\ \pi_\Lambda(x,dy_2)d\mu(x)=0. \tag{2.14}$$

But by (2.13) and the fact hat $\mu \ll \mu\pi_{\Lambda'}$ we get

$$E_\Lambda = E_{\Lambda'}\quad \mu\pi_\Lambda\text{-a.e.}\quad \text{whenever}\quad \Lambda \le \Lambda'\ ; \tag{2.15}$$

thus (2.14) gives

$$\iint_{E_\Lambda}\int |f_\Lambda(y_1)-\xi_{\Lambda'}(y_1,y_2)f_\Lambda(y_2)|\pi_\Lambda(x,dy_1)\pi_\Lambda(x,dy_2)d\mu(x) = 0 \tag{2.16}$$

for all $\Lambda \leq \Lambda'$. From (2.15) we have $E_\Lambda = E$ $\mu\pi_\Lambda$-a.e. for each $\Lambda \in I$, hence by (2.16) we get

$$\iint_E\int |f_\Lambda(y_1)-G(y_1,y_2)f_\Lambda(y_2)|\pi_\Lambda(x,dy_1)\pi_\Lambda(x,dy_2)d\mu(x) = 0. \tag{2.17}$$

Now $\int f_\Lambda(y_1)\pi_\Lambda(\cdot,dy_1) = 1$ μ-a.e., thus from (2.17) we have

$$\int_E |1- \int G(y,x)\ \pi_\Lambda(x,dy)\ f_\Lambda(x)|\ d(\mu\pi_\Lambda)(x) = 0$$

and therefore $f_\Lambda(x) = h_\Lambda(x,x)$ for $\mu\pi_\Lambda$-a.e. $x \in E$. Let

$$d(y_1,y_2) = \begin{cases} 1 & \text{if } G(y_1,y_2) > 0, \\ 0 & \text{otherwise;} \end{cases}$$

from (2.15) and (2.16) we get

$$\iint |d(y_1,y_2) - \chi_E(y_1)\ \chi_E(y_2)|\pi_\Lambda(x,dy_1)\ \pi_\Lambda(x,dy_2) = 0,$$

and this gives

$$\int_F d(z,x)\ \pi_\Lambda(x,dz) = \pi_\Lambda(x,F\cap E)\quad \chi_E(x) \quad \text{for } \mu\pi_\Lambda\text{-a.e.} \quad x \in X, \tag{2.18}$$

for each $F \in \mathcal{F}$. (2.18), together with the fact that $\pi_\Lambda(x,E) > 0$ for $\mu\pi_\Lambda$-a.e. $x \in E$ (which holds because $E = E_\Lambda$ $\mu\pi_\Lambda$-a.e.), implies that $\Delta_\Lambda(x,x) = \chi_E(x)$ for $\mu\pi_\Lambda$-a.e. $x \in X$, and hence $f_\Lambda = w_\Lambda$ $\mu\pi_\Lambda$-a.e. □

3. Construction of Specifications, Continued

In this section we prove that the method of constructing specifications which was described in Section 2 actually works. Let $G : X \times X \to \mathbb{R}^+$ be $\mathcal{F} \times \mathcal{F}$-measurable and satisfy:

$$G(x_1,x_2) > 0 \quad \text{whenever} \quad G(x_2,x_1) > 0, \tag{3.1}$$

$$G(x_1,x_3) = G(x_1,x_2)\ G(x_2,x_3) \quad \text{whenever } G(x_1,x_2)\ G(x_2,x_3) > 0. \tag{3.2}$$

Note that because of (3.1) we could replace (3.2) by the stronger condition

$$G(x_1,x_3) = G(x_1,x_2)\ G(x_2,x_3) \quad \text{whenever} \quad G(x_2,x_3) > 0. \tag{3.3}$$

Define $h_\Lambda : X \to [0, +\infty]$ by

$$h_\Lambda(x,y) = \left[\int G(z,y)\ \pi_\Lambda(x,dz)\right]^{-1}. \tag{3.4}$$

For some applications it is convenient to slightly generalize the set-up considered in Section 2. Therefore instead of taking a single set $E \in F$ we suppose that for each $\Lambda \in I$ we are given $D_\Lambda \in F_\Lambda \times F$. (The situation of Section 2 will be the special case when $D_\Lambda = X \times E$ for all $\Lambda \in I$.) For $\Lambda \in I$, $x \in X$ let $D_\Lambda(x) = \{y : (x,y) \in D_\Lambda\}$, and define $\Delta_\Lambda : X \times X \to \mathbb{R}$ by

$$\Delta_\Lambda(x,y) = \begin{cases} 1 & \text{if } \pi_\Lambda(x,D_\Lambda(x)) > 0 \text{ and } \{z : G(z,y) > 0\} = D_\Lambda(x) \\ & \qquad \pi_\Lambda(x,\cdot)\text{-a.e.}, \\ 0 & \text{otherwise.} \end{cases} \tag{3.5}$$

If $\Delta_\Lambda(x,y) = 1$ then $h_\Lambda(x,y) < +\infty$; let $g_\Lambda(x,y) = \Delta_\Lambda(x,y)\ h_\Lambda(x,y)$ and $w_\Lambda(x) = g_\Lambda(x,x)$.

<u>Theorem 3.1:</u> Suppose

$$\text{if } \Lambda \le \Lambda' \text{ then } D_\Lambda(x) = D_{\Lambda'}(x)\ \pi_\Lambda(x,\cdot)\text{-a.e. for each } x \in X, \tag{3.6}$$

$$\text{if } \Lambda \le \Lambda' \text{ and } \Delta_{\Lambda'}(x,y) = 1 \text{ then either } \Delta_\Lambda(x,y) = 1 \text{ or } \Delta_\Lambda(x,z) = 0 \text{ for all } z \in X. \tag{3.7}$$

Then $\{w_\Lambda\}_{\Lambda\in I}$ satisfy (2.2) and (2.3); thus defining $\pi_\Lambda^{(w)}$ by (2.1) gives us an $\mathbb{F}$-specification $V_w = \{\pi_\Lambda^{(w)}\}_{\Lambda\in I}$. Furthermore (2.5) holds, and so we in fact have

$$\pi_\Lambda^{(w)}(x,F) = \int_F g_\Lambda(x,y)\pi_\Lambda(x,dy) \quad \text{for all } F \in F,\ x \in I. \tag{3.8}$$

<u>Proof:</u> At the end of the section. □

<u>Remarks:</u>

(1) If $D_\Lambda = X \times E$ for all $\Lambda \in I$ (the situation of Section 2) then $D_\Lambda(x) = E$ for all $\Lambda \in I$, $x \in X$, and so (3.6) is clearly true in this case.

(2) Let E_0 consists of those $F \in F$ for which $\pi_\Lambda(x,F) = 0$ whenever $\Lambda \leq \Lambda'$, $\pi_{\Lambda'}(x,X) = 1$ and $\pi_{\Lambda'}(x,F) = 0$; put $E = \{F \in E_0 : X - F \in E_0\}$. It is easily checked that E is a σ-algebra and that $E \supset F_\infty$ (where $F_\infty = \bigcap_{\Lambda \in I} F_\Lambda$). Assuming (3.6) holds, a sufficient condition for (3.7) to be true is that $D_\Lambda(x) \in E$ for all $x \in X$, $\Lambda \in I$ and that $G(\cdot,y)$ is E-measurable for each $y \in X$.

(3) Suppose we have $X = T^S$ with S and T countable, and that F is the product σ-algebra on X. Let I denote the non-empty finite subsets of S, ordered by inclusion, and let F_Λ be the sub-σ-algebra of F consisting of the events which are determined by what happens outside Λ. In this case it is a simple matter to choose $V = \{\pi_\Lambda\}_{\Lambda \in I}$ so that $E = F$ (where E is given by (2)) and also so that $\mu \ll \mu\pi_\Lambda$ for all probability measures on T; then define $\pi_\Lambda(x,F)$ to be the obvious version of $E_\lambda(\chi_F | F_\Lambda))$. Therefore Theorem 3.1 and Proposition 2.2 give us that if μ is any probability measure on (X,F) then there exists an $\mathbb{F}$-specification W with $\mu \in G(W)$. This generalizes a result of Goldstein (1978).

We now consider a simple case which occurs quite often in practice. Let $G : X \times X \to \mathbb{R}^+$ be $F \times F$-measurable and satisfy (3.1) and (3.2); put $D = \{(x,y) : G(x,y) > 0\}$. Suppose there exists $M \in F$ and for each $\Lambda \in I$ $C_\Lambda \in F_\Lambda \times F_\Lambda$ such that $D = (M \times M) \cap C_\Lambda$. Put $B = \{x : G(x,x) > 0\}$ and let $D_\Lambda = X \times M$ for all $\Lambda \in I$. Let $\Lambda \leq \Lambda'$, $y \in M$; then $\{z\ G(z,y) > 0\} = M\ \pi_\Lambda(x,\cdot)$-a.e. is the same as $\pi_\Lambda(x : \{z : C_{\Lambda'}(z,y) > 0\} \cap M\} = \pi_\Lambda(x,M)$.
But $\pi_\Lambda(x : \{z : C_{\Lambda'}(z,y) > 0\} \cap M) = \chi_{C_{\Lambda'}(\cdot,y)}(x)\ \pi_\Lambda(x,M)$, and thus $\Delta_\Lambda(x,y) = 1$ if and only if $\pi_\Lambda(x,M) > 0$, $y \in M$ and $C_{\Lambda'}(x,y) > 0$. From this it is clear that (3.6) and (3.7) hold. Also, we have $\Delta_\Lambda(y,y) = 1$ if and only if $\pi_\Lambda(y,M) > 0$ and $y \in B$, because $B = M \cap \{y : C_\Lambda(y,y) > 0\}$. Therefore as a corollary to Theorem

3.1 we get

Theorem 3.2:

Let $G : X \times X \to \mathbb{R}^+$ be $F \times F$-measurable and satisfy (3.1) and (3.2); put $D = \{(x,y) : G(x,y) > 0\}$ and $B = \{x : G(x,x) > 0\}$. Suppose that there exists $M \in F$ and for each $\Lambda \in I$ $C_\Lambda \in F_\Lambda \times F_\Lambda$ such that $D = (M \times M) \cap C_\Lambda$. Then for each $\Lambda \in I$, $x \in X$ $\int_B [\int G(z,y)\ \pi_\Lambda(y,dz)]^{-1}\ \pi_\Lambda(x,dy)$ is either 0 or 1, and if we define $\pi_\Lambda^G : X \times F \to \mathbb{R}^+$ by

$$\pi_\Lambda^G(x,F) = \int_{F \cap B} [\int G(z,y)\ \pi_\Lambda(y,dz)]^{-1}\ \pi_\Lambda(x,dy)$$

then $V_G = \{\pi_\Lambda^G\}_{\Lambda \in I}$ is an $\mathbb{F}$-specification.

The simplest case of Theorem 3.2 is when we have $D \in F_\Lambda \times F_\Lambda$ for all $\Lambda \in I$.

Proof of Theorem 3.1:

This will be carried out via a number of lemmas. First note that since we have $w_\Lambda(x) = g_\Lambda(x,x)$ and g_Λ is $F_\Lambda \times F$-measurable a sufficient condition for (2.2) to hold is:

$$g_{\Lambda'}(x,y_1)\ g_\Lambda(x,y_2) = g_{\Lambda'}(x,y_2)\ g_\Lambda(x,y_1) \quad \text{for all } x,y_1 y_2 \in X \tag{3.9}$$

(because if $g : X \times X \to \mathbb{R}^+$ is $F_\Lambda \times F$-measurable then

$$\int g(y,y)\ \pi_\Lambda(x,dy) = \int g(x,y)\ \pi_\Lambda(y,dy) \quad \text{for all } x \in X).$$

Lemma 3.1:

If $G(y_1,y_2) > 0$ then

$$h_\Lambda(x,y_1) = G(y_1,y_2)\ h_\Lambda(x,y_2) \quad \text{for all } \Lambda \in I,\ x \in X. \tag{3.10}$$

Proof:

From (3.3) we have $G(z,y_2) = G(y_1,y_2)\ G(z,y_1)$ for all $z \in X$, and thus

$$\int G\ (z,y_2)\ \pi_\Lambda(x,dz) = G(y_1,y_2) \int G(z,y_1)\ \pi_\Lambda(x,dz);$$

(3.10) follows immediately. □

For each $y \in X$ put $S_y = \{z : G(z,y) > 0\}$; (3.1) and (3.2) give us that for each $y,z \in Y$ either $S_y = S_z$ or $S_y \cap S_z = \emptyset$ (although $S_y = \emptyset$ is of course possible).

Lemma 3.2:

If $0 < h_\Lambda(x,y) < +\infty$ then $0 < h_\Lambda(x,z) < +\infty$ for all $z \in S_y$ and $\int_{S_y} h_\Lambda(x,z)\ \pi_\Lambda(x,dz) = 1$.

Proof:

Suppose $0 < h_\Lambda(x,y) < +\infty$; then (3.10) gives us $0 < h_\Lambda(x,z) < +\infty$ for all $z \in S_y$ and

$$\int_{S_y} h_\Lambda(x,z)\ \pi_\Lambda(x,dz) = \int_{S_y} G(z,y)\ \pi_\Lambda(x,dz)\ h_\Lambda(x,y)$$

$$= \int G(z,y)\ \pi_\Lambda(x,dz)\ h_\Lambda(x,y) = 1. \qquad \square$$

Lemma 3.3:

Either $g_\Lambda(x,y) = 0$ for all $y \in X$ or $g_\Lambda(x,\cdot) > 0$ $\pi_\Lambda(x,\cdot)$-a.e. on $D_\Lambda(x)$ and $\int g_\Lambda(x,z)\ \pi_\Lambda(x,dz) = 1$.

Proof:

Suppose $y \in X$ exists with $g_\Lambda(x,y) > 0$, so we have $\Delta_\Lambda(x,y) = 1$ and $0 < h_\Lambda(x,y) < +\infty$. If $z \in S_y$ then $S_z = S_y$ and hence $\Delta_\Lambda(x,z)=1$. Conversely, if $\Delta_\Lambda(x,z) = 1$ then $S_z = D_\Lambda(x)$ $\pi_\Lambda(x,\cdot)$-a.e. and so $S_z \cap S_y \neq \emptyset$ (since $\pi_\Lambda(x,D_\Lambda(x)) > 0$), hence $z \in S_y$. Therefore $\{z : \Delta_\Lambda(x,z) = 1\} = S_y$, and so Lemma 2.2 gives us what we need. □

Lemma 3.3 shows that (2.5) holds.

Lemma 3.4:

If $\Lambda \le \Lambda'$ and $h_{\Lambda'}(x,y) > 0$ then $h_\Lambda(\cdot,y) > 0$ $\pi_{\Lambda'}(x,\cdot)$-a.e.

Proof:

$$[h_{\Lambda'}(x,y)]^{-1} = \int G(z,y)\ \pi_{\Lambda'}(x,dz) = \iint G(v,y)\ \pi_\Lambda(z,dv)\pi_{\Lambda'}(x,dz)$$

$$= \int_\Lambda [h\ (z,y)]^{-1}\ \pi_{\Lambda'}(x,dz). \qquad \square$$

Lemma 3.5:

If $\Lambda \le \Lambda'$ and $\Delta_{\Lambda'}(x,y) = 1$ then $\Delta_\Lambda(\cdot,y) = 1$ $\pi_{\Lambda'}(x, \ )$-a.e. on $D_{\Lambda'}(x)$.

Proof:

Write $C_\Lambda(x) = X - D_\Lambda(x)$ and let $d : X \times X \to \mathbb{R}^+$ be given by

$$d(x_1,x_2) = \begin{cases} 1 & \text{if } G(x_1,x_2) > 0, \\ 0 & \text{otherwise ;} \end{cases}$$

thus $\Delta_\Lambda(x,y) = 1$ if and only if $\pi_\Lambda(x,D_\Lambda(x)) > 0$,

$$\int_{C_\Lambda(x)} d(z,y)\ \pi_\Lambda(x,dz) = 0 \quad \text{and} \quad \int_{D_\Lambda(x)} d(z,y)\ \pi_\Lambda(x,dz) = \pi_\Lambda(x,D_\Lambda(x)).$$

Define $m_\Lambda : X \times X \to \mathbb{R}^+$ by

$$m_\Lambda(x,y) = \begin{cases} [\pi_\Lambda(x,D_\Lambda(x)]^{-1} \displaystyle\int_{D_\Lambda(x)} d(z,y)\ \pi_\Lambda(x,dz) & \text{if } \pi_\Lambda(x,D_\Lambda(x))>0, \\ 0 & \text{otherwise;} \end{cases}$$

thus $0 \le m_\Lambda(x,y) \le 1$ and $\Delta_\Lambda(x,y) = 1$ if and only if $m_\Lambda(x,y) = 1$ and $\int_{C_\Lambda(x)} d(z,y)\ \pi_\Lambda(x,dz) = 0$. Now suppose $\Delta_{\Lambda'}(x,y) = 1$; we have (using (3.6))

$$\pi_{\Lambda'}(x,D_{\Lambda'}(x)) = \int_{D_{\Lambda'}(x)} d(z,y)\pi_{\Lambda'}(x,dz) = \int\int_{D_{\Lambda'}(x)} d(v,y)\pi_\Lambda(z,dv)\pi_{\Lambda'}(x,dz)$$

$$= \int\int_{D_{\Lambda'}(x)} d(v,y)\pi_\Lambda(z,dv)\pi_{\Lambda'}(x,dz) = \int\int_{D_\Lambda(z)} d(v,y)\pi_\Lambda(z,dv)\pi_{\Lambda'}(x,dz)$$

$$= \int\pi_\Lambda(z,D_\Lambda(z))m_\Lambda(z,y)\pi_{\Lambda'}(x,dz) = \int\int_{D_\Lambda(z)} m_\Lambda(v,y)\pi_\Lambda(z,dv)\pi_{\Lambda'}(x,dz)$$

$$= \int\int_{D_{\Lambda'}(z)} m_\Lambda(v,y)\pi_\Lambda(z,dv)\pi_{\Lambda'}(x,dz) = \int\int_{D_{\Lambda'}(x)} m_\Lambda(v,y)\pi_\Lambda(z,dv)\pi_{\Lambda'}(x,dz)$$

$$= \int_{D_{\Lambda'}(x)} m_\Lambda(z,y)\pi_{\Lambda'}(x,dz),$$

and thus $m_\Lambda(\cdot,y) = 1$ $\pi_{\Lambda'}(x,\cdot)$-a.e. on $D_{\Lambda'}(x)$. Similarly we have

$$0 = \int_{C_{\Lambda'}(x)} d(z,y)\pi_{\Lambda'}(x,dz) = \int\int_{C_{\Lambda'}(x)} d(v,y)\pi_{\Lambda}(z,dv)\pi_{\Lambda'}(x,dz)$$

$$= \int\int_{C_{\Lambda'}(x)} d(v,y)\pi_{\Lambda}(z,dv)\pi_{\Lambda'}(x,dz) = \int\int_{C_{\Lambda}(z)} d(v,y)\pi_{\Lambda}(z,dv)\pi_{\Lambda'}(x,dz),$$

and so $\int_{C_{\Lambda}(\cdot)} d(v,y)\pi_{\Lambda}(\cdot,dv) = 0$ $\pi_{\Lambda'}(x,\cdot)$-a.e. Therefore $\Delta_{\Lambda}(\cdot,y) = 1$ $\pi_{\Lambda'}(x,\cdot)$-a.e. on $D_{\Lambda'}(x)$. □

Lemma 3.6:

If $\Lambda \le \Lambda'$ and $\int g_{\Lambda'}(x,y)\pi_{\Lambda'}(x,dy) = 1$ then $\int g_{\Lambda}(\cdot,z)\pi_{\Lambda}(\cdot,dz) = 1$ $\pi_{\Lambda'}(x,\cdot)$-a.e. on $D_{\Lambda'}(x)$.

Proof:

We can find $y \in X$ with $g_{\Lambda'}(x,y) > 0$; thus $0 < h_{\Lambda'}(x,y) < +\infty$ and $\Delta_{\Lambda'}(x,y) = 1$. By Lemmas 3.4 and 3.5 this gives us $g_{\Lambda}(\cdot,y) > 0$ $\pi_{\Lambda'}(x,\cdot)$-a.e. on $D_{\Lambda'}(x)$, and so from Lemma 2.3 we get $\int g_{\Lambda}(\cdot,z)\pi_{\Lambda}(\cdot,dz) = 1$ $\pi_{\Lambda'}(x,\cdot)$-a.e. on $D_{\Lambda'}(x)$.

Lemma 3.7:

(2.3) holds for all $\Lambda \le \Lambda'$.

Proof:

If $0 < Z_{\Lambda'}^{(w)}(x) < +\infty$ then by Lemma 3.3 we have $Z_{\Lambda'}^{(w)}(x) = 1$, i.e. $\int g_{\Lambda'}(x,y)\pi_{\Lambda'}(x,dy) = 1$. Therefore by Lemma 3.6 we get $Z_{\Lambda}^{(w)}(\cdot) = 1$ $\pi_{\Lambda'}(x,\cdot)$-a.e. on $D_{\Lambda'}(x)$. Thus

$$\int_{A_{\Lambda}} w_{\Lambda'}(y)\pi_{\Lambda'}(x,dy) = \int_{A_{\Lambda}} g_{\Lambda'}(x,y)\pi_{\Lambda'}(x,dy) = \int w_{\Lambda'}(y)\pi_{\Lambda'}(x,dy)$$

because $\{y : g_{\Lambda'}(x,y) > 0\} = D_{\Lambda'}(x)$ $\pi_{\Lambda'}(x,\cdot)$-a.e. □

We complete the proof of Theorem 2.1 with:

Lemma 3.8:

(3.9) holds for all $\Lambda \le \Lambda'$.

Proof:

If $G(y_1,y_2) > 0$ then from (3.10) and the definition of Δ_{Λ} we get

$g_\Lambda(x,y_1) = G(y_1,y_2)\ g_\Lambda(x,y_2)$, and so (3.9) holds whenever $G(y_1,y_2) > 0$. Therefore it suffices to show $\Delta_{\Lambda'}(x,y_1)\Delta_\Lambda(x,y_2) = \Delta_{\Lambda'}(x,y_2)\Delta_\Lambda(x,y_1) = 0$ when $G(y_1,y_2) = 0$. If $\Delta_{\Lambda'}(x,y_1)\Delta_\Lambda(x,y_2) = 1$ then by (3.7) we have $\Delta_\Lambda(x,y_1 = 1$, and if both $\Delta_\Lambda(x,y_1)$ and $\Delta_\Lambda(x,y_2)$ are 1 then $G(y_1,y_2) > 0$. □

4. Examples: A General Lattice Model

We now apply the construction given in Section 3 to same examples. These examples will be given in the framework of a "lattice" model, but the set-up will be sufficiently general to also include many continuous models.

Let X be a set and S a countable set; suppose for each $A \subset S$ that we have a σ-algebra $F(A)$ of subsets of X with

$$F(A) \subset F(B) \quad \text{whenever } A \subset B. \tag{4.1}$$

Put $F = F(S)$, and let I denote the non-empty finite subsets of S, ordered by inclusion. The basic collection of σ-algebras $\mathbb{F} = \{F_\Lambda\}_{\Lambda\in I}$ is defined by $F_\Lambda = F(S - \Lambda)$.

Suppose we have a physical system with phase space (or configuration space) (X,F). The interaction for such a system will give us a collection of functions $\Phi = \{\Phi\ \}_{\Lambda\in I}$. where $\Phi_\Lambda : X \to \mathbb{R}$ is $F(\Lambda)$-measurable. For each $\Lambda \in I$ let $U_\Lambda = \sum_{A\subset\Lambda}$; U_Λ is the potenial energy for the region Λ and the interaction Φ.

Suppose we also have an $\mathbb{F}$-specification $V = \{\pi_\Lambda\}_{\Lambda\in I}$ which corresponds to the same system as above but with no interaction. Then there is a fairly standard way of defining the $\mathbb{F}$-specification V_Φ for the system with the interaction Φ. V_Φ is given in terms of V and the conditional energy corresponding to Φ. We now define V_Φ and then show that it can also be obtained using the construction given in Section 3 by defining G in terms of the functions $\{f_\Lambda\}_{\Lambda\in I}$ (as in Section 1) with $f_\Lambda = \exp(-U_\Lambda)$.

The conditional energy for the region Λ and the interaction should be given by $E_\Lambda = \sum_{\substack{A\in I\\ A\cap\Lambda\neq\emptyset}} \Phi_A$. However, since this does not always make sense we must be careful about the domain on which E_Λ is defined. For $t \in S$ let $M(t) = \{x : \sum_{\substack{A \in I \\ t \in A}} |\Phi_A(x)| > \infty\}$, and put $M = \bigcap_{t \in S} M(t)$. If $x \in M$ and $\Lambda \in I$ then

$$\sum_{\substack{A \in I \\ A \cap \Lambda \neq \emptyset}} |\Phi_A(x)| = \sum_{t \in \Lambda} \sum_{\substack{A \in I \\ t \in A}} \frac{1}{|A \cap \Lambda|} |\Phi_A(x)| \leq \sum_{t \in \Lambda} \sum_{\substack{A \in I \\ t \in A}} |\Phi_A(x)| < +\infty,$$

and therefore we can define $E_\Lambda : M \to \mathbb{R}$ by $E_\Lambda(x) = \sum_{\substack{A \in I \\ A \cap \Lambda \neq \emptyset}} \Phi_A(x)$.

$\Lambda \in I$ define $v_\Lambda : X \to \mathbb{R}^+$ by

$$v_\Lambda(x) = \begin{cases} \exp(-E_\Lambda(x)) & \text{if } x \in M, \\ 0 & \text{otherwise.} \end{cases}$$

<u>Proposition 4.1:</u>

The collection $v = \{v_\Lambda\}_{\Lambda \in I}$ satisfies (2.2) and (2.3); thus defining $\pi_\Lambda^\Phi : X \times F \to \mathbb{R}^+$ by

$$\pi_\Lambda^\Phi(x,F) = \begin{cases} [Z_\Lambda^\Phi(x)]^{-1} \int_{F \cap M} \exp(-E_\Lambda(y))\, \pi_\Lambda(x,dy) & \text{if } 0 < Z_\Lambda^\Phi(x) < +\infty, \\ 0 & \text{otherwise,} \end{cases}$$

with $Z_\Lambda^\Phi(x) = \int_M \exp(-E_\Lambda(y))\, \pi_\Lambda(x,dy)$, gives us an $\mathbb{F}$-specification $V_\Phi = \{\pi_\Lambda^\Phi\}_{\Lambda \in I}$.

<u>Proof:</u>

For $\Lambda \subset \Lambda' \in I$ define $v_{\Lambda',\Lambda} : X \to \mathbb{R}^+$ by

$$v_{\Lambda',\Lambda}(x) = \begin{cases} \exp(-\sum_{\substack{A \cap \Lambda' \neq \emptyset \\ A \cap \Lambda = \emptyset}} \Phi_A(x)) & \text{if } \sum_{\substack{A \cap \Lambda' \neq \emptyset \\ A \cap \Lambda = \emptyset}} |\Phi_A(x)| < +\infty, \\ 0 & \text{otherwise.} \end{cases}$$

Then $v_{\Lambda',\Lambda}$ is F_Λ-measurable and a simple computation showhs that $v_{\Lambda'} = v_{\Lambda',\Lambda}\, v_\Lambda$. From this it easily follows that (2.2) and (2.3) hold. □

We now show how to obtain V_Φ using the construction of the previous sections. Choose an increasing sequence $\{\Lambda_n\}_{n\geq 1}$ with $S = \bigcap_{n\geq 1} \Lambda_n$ and define $G : X \times X \to \mathbb{R}^+$ by

$$G(x,y) = \begin{cases} \lim_{n\to\infty} \exp[U_{\Lambda_n}(y) - U_{\Lambda_n}(x)] & \text{if } x,y \in M \text{ and the limit exists,} \\ 0 & \text{otherwise.} \end{cases}$$

Then G is $F \times F$-measurable and satisfies (3.1) and (3.2).

Proposition 4.2: G satisfies the hypotheses of Theorem 3.2.

Proof: For each $\Lambda \in I$ define $G_\Lambda : X \times X \to \mathbb{R}^+$ by

$$G_\Lambda(x,y) = \begin{cases} \lim_{n\to\infty} \exp\left[\sum_{\substack{A\subset\Lambda_n \\ A\cap\Lambda=\emptyset}} \Phi_A(y) - \sum_{\substack{A\subset\Lambda_n \\ A\cap\Lambda=\emptyset}} \Phi_A(x)\right] & \text{if the limit exists,} \\ 0 & \text{otherwise.} \end{cases}$$

Then G_Λ is $F_\Lambda \times F_\Lambda$-measurable and it is easily checked that we have $v_\Lambda(y)\ G(x,y) = v_\Lambda(x)\ G_\Lambda(x,y)$ for all $x,y \in M$. In particular this gives us $\{(x,y) : G(x,y) > 0\} = (M \times M) \cap C_\Lambda$, where $C_\Lambda = \{(x,y)\ G\Lambda(x,y) > 0\}$, and of course $C_\Lambda \in F_\Lambda \times F_\Lambda$. □

From Proposition 4.2 we get an $\mathbb{F}$-specification $V_G = \{\pi_\Lambda^G\}_{\Lambda\in I}$, where V_G is defined as in Theorem 3.2.

Proposition 4.3: $V_G = V_\Phi$.

Proof:

Let G_Λ be as in the proof of Proposition 4.2. Then we have $v_\Lambda(y)\ G(z,y) = v_\Lambda(z)\ G_\Lambda(z,y)$ for all $z,y \in M$, and thus for all $z \in X$, $y \in M$. Hence

$$v_\Lambda(y) \int G(z,y)\ \pi_\Lambda(x,dz) = \int v_\Lambda(z)\ G_\Lambda(z,y)\ \pi_\Lambda(x,dz)$$

$$= G_\Lambda(x,y) \int v_\Lambda(z)\ \pi_\Lambda(x,dz) \quad \text{for all } x \in X,\ y \in M.$$

Therefore $[\int G(z,y)\ \pi_\Lambda(x,dz)]^{-1} = v_\Lambda(y)\ [G_\Lambda(x,y)\int v_\Lambda(z)\ \pi_\Lambda(x,dz)]^{-1}$ for all $x \in X$, $y \in M$. But $B = \{x : G(x,x) > 0\} = M$ and $G_\Lambda(y,y) = 1$ for all $y \in X$, thus we get

$$\pi_\Lambda^G(x,F) = \int_{F \cap B} [\int G(z,y)\ \pi_\Lambda(y,dz)]^{-1}\ \pi_\Lambda(x,dy)$$

$$= \int_{F \cap M} v_\Lambda(y)\ [\int v_\Lambda(z)\ \pi_\Lambda(y,dz)]^{-1}\ \pi_\Lambda(x,dy) = \pi_\Lambda^\Phi(x,F). \qquad \square$$

5. Invariant Specifications

In many examples we will have a group of measurable transformations acting on the basic space $(X,\mathcal{F})$, and we are interested in measures and functions which are invariant under the action of this group.

Let $\Psi : X \to X$ be a bijection with both Ψ and Ψ^{-1} $\mathcal{F}$-measurable. If $\mathcal{B}$ is a sub-σ-algebra of $\mathcal{F}$ then let $\Psi(\mathcal{B}) = \{B \in \mathcal{F} : \Psi^{-1}(B) \in \mathcal{B}\}$; if $\mu \in P(X,\mathcal{F})$ then $\Psi\mu \in P(X,\mathcal{F})$ is given by $(\Psi\mu)(F) = \mu(\Psi^{-1}(F))$. For a kernel $\pi : X \times \mathcal{F} \to \mathbb{R}^+$ we define the transformed kernel $\Psi\pi : X \times \mathcal{F} \to \mathbb{R}^+$ by $\Psi\pi(x,F) = \pi(\Psi^{-1}(x), \Psi^{-1}(F))$.

Let $V = \{\pi_\Lambda\}_{\Lambda \in I}$ be an $\mathbb{F}$-specification; we say that V is <u>Ψ-invariant</u> if for each $\Lambda \in I$ there exists $\Lambda' \in I$ with $\Psi(\mathcal{F}_{\Lambda'}) \subset \mathcal{F}_\Lambda$ and $(\Psi\pi_{\Lambda'})\pi_\Lambda = \Psi\pi_{\Lambda'}$. The justification for this definition is that if V is Ψ-invariant and $\mu \in G(V)$ then we also have $\Psi\mu \in G(V)$ (see Preston (1976), Section 4).

In this section we consider the question: If V is Ψ-invariant then when does the construction of Section 2 and 3 (in terms of V) also give specifications which are Ψ-invariant? For simplicity we look at the situation of Theorem 3.2. Let $G : X \times X \to \mathbb{R}^+$ be $\mathcal{F}\times\mathcal{F}$-measurable and satisfy (3.1) and (3.2); let $D = \{(x,y) : G(x,y) > 0\}$ and $B = \{x : G(x,x) > 0\}$. We thus suppose there exists $M \in \mathcal{F}$ and for each $\Lambda \in I$ $C_\Lambda \in \mathcal{F}_\Lambda \times \mathcal{F}_\Lambda$ such that $D = (M \times M) \cap C_\Lambda$. Theorem 3.2 gives us that

$\int_B [\int G(z,y)\ \pi_\Lambda(y,dz)]^{-1}\ \pi_\Lambda(x,dy)$ is either 0 or 1 for each $\Lambda \in I$, $x \in X$, and that we define $\pi_\Lambda^G : X \times F \to \mathbb{R}^+$ by

$$\pi_\Lambda^G(x,F) = \int_{F\cap B} [\int G(z,y)\ \pi_\Lambda(y,dz)]^{-1}\ \pi_\Lambda(x,dy)$$

then $V_G = \{\pi_\Lambda^G\}_{\Lambda\in I}$ is an $\mathbb{F}$-specification.

Theorem 5.1: Suppose $V = \{\pi_\Lambda\}_{\Lambda\in I}$ is Ψ-invariant, and

$$G(\Psi(x),\ \Psi(y)) = G(x,y) \quad \text{for all } x,y \in X. \tag{5.1}$$

Then V_G is Ψ-invariant.

Proof:

First note that if π and π' are kernels then having $(\Psi\pi')\pi = \Psi\pi'$ is the same as having $\pi'(\Psi^{-1}\pi) = \pi'$. Let $\Lambda \in I$, we thus need to find $\Lambda' \in I$ with $\Psi(F_{\Lambda'}) \subset F_\Lambda$ and $\pi_{\Lambda'}^G(\Psi^{-1}\ \pi_\Lambda^G) = \pi_{\Lambda'}^G$. Since V is Ψ-invariant we can find $\Lambda' \in I$ with $\Psi(F_{\Lambda'}) \subset F_\Lambda$ and $\pi_{\Lambda'}(\Psi^{-1}\pi_\Lambda) = \pi_{\Lambda'}$. Applying Theorem 3.2 to the kernels $\pi_{\Lambda'}$ and $\Psi^{-1}\pi_\Lambda$ we get $\pi_\Lambda^G(\Psi^{-1}\pi_\Lambda)^G = \pi_\Lambda^G$, where

$$(\Psi^{-1}\pi_\Lambda)^G(x,F) = \int_{F\cap B}\left[\int G(z,y)\Psi^{-1}\ \pi_\Lambda(x,dz)\right]^{-1}\Psi^{-1}\ \pi_\Lambda(x,dy).$$

It thus suffices to show that $\Psi^{-1}\pi_\Lambda^G = (\Psi^{-1}\pi_\Lambda)^G$. But

$$\Psi^{-1}\pi_\Lambda^G(x,F) = \pi_\Lambda^G(\Psi(x),\ \Psi(F))$$

$$= \int_{\Psi(F)\cap B} [\int G(z,y)\ \pi_\Lambda(\Psi(x),dz)]^{-1}\ \pi_\Lambda(\Psi(x),dy)$$

$$= \int_{\Psi(F\cap B)} [\int G(\Psi(z),y)\Psi^{-1}\ \pi_\Lambda(x,dz)]^{-1}\ \pi_\Lambda(\Psi(x),dy)$$

$$= \int_{F\cap B} [\int G(\Psi(z),\Psi(y))\Psi^{-1}\ \pi_\Lambda(x,dz)]^{-1}\ \Psi^{-1}\pi_\Lambda(x,dy)$$

$$= \int_{F\cap B} [\int G(z,y)\Psi^{-1}\ \pi_\Lambda(x,dz)]^{-1}\ \Psi^{-1}\pi_\Lambda(x,dy) = (\Psi^{-1}\pi_\Lambda)^G(x,F) \square$$

Let H be a group of measurable transformations acting on (X,F), thus each $\Psi \in H$ is a bijection with both Ψ and Ψ^{-1} $\mathcal{F}$-measurable. $\mathcal{V}$ is called <u>H-invariant</u> if it is Ψ-invariant for all $\Psi \in H$. Theorem 5.1 thus gives us that if $\mathcal{V}$ is H-invariant then $\mathcal{V}_G$ is also H-invariant provided (5.1) holds for all $\Psi \in H$.

Let us look again at the lattice model of Section 4, but now with S being a countable group, (usually we will have $S = \mathbb{Z}^d$). The group operation will be written additively, but we do not assume S is abelian. We suppose we have a group of transformations $\{\Psi_t\}_{t\in S}$ acting on $(X,\mathcal{F})$ such that

$$\mathcal{F}(A) = \Psi_t^{-1}(\mathcal{F}(A+t)) \quad \text{for all } A \subset S,\ t \in S, \tag{5.2}$$

where $A + t = \{s + t : s \in A\}$. Let $\mathcal{V} = \{\pi_\Lambda\}_{\Lambda\in I}$ be an $\mathbb{F}$-specification such that

$$\Psi_t^{-1}\pi_{\Lambda+t} = \pi_\Lambda \qquad \text{for all } \Lambda \in I,\ t \in S; \tag{5.3}$$

then $\mathcal{V}$ is clearly S-invariant. We also suppose that the interaction $\Phi = \{\Phi_\Lambda\}_{\Lambda\in I}$ is S-invariant in that

$$\Phi_{\Lambda+t}(\Psi_t(x)) = \Phi_\Lambda(x) \quad \text{for all } \Lambda \in I,\ t \in S,\ x \in X. \tag{5.4}$$

Clearly we have $\Psi_t^{-1}(M) = M$ for all $t \in S$, i.e. M is S-invariant (where M is as in Section 4). Also it is easily checked that $v_{\Lambda+t}(\Psi_t(x)) = v_\Lambda(x)$ for all $\Lambda \in I$, $t \in S$, $x \in X$, and from this it follows that $\Psi_t^{-1}\pi^\Phi_{\Lambda+t} = \pi^\Phi_\Lambda$ for all $\Lambda \in I$, $t \in S$; so $\mathcal{V}_\Phi$ is S-invariant.

Let G be defined as in Section 4, and let $G_0 : X \times X \to \mathbb{R}^+$ be given by

$$G_0(x,y) = \begin{cases} G(x,y) & \text{if } G(x,y) = G(\Psi_t(x),\Psi_t(y)) \text{ for all } t \in S, \\ 0 & \text{otherwise.} \end{cases}$$

Clearly G_0 is $\mathcal{F} \times \mathcal{F}$-measurable and satisfies (3.1), (3.2) and also $G_0(\Psi_t(x),\Psi_t(y)) = G_0(x,y)$ for all $x,y \in X$, $t \in S$. Let G_Λ be as in Section 4; for $x,y \in M$ we have

$$G(\Psi_t(x),\Psi_t(y)) = \frac{v_{\Lambda+t}(\Psi_t(x))}{v_{\Lambda+t}(\Psi_t(y))} G_{\Lambda+t}(\Psi_t(x),\Psi_t(y))$$

$$= \frac{v_\Lambda(x)}{v_\Lambda(y)} G_{\Lambda+t}(\Psi_t(x),\Psi_t(y)),$$

and thus $(x,y) : G_O(x,y) > 0\} = (M \times M) \cap C_\Lambda$, where $C_\Lambda = \{(x,y) : G_\Lambda(x,y) > 0$ and $G_\Lambda(x,y) = G^t_\Lambda(x,y)$ for all $t \in S$, with $G^t_\Lambda(x,y) = G_{\Lambda+t}(\Psi_t(x),\Psi_t(y))$. But G^t_Λ is $F_\Lambda \times F_\Lambda$-measurable for each $t \in S$, and thus $C_\Lambda \in F_\Lambda \times F_\Lambda$. Therefore, as at the beginning of this section, we can define an $\mathbb{F}$-specification $V_{G_O} = \{\pi_\Lambda^{G_O}\}_{\Lambda \in I}$ and Theorem 5.1 gives us that V_{G_O} is S-invariant. (In fact the proof of Theorem 5.1 shows that (5.3) holds for V_{G_O}). Finally, exactly as in the proof of Proposition 4.3, we get $V_{G_O} = V_\Phi$.

References

Dobrushin, R.L. (1968a). Description of a random field by means of conditional probabilities and the conditions governing its regularity. Theory of Prob. and its Appl., 13, 197-224.

Dobrushin, R.L. (1968b). Gibbsian random fields for lattice systems with pairwise interactions. Funct. Anal. and its Appl., 2, 292-301.

Dobrushin, R.L. (1968 c). The problem of uniqueness of a Gibbsian random field and the problem of phase transition. Funct. Anal. and its Appl., 2, 302-312.

Dobrushin, R.L. (1969). Gibbsian random fields. The general case. Funct. Anal. and its Appl. , 3, 22-28.

Föllmer, H. (1975 a). Phase transition and Martin boundary. Sém. de Probabilités, IX, Springer Lecture Notes in Math., Vol. 465.

Föllmer, H. (1975 b). On the potential theory of stochastic fields. Invited address: 40th Session ISI, Warsaw, Poland.

Goldstein, S. (1978). A note on specifications. Preprint.

Lanford, O.E. and Ruelle, D. (1969). Observables at infinity and states with short range correlations in statistical mechanics. Comm. Math. Phys., 13, 194-215.

Preston, C.J. (1975). Random fields. Springer Lecture Notes in Math., Vol. 534.

Preston, C.J. (1979). Canonical and microcanonical Gibbs states. Zeitschrift für Wahrscheinlichkeitstheorie, 46, 125-158.

Ruelle, D. (1970). Superstable interactions in statistical mechanics. Comm. Math. Phys., 18, 127-159.

On the Global Markov Property

Hans Föllmer

Mathematikdepartement, ETH-Zentrum
Zürich, Switzerland

Abstract

In a lattice model with nearest neighbor interaction, the Gibbs measures P have the local but not necessarily the global Markov property. We show that the global Markov property does hold in the following two cases: (1) The interaction satisfies Dobrushin's uniqueness condition, or (2) the interaction is attractive, and P is a "high density state".

1. Introduction

Consider a lattice model with countable parameter set T, state space E, configuration space $\Omega = E^T$ and a nearest neighbor interaction with respect to some graph structure on T. Any corresponding Gibbs measure P on Ω has the local Markov property, i.e., for each finite subset V of T we have

$$P[A|\underline{\underline{F}}_{T-V}] = P[A|\underline{\underline{F}}_{\partial V}] \qquad (A \in \underline{\underline{F}}_V), \tag{1.1}$$

where ∂V is the set of neighbors of V, $\underline{\underline{F}}_S$ is the σ-field of events depending on the situation on $S \subseteq T$ and $P[\cdot|\underline{\underline{F}}_S]$ denotes the conditional probability with respect to P and $\underline{\underline{F}}_S$. Let us say that P has the global Markov property if (1.1) holds for arbitrary subsets V of T.

This note contains two remarks concerning the passage from the local to the global Markov property. They are formulated for lattice systems, but see (4.11) for extensions. The first is that Dobrushin's uniqueness condition on the interaction implies the global Markov property for the corresponding Gibbs measure. A very similar argument appears in the work of Albeverio and Høegh-Krohn [1]. Now suppose that the interaction is attractive and admits a "high density state" P^+ resp. a "low density state" P^-. The second remark is that P^+ resp. P^- has the global Markov property, and this settles the question if there is no phase transition.

Both remarks came up during a visit at Cornell, and thanks are due to E.B. Dynkin, H. Kesten and F. Spitzer for a stimulating discussion.

2. Markov Specification

Let T be a countable set of sites, $(E,\underline{\underline{E}})$ a standard Borel space and $\Omega = E^T$ the set of all configurations $\omega: T \to E$. For $S \subseteq T$ we denote by $\underline{\underline{F}}_S$ the σ-field on Ω generated by the maps $\omega \to \omega(s)$ $(s \in S)$, and by $\underline{\underline{F}}_S^b$ the class of bounded $\underline{\underline{F}}_S$-measurable functions on Ω.

Let $\underline{\underline{V}}$ be the class of finite subsets V of T, and let $\pi = (\pi_V)_{V \in \underline{\underline{V}}}$ be a local specification with respect to the σ-fields $(\underline{\underline{F}}_{T-V})_{V \in \underline{\underline{V}}}$. This means that for each $V \in \underline{\underline{V}}$ we have a kernel $\pi_V(\omega, A)$ $(\omega \in \Omega, A \in \underline{\underline{F}}_T)$ such that

$$\pi_V(\omega,.) \text{ is a probability measure on } \underline{\underline{F}}_T \text{ which coincides with } \delta_\omega \text{ on } \underline{\underline{F}}_{T-V} \quad (\omega \in \Omega), \tag{2.1}$$

$$\pi_V(.,A) \text{ is } \underline{\underline{F}}_{T-V}\text{-measurable} \quad (A \in \underline{\underline{F}}_T), \tag{2.2}$$

and these kernels are consistent in the sense that

$$\pi_W \pi_V = \pi_W \quad (V \subseteq W \in \underline{\underline{V}}). \tag{2.3}$$

Let $G(\pi)$ be the class of Gibbs states with respect to π, i.e., all probability measures P on $(\Omega, \underline{\underline{F}}_T)$ such that

$$P[A|\underline{\underline{F}}_{T-V}](.) = \pi_V(.,A) \quad P\text{-a.s.} \quad (A \in \underline{\underline{F}}_T, V \in \underline{\underline{V}}). \tag{2.4}$$

This is equivalent to

$$P\pi_V = P \quad (V \in \underline{\underline{V}}) \tag{2.5}$$

where $P\pi_V$ denotes the measure $\int P(d\omega)\, \pi_V(\omega,.)$; see [7]. Throughout this paper we assume $G(\pi) \neq \emptyset$.

Now suppose that T carries a locally finite graph structure, i.e., for each $S \subseteq T$ there is a set $\partial S \subseteq T-S$ of neighbors of S, with $\partial V \in \underline{\underline{V}}$ for each $V \in \underline{\underline{V}}$. Let us assume that π has the following Markov property:

$$V \in \underline{\underline{V}}, \quad A \in \underline{\underline{F}}_V \Rightarrow \pi_V(.,A) \quad \text{is } \underline{\underline{F}}_{\partial V}\text{-measurable}. \tag{2.6}$$

Then each $P \in G(\pi)$ has the local Markov property in the sense that

$$V \in \underline{\underline{V}}, \quad A \in \underline{\underline{F}}_V \Rightarrow P[A|\underline{\underline{F}}_{T-V}] = P[A|\underline{\underline{F}}_{\partial V}]. \tag{2.7}$$

We say that P has the global Markov property if in (2.7) we can replace $V \in \underline{\underline{V}}$ by an arbitrary subset S of T. The problem now is to find conditions on π resp. on $P \in G(\pi)$ which guarantee the global Markov property.

Remark (2.8)

The situation is completely understood in the one-dimensional case $T = Z^1$ with countable state space E. In this case a measure $P \in G(\pi)$ has the global Markov property if and only if it is a Markov chain. If $G(\pi) \neq \emptyset$ then there exist extreme points in the convex set $G(\pi)$, and any such extreme point is a Markov chain. Fixing such a reference chain P with its transition laws $P_n(x,y)$ $(n \in Z,\ x,y \in E)$, and assuming $P_n(.,.) > 0$ for simplicity, one can obtain an explicit integral representation of $G(\pi)$ in terms of the exit boundary B and the entrance boundary B* associated to (P_n). The Markov chains in $G(\pi)$ can then be characterized by a "splitting property" of their representing measures on $B \times B^*$. See [6,8] for the details, and also [2] for an extension to the continuous case.

3. Strong Uniqueness implies the Global Markov Property

For any $S \subseteq T$ and any $\omega \in \Omega$ let us define the conditional specification $\pi^S_\omega = (\pi^S_{\omega,V})_{V \in \underline{\underline{V}}, V \subseteq S}$ with

$$\pi^S_{\omega,V}(\eta,.) = \pi_V(\eta^S_\omega,.) \tag{3.1}$$

where η^S_ω is the configuration which coincides with ω on T-S and with η on S. π^S_ω is a local specification with respect to the σ-fields $\underline{\underline{F}}_{S-V}$ $(V \in \underline{\underline{V}},\ V \subseteq S)$. Now take $P \in G(\pi)$. Since $(\Omega, \underline{\underline{F}}_T)$ is standard Borel, we can choose conditional probability distributions

$$P^S_\omega = P[.|\underline{\underline{F}}_{T-S}](\omega), \qquad P^{\partial S}_\omega = P[.|\underline{\underline{F}}_{\partial S}](\omega).$$

For $f \in \underline{\underline{F}}^b_{T-S}$, $g \in \underline{\underline{F}}^b_S$ and finite $V \subseteq S$ we have

$$E[f(\omega)\int g dP_\omega^S] = E[fg] = E[f\pi_V g] = E[f(\omega)\int \pi_V g dP_\omega^S]$$

where $\pi_V g$ denotes the function $\pi_V g(\omega) = \int \pi_V(\omega, d\eta) g(\eta)$ $(\omega \in \Omega)$. This implies, via (2.4),

$$P_\omega^S \in C(\pi_\omega^S) \quad \text{for P-almost all } \omega \in \Omega. \tag{3.2}$$

The analogous argument for $P_\omega^{\partial S}$ and functions $f \in \underline{\underline{F}}^b_{\partial S}$ yields

$$P_\omega^{\partial S} \in G(\pi_\omega^S) \quad \text{for P-almost all } \omega \in \Omega. \tag{3.3}$$

But (3.2) and (3.3) imply $P_\omega^{\partial S} = P_\omega^S$ P-a.s. - and this is the Markov property (2.7) with respect to S - as soon as we have <u>strong uniqueness</u> in the sense that

$$|G(\pi_\omega^S)| \leq 1 \quad (\omega \in \Omega). \tag{3.4}$$

Let us now recall Dobrushin's uniqueness condition. Define

$$a_{st} = \sup \frac{1}{2} \| \pi_{\{t\}}(\omega,.) - \pi_{\{t\}}(\eta,.) \| \quad (s,t \in T)$$

where the sup is taken over all configurations ω,η which coincide everywhere except in s ($\|\cdot\|$ = total variation). The Dobrushin condition

$$\sup_t \sum_s a_{st} < 1 \tag{D}$$

implies $|G(\pi)| = 1$; see [3]. But it is easy to check that it also implies the corresponding Dobrushin condition (D_ω^S) for each conditional specification π_ω^S $(S \subseteq T,\ \omega \in \Omega)$, hence (3.4). We have thus shown:

<u>Theorem</u> (3.5)

Under (D) the unique Gibbs state $P \in G(\pi)$ has the global Markov property.

<u>Remark</u> (3.6)

The same argument holds if we define a_{st} in terms of a Wasserstein distance and replace (D) by conditions (5.2), (5.4), (5.5) of [4].

4. Attractive Interactions

Suppose we have a measurable partial order on E. We denote by $\leq$ the induced product order on Ω, and by $\underline{\underline{M}}$ the class of bounded measurable functions on Ω which are increasing with respect to $\leq$. Let us assume that there is an increasing sequence (x_n) in E such that each $x \in E$ is dominated by some x_n, and that $\underline{\underline{M}}$ determines probability measures on $(\Omega, \underline{\underline{F}}_T)$ in the sense that for each $V \in \underline{\underline{V}}$

$$\int f dP = \int f dQ \quad (f \in \underline{\underline{M}} \cap \underline{\underline{F}}_V^b) \Rightarrow P = Q \text{ on } \underline{\underline{F}}_V. \tag{4.1}$$

Now suppose that π is attractive, i.e.,

$$f \in \underline{\underline{M}} \Rightarrow \pi_V f \in \underline{\underline{M}} \quad (V \in \underline{\underline{V}}). \tag{4.2}$$

Under mild compactness assumptions on π (see [7] pp. 163, 182) there is a "high density state", i.e., a unique extreme point P^+ of $G(\pi)$ such that

$$E^+[h] = \inf_{V \in \underline{\underline{V}}} \sup_{\eta} \pi_V h(\eta) \quad (h \in \underline{\underline{H}})$$

with $\underline{\underline{H}} = \underline{\underline{M}} \cap \bigcup_{V \in \underline{\underline{V}}} \underline{\underline{F}}_V^b$.

Example (4.3)

If E has a maximum element x^+ and π is defined in terms of a ferromagnetic nearest neighbor potential, then π is attractive, and the high density state is given by $E^+[f] = \lim_{V \uparrow T} \pi_V f(\eta^+)$ $(f \in \underline{\underline{M}})$ with $\eta^+(t) = x^+$ for each $t \in T$.

Theorem (4.4)

P^+ has the global Markov property.

Proof

Let S be an arbitrary subset of T. Take $V_o \in \underline{\underline{V}}$ and $g \in \underline{\underline{M}} \cap \underline{\underline{F}}^b_{V_o \cap S}$. It is enough to show that the conditional expectation of g with respect to P^+ and $\underline{\underline{F}}_{T-S}$ is given by $g_\infty = \inf_{V \in \underline{\underline{V}}} g_V$ where

$$\begin{aligned} g_V(\omega) &= \sup_{\eta} \pi_{V\cap S}\, g(\eta^S_\omega) \\ &= \lim_m \pi_{V\cap S}\, g((\eta_m)^S_\omega) \end{aligned} \tag{4.5}$$

(define $\eta_m(t) = x_m$ $(t \in T)$ and recall the notation of (3.1)) is $\underline{\underline{F}}_{\partial S \cap V}$-measurable and decreasing in V due to (4.1).

1) For each $V \in \underline{\underline{V}}$ and each $\eta \in \Omega$ we have

$$E_V^\eta[g|\underline{\underline{F}}_{T-S}](\omega) = \pi_{V\cap S}\, g(\eta_\omega^S), \tag{4.6}$$

where E_V^η denotes expectation with respect to the measure $\pi_V(\eta,.)$. In fact:

$$E_V^\eta[fg] = \int \pi_V(\eta,d\omega)\ \pi_{V\cap S}(fg)(\omega)$$

$$= \int \pi_V(\eta,d\omega) f(\omega)\ \pi_{V\cap S} g(\omega)$$

$$= E_V^\eta[f(\omega)\ \pi_{V\cap S} g(\eta_\omega^S)] \qquad (f \in \underline{\underline{F}}_{T-S}^b)$$

due to (2.1) and (2.3).

2) We want to show

$$E^+[fg] = E^+[fg_\infty] \tag{4.7}$$

for any $f \in \underline{\underline{F}}_{T-S}^b$, and we may assume $f \in \underline{\underline{M}} \cap \underline{\underline{F}}_W^b$ for some finite subset W of T-S due to (4.1). Fix a sequence $(V_n)_{n=0,1,\ldots} \subseteq \underline{\underline{V}}$ increasing to T and write $g_0 = g$, $g_n = g_{V_n}$ $(n=1,2,\ldots)$. Since P^+ is an extreme point of $G(\pi)$, it is 0-1 on the tail field (see [7] p. 24), and by backward martingale convergence we have, for P^+-almost all η,

$$E^+[h] = \lim \pi_{V_n}\ h(\eta) \tag{4.8}$$

for each $h \in \underline{\underline{H}}_0 = \{fg_n | n=0,1,\ldots,\infty\}$. We choose such an η and call it $\bar\eta$ (in the situation of (4.3) it would simplify the proof to take $\bar\eta = \eta^+$). Then

$$E^+[fg] = \lim\ \pi_{V_n}(fg)(\bar\eta)$$

$$= \lim\ E_{V_n}^{\bar\eta}\ [f(\omega)\ \pi_{V_n\cap S} g(\bar\eta_\omega^S)]$$

due to (4.6). But $\pi_{V_n\cap S} g(\eta_\omega^S) \le g_n(\omega) \le g_m(\omega)$ $(n \ge m)$ so that

$$E^+[fg] \le \lim \pi_{V_n}(fg_m)(\bar{\eta}) = E^+[fg_m] \quad (m=1,2,\dots),$$

and since g_m decreases to g_∞ we obtain

$$E^+[fg] \ge E^+[fg_\infty] \tag{4.9}$$

by monotone integration.

3) In order to get the converse inequality, let us first note that for any sequence (η_n) with $\eta_n \ge \bar{\eta}$

$$E^+[h] = \lim \pi_{V_n} h(\eta_n) \quad (h \in \underline{\underline{H}}_o \cap \underline{\underline{H}}) \tag{4.10}$$

since

$$\begin{aligned} E^+[h] &= \lim \pi_{V_n} h(\bar{\eta}) \\ &\le \lim\inf \pi_{V_n} h(\eta_n) \le \lim\sup \pi_{V_n} h(\eta_n) \\ &\le \lim_n \sup_\eta \pi_{V_n} h(\eta) = E^+[h] \end{aligned}$$

due to (4.2) and (4.8). Let us now fix $\varepsilon > 0$ and take $\eta_n \ge \bar{\eta}$ such that

$$\begin{aligned} E^{\bar{\eta}}_{V_n}[f(\omega)\, \pi_{V_n \cap S} g((\eta_n)^S_\omega)] &\ge E^{\bar{\eta}}_{V_n}[f(\omega) g_n(\omega)] - \varepsilon \\ &\ge \pi_{V_n}(fg_\infty)(\bar{\eta}) - \varepsilon \end{aligned}$$

(via (4.5) and monotone integration). From (4.10) we get

$$E^+[fg] = \lim \pi_{V_n}(fg)(\eta_n)$$

with

$$\begin{aligned} \pi_{V_n}(fg)(\eta_n) &= E^{\eta_n}_{V_n}[f(\omega)\, \pi_{V_n \cap S} g((\eta_n)^S_\omega)] \\ &\ge E^{\bar{\eta}}_{V_n}[f(\omega)\, \pi_{V_n \cap S} g((\eta_n)^S_\omega)] \\ &\ge \pi_{V_n}(fg_\infty)(\bar{\eta}) - \varepsilon \end{aligned}$$

so that

$$E^{+}[fg] \geq \lim \pi_{V_n}(fg_\infty)(\bar{\eta}) - \varepsilon = E^{+}[fg_\infty] - \varepsilon.$$

This yields (4.7).

Remarks (4.11)

1) If π admits no phase transition, i.e., if $|G(\pi)| = 1$, then the unique Gibbs state in $G(\pi)$ has the global Markov property. For an attractive interaction we have thus strengthened (3.5).

2) The proof shows that for $S \subseteq T$ the high density states P_ω^{+} for the conditional specifications π_ω^{S} ($\omega \in \Omega$) form a version of the conditional probability distribution $P^{+}[\cdot|\underline{\underline{F}}_{T-S}]$. This argument does not involve the Markov property. But if π is Markov, then the map $\omega \to P_\omega^{+}$ is $\underline{\underline{F}}_{\partial S}$-measurable, and this yields (4.4).

3) Instead of working with a lattice system, we could have formulated the arguments in this and the previous section (up to 3.4) for a general Markov specification as defined in [5]. But it would require some additional structure: we need an operation $(\omega,\eta) \to \eta_\omega^{S}$ which corresponds to "patching together" two configurations ω (restricted to the exterior of S) and η (restricted to the interior of S).

References

[1] Albeverio, S., Høegh-Krohn, R., Olsen, G.: The global Markov property for lattice systems, Preprint Bielefeld (Nov. 1978)

[2] Dang-Ngoc, N., Yor, M.: Champs markoviens et mesures de Gibbs sur R. Ann. Scient. Ec. Norm. Sup. 11, 29-69 (1978)

[3] Dobrushin, R.L.: The description of a random field by means of conditional probabilities and conditions for its regularity. Th.Prob.Appl. 13, 197-224 (1968)

[4] Dobrushin, R.L.: Prescribing a system of random variables by conditional distributions. Th.Prob.Appl. 15, 459-486 (1970)

[5] Föllmer, H.: Phase transition and Martin boundary. Sém. Prob. Strasbourg IX, Lecture Notes Math. 465, 305-317 (1975)

[6] Föllmer, H.: On the Potential theory of stochastic fields. Proc. 40th session ISI, Warsaw (1975)

[7] Preston, C.: Random fields. Lecture Notes Math. 534 (1976)

[8] Spitzer, F.: Phase transition in one dimensional nearest neighbor systems. J. Funct. Anal. 20, 240-254 (1975)

Uniqueness and Global Markov Property for Euclidean Fields and Lattice Systems

Sergio Albeverio and Raphael Høegh-Krohn

Fakultät für Mathematik, Universität Bielefeld
General Republic of Germany

Matematisk Institutt, Universitetet i Oslo
Norway

Abstract

We give the results and the ideas of the proof concerning the uniqueness of the Gibbs states, in the class of regular generalized random fields, for the Euclidean Sine-Gordon fields in two dimensions. We also discuss in the same way the proof of the global Markov property of these fields. Consequences and applications of these properties in the theory of random fields and quantum field theory are also shortly discussed. Finally we show how the same basic ideas give a proof of the global Markov property for lattice systems.

1. Introduction

Markov processes occupy a central position in the theory of stochastic processes. Their uses in connection with physical problems are also manifold (diffusion problems, heat equation, quantum mechanics, e.g.). The interest in extending the theory of stochastic processes (where the time parameter is either a discrete or some suitable continuous subset of $\mathbb{R}$) to stochastic (i.e. random) fields (where the time parameter is either a discrete subset of $\mathbb{R}^n$ or a suitable continuous subset of $\mathbb{R}^n$) have arisen both within probability itself (e.g. a line initiated by P. Lévy, McKean and others, looking at Markov fields, and another line pursued especially by R. Cairoli, I.B. Walsh and others, looking at many parameters martingales) as well as more directly from problems which arise naturally in such domains as statistical mechanics, quantum field theory, information theory, mathematical biology. It is also expected that Markov fields might play an important role in the general theory of stochastic fields. For Markov processes the class of stationary processes is particularly important (due in particular to the large range of applications on one hand and or the other hand to the amount of techniques available for its study, from harmonic analysis to ergodic theory). Similarly stationary Markov fields are expected to play a major role, where stationarity here means invariance under the translations in $\mathbb{R}^n$ or, even more particularly important, homogeneity with respect to the whole Euclidean group in $\mathbb{R}^n$.

It appears that in $\mathbb{R}^n$ two natural concepts of Markov property arise, a local one and a global one. They are best illustrated in the case $n = 2$ (and this is anyhow the only case where we have results (in the continuous case) including non Gaussian homogeneous fields). Let C be a piecewise smooth curve in $\mathbb{R}^2$ such that $\mathbb{R}^2 - C$ has two components Ω_+ and Ω_-. Suppose that ξ is a random field such that it has meaning to restrict ξ t. C and to consider the σ-algebra generated by this restriction. Let $E(\cdot|C)$ be the conditional expectation with respect to this σ-algebra. Let $f_\pm$ be functions measurable with respect to the random variables associated with $\Omega^\pm$ respectively. If we have

$$E(f_+ f_-|C) = E(f_+ |C) E(f_- |C) \tag{1.1}$$

for any arbitrary (piecewise smooth, but possibly extending to infinity) curve C we say that the random field ξ has the global Markov property. If (1.1) holds only for C such that one of the components $\Omega_\pm$ is bounded we say that ξ has the local Markov property. If ξ is a global Markov field one may take C to run through a family of parallel straight lines. One can then see from (1.1) that ξ is then a Markov process with respect to translations in the direction orthogonal to these straight lines (this is the Markov property with respect to half-planes of Ref. [1], [2], [3]).

Examples of global Markov fields are the free Euclidean fields [1] these are the only Gaussian homogeneous global Markov fields satisfying some regularity condition). An example of (non-homogeneous, non Gaussian) global Markov fields are the random fields obtained by the construction of multiplicative functionals of the free Euclidean fields [4]. These are the so called space-time cut-off quantum field models with polynomial [5], exponential [6] or trigonometric [7] interaction. In quantum field theory (but not only in this domain, as we remarked above) the main interest is in the homogeneous (i.e. Euclidean invariant) random fields which arise by studying the thermodynamic limit (removal of the space-time cut-off) of above models. The problem of whether these homogeneous random fields (constructed in [5], [6], [7]) have the global Markov property has been open for quite some time. Physically this problem corresponds to the one of knowing whether or not the corresponding quantum fields are canonical in the sense that the whole "Schrödinger representation" (of Friedrichs, Segal and Araki, Coester and Haag) holds (see [8] - [12]).

In this lecture we shortly indicate how we proved the uniqueness and global Markov property for the homogeneous fields (i.e. Euclidean quantum fields) with a weak trigonometric interaction, and hence exhibited the first class of non Gaussian homogeneous random fields with the global Markov property. The detailed proof is contained in [13]. We also shortly mention, as a further illustration of the method, the corresponding result for the case of a spin system on $\mathbb{Z}^d$, at high temperature (the uniqueness result was proven by Dobrushin [14] in his fundamental work on this entire circle of problems; what we remarked in [15], in collaboration with G. Olsen, is that this uniqueness result is actually strong enough as to yield the global Markov property at high temperature).

2. The Free Euclidean Field

Let μ_o be the probability measure on Schwartz space of tempered distributions $S'(\mathbb{R}^2)$ which gives the free Euclidean field $\xi(x)$, $x \in \mathbb{R}^2$, as a Gaussian homogeneous generalized random field, by

$$E_o(e^{i\int\xi(x)\phi(x)dx}) = e^{-\frac{1}{2}\iint G(x-y)\phi(x)\phi(y)dxdy}, \tag{2.1}$$

where E_o means expectation with respect to μ_o, $G(x-y)$ is the kernel of the operator $(-\Delta + m_o^2)^{-1}$, $m_o > 0$ in $L^2(\mathbb{R}^2)$, and $\phi \in S(\mathbb{R}^2)$. We have that $\phi \to \langle\xi,\phi\rangle \equiv \int\xi(x)\phi(x)dx$ (distributional notation) extends by uniform continuity from $S(\mathbb{R}^2)$ to the Sobolev space $H_{-1}(\mathbb{R}^2)$ (the Hilbert space of tempered distributions η whose Fourier transform $\hat{\eta}$ are measurable functions satisfying $(\hat{\eta},\hat{\eta})_{-1} \equiv \int_{\mathbb{R}^2} \frac{|\hat{\eta}(p)|^2}{p^2+m_o^2}\, dp < \infty$, $(,)_{-1}$ being the scalar product in $H_{-1}(\mathbb{R}^2)$). For $\xi \in \text{supp}\,\mu_o$ the map $\phi \to \langle\xi,\phi\rangle$ gives in fact an isometric imbedding of $H_{-1}(\mathbb{R}^2)$ into $L^2(S'(\mathbb{R}^2), d\mu_o)$. In particular if ρ is a Borel measure on $\mathbb{R}^2$ of finite energy in the sense that

$$\int G(x-y)\, d\rho(x)d\rho(y) < \infty \tag{2.2}$$

(in which case ρ is identifed with an element of $H_{-1}(\mathbb{R}^2)$) we have that $\langle\xi,\rho\rangle \equiv \int\xi(x)d\rho(x)$ is μ_o-measurable (in fact in $L^2(d\mu_o)$).

For any Borel subset Λ of $\mathbb{R}^2$ let B_Λ be the σ-algebra generated by all the linear functions $\xi \to \int\xi(x)d\rho(x)$ with $\xi \in \text{supp}\,\mu_o$, supp $\rho \subset \Lambda$.

Let now C be a piecewise C^1-curve and let $g \in C_o(\mathbb{R}^2)$ (where $C_o(\mathbb{R}^2)$ is the space of continuous functions of compact support in $\mathbb{R}^2$). Consider the Dirichlet problem

$$\begin{cases} (-\Delta + m_o^2)\ \psi_g^C\ (x) = 0 & \text{for } x \in \mathbb{R}^2\text{-}C \\ \psi_g^C(z) = g(z) & \text{for } z \in C. \end{cases} \tag{2.3}$$

It is well known that the function $\psi_g^C(x)$ is uniquely determined by (2.3), in fact for all $x \in \mathbb{R}^2$

$$\psi_g^C(x) = \int_C P_C(x,dz)\ g(z)\ , \tag{2.4}$$

where $P_C(x,dz)$ is the Poisson kernel for the region $\mathbb{R}^2$-C. $P_C(x,\cdot)$ is for any $x \in \mathbb{R}^2 - C$ a finite measure on C which is in $H_{-1}(\mathbb{R}^2)$, in fact we have

$$\begin{aligned} &\int_C \int_C P_C(x,dz_1)\ G(z_1 - z_2)\ P_C(y,dz_2) \\ &= G(x-y) - G^C(x,y) \equiv K_C(x,y) \end{aligned} \tag{2.5}$$

where $G^C(x,y)$ is the kernel of $(-\Delta_C + m_o^2)^{-1}$, Δ_C being the Laplacian on $L^2(\mathbb{R}^2)$, with Dirichlet boundary conditions on C.

Since $P_C(x,\cdot) \in H_{-1}\ (\mathbb{R}^2)$ for any $x \notin C$ we can consider

$$\xi \to \psi_\xi^C(x) \equiv \int_C P_C(x,dz)\xi(z) \equiv \langle\xi, P_C(x,\cdot)\rangle\ , \tag{2.6}$$

for $\xi \in$ supp μ_o, which is a μ_o-measurable function (in fact an element of $L^2(d\mu_o)$), for any $x \notin C$. Moreover $\xi \to \psi_\xi{}^C(x)$ is B_C-measurable. Note that for $x \in C$ we have $\psi_\xi^C(x) = \xi(x)$, (in the sense of distributions), thus $\psi_\xi{}^C(x)$ can be interpreted as the (generalized) solution of the Dirichlet problem (2.3) with $g(z) = \xi(z)$, $\xi \in$ supp μ_o.

For $x \notin C$ we have thus in particular that $\psi_\xi{}^C(x)$ is an m_o^2-harmonic function, thus it is in particular C^∞. From (2.5) and (2.6) we see that

$$E_o(\psi_\xi{}^C(x)\ \psi_\xi{}^C(y)) = K_C(x,y). \tag{2.7}$$

In particular we get from (2.7), using the properties of the Green's functions G, G_C:

$$E_o(|\psi_\xi{}^C(x)|^2) = K_C(x,x) \le a|\ln d(x,C)|e^{-m_o d(x,C)} \tag{2.8}$$

for some constant a.

Let now $\xi^C(x)$ be the distribution defined by

$$\xi^C(x) \equiv \xi(x) - \psi_\xi{}^C(x) \tag{2.9}$$

$\xi \in \operatorname{supp} \mu_o$. From (2.7) we then have that

$$E_o(\xi^C(x)\ \ \xi^C(y)) = G_C(x,y) \tag{2.10}$$

and that

$$E_o(\xi^C(x)\ \psi_\xi{}^C(y)) = 0\ , \tag{2.11}$$

hence ξ^C and $\psi_\xi{}^C$ are μ_o-independent Gaussian fields. Because of (2.10) ξ^C is the free field with Dirichlet boundary conditions on C.

Let now C be such that $\mathbb{R}^2 - C$ has two components Ω_+ and Ω_- and let us define the distributions $\xi_\pm{}^C(x)$ by

$$\xi_\pm{}^C(x) \equiv \xi^C(x) \quad \text{if } x \in \Omega_\pm \quad \text{and } \xi_\pm{}^C(x) = 0$$

if $x \notin \Omega_\pm$. Since $G_C(x,y) = 0$ for $x \in \Omega_+$ and $y \in \Omega_-$ we have from (2.10) that $\xi_+{}^C$ and $\xi_-{}^C$ are independent Gaussian fields.

Thus the splitting

$$\xi(x) = \xi_+{}^C(x) + \xi_-{}^C(x) + \psi_\xi{}^C(x) \tag{2.12}$$

is a splitting of the Gaussian field $\xi \in \operatorname{supp} \mu_o$ into three mutually independent Gaussian fields. From the fact that $\psi_\xi{}^C$ is B_C-measurable and the definitions we have that $\xi_\pm{}^C(x)$ are $B_{\Omega_\pm}$-measurable. Thus (2.12) gives a splitting of the Euclidean free field ξ into three mutually independent Gaussian fields measurable respectively with respect to the B_{Ω_+}, B_{Ω_-} and B_C-σ-algebras. This is clearly equivalent with the Markov property (1.1) with respect to C i.e.

$$E_o(f_+ f_- \mid C) = E_o(f_+ \mid C)\, E_o(f_- \mid C)$$

for all $f_\pm \in L^1(d\mu_o)$ (or positive) which are $B_{\Omega_\pm}$-measurable, and where $E_o(\cdot|C)$ means conditional expectation with respect to the σ-algebra B_C. Since C was an arbitrary (piecewise C^1, not necessarily bounded) curve we have that the free Euclidean field ξ has the global Markov property (this fact was known since the work [1], [16]). Moreover we remark that (2.12) implies the basic property

$$E_o(f(\cdot)|C)\ (\eta) = E_o{}^C(f(\cdot + \psi_\xi{}^C)), \tag{2.13}$$

for any $f \in L^1(d\mu_o)$, where E_o^C is the expectation with respect to the Gaussian measure μ_o^C such that

$$E_o^C(e^{i \int \xi^C(x)\phi(x)dx}) = e^{-\frac{1}{2}\int G_C(x,y)\phi(x)\phi(y)\,dx\,dy} \tag{2.14}$$

i.e. the Gaussian measure of the free Euclidean field with Dirichlet boundary condition on C.

3. Interactions and their Gibbs Measures

Let $\Lambda \to U_\Lambda$ be a map from the bounded Borel subsets Λ of $\mathbb{R}^2$ into $L^1(S'(\mathbb{R}^2), d\mu_o)$ such that U_Λ is B_Λ-measurable and $\Lambda \to U_\Lambda$ is an additive and Euclidean covariant set function in the sense that

$$U_{\Lambda_1 \cup \Lambda_2} = U_{\Lambda_1} + U_{\Lambda_2} \quad \text{if } \Lambda_1 \cap \Lambda_2 = \emptyset$$

and (3.1)

$$U_\Lambda\ (\xi) = U_{T\Lambda}\ (T\xi)$$

for any Euclidean transformation T of $\mathbb{R}^2$. If moreover we have $e^{-U_\Lambda} \in L^1(d\mu_o)$ then we say that the system U_Λ, for Λ bounded

Borel in $\mathbb{R}^2$ is an <u>interaction</u>. We denote by U the system all (U_Λ, Λ). Set

$$d\mu_\Lambda = E_o(e^{-U_\Lambda}) \, e^{-U_\Lambda} \, d\mu_o. \tag{3.2}$$

It is easy to see, using the global Markov property of the free Euclidean field ξ and the formula (2.13), that the properties (3.1) imply the global Markov property for the field associated with μ_Λ. In fact let C be a piecewise C^1 curve such that $R^2 - C = \Omega_+ \cup \Omega_-$, $\Omega_+ \cap \Omega_- = \emptyset$. Then if $f_\pm \in L^1(d\mu_o)$ and $f_\pm$ are $B_{\Omega_\pm}$-measurable we have from (3.2) and (3.1), with $\Lambda_\pm \equiv \Lambda \cap \Omega_\pm$:

$$E\,(f_+ \, f_- \mid C) = E_o(e^{-U_\Lambda} \mid C)^{-1} \, E_o(e^{-U_\Lambda} \, f_+ \, f_- \mid C)$$

$$= E_o(e^{-U_{\Lambda_-}} \mid C)^{-1} \, E_o(e^{-U_{\Lambda_+}} \mid C)^{-1} \, E_o(e^{-U_{\Lambda_-}} f_- \mid C) \tag{3.3}$$

$$E_o(e^{-U_{\Lambda_+}} \, f_+ \mid C) = E_{\Lambda_+}(f_+ \mid C) \, E_{\Lambda_-}(f_- \mid C) \,,$$

where $E_\Lambda(\cdot \mid C)$ is the conditional expectation with respect to μ_Λ and B_C. Now observing that,

$$E_\Lambda(f_\pm \mid C) = E_{\Lambda_\pm}(f_\pm \mid C), \tag{3.4}$$

which follows from (2.12) and the definition of U_Λ, we get from (3.3)

$$E_\Lambda(f_+ \, f_- \mid C) = E_\Lambda(f_+ \mid C) \, E_\Lambda(f_- \mid C), \tag{3.5}$$

which is the global Markov property of the measure μ_Λ (or its associated random field).

Let now Ω be a bounded measurable subset of $\mathbb{R}^2$ with piecewise C^1 boundary $\partial\Omega$ and let f be B_Ω-measurable and in $L^1(d\mu_\Lambda)$ or $\geq o$. From (3.4) we have

$$E_\Lambda(f \mid \partial\Omega) = E_{\Lambda\cap\Omega}(f \mid \partial\Omega). \tag{3.6}$$

In particular we have, if $\Lambda \supset \Omega$:

$$E_\Lambda(f \mid \partial\Omega) = E_\Omega(f \mid \partial\Omega). \tag{3.7}$$

We shall now introduce the following

<u>Definition:</u> A measure μ on $S'(\mathbb{R}^2)$ is a <u>Gibbs measure</u> for the interaction U if it is locally absolutely continuous with respect to the free Euclidean field μ_o (in the sense that $\mu \upharpoonright \mathcal{B}_\Lambda << \mu_o \upharpoonright \mathcal{B}_\Lambda$ for any bounded Borel measurable subset Λ of $\mathbb{R}^2$) and for any bounded domain Ω with piecewise C^1 boundary $\partial\Omega$ we have

$$E(f \mid \partial\Omega) = E_\Omega(f \mid \partial\Omega), \tag{3.8}$$

for any $\mathcal{B}_\Omega$-measurable f which is nonnegative or in $L^1(d\mu)$.

One shows easily that any weak limit point of μ_Λ as $\Lambda \uparrow \mathbb{R}^2$ is a Gibbs measure. The local Markov property mentioned in the introduction may be now defined as the property for a measure μ on $S'(\mathbb{R}^2)$ to satisfy

$$E(f\ g \mid \partial\Omega) = E(f \mid \partial\Omega)\ E(g \mid \partial\Omega) \tag{3.9}$$

for arbitrary bounded domains Ω with piecewise C^1 boundary and arbitrary $\mathcal{B}_\Omega$ resp. $\mathcal{B}_{\mathbb{R}^2-\Omega}$ f resp. g which are non negative or in $L^2(d\mu)$. We also note that (3.9) is equivalent with

$$E(f \mid \mathbb{R}^2 - \Omega) = E(f \mid \partial\Omega) \tag{3.10}$$

for arbitrary bounded domains Ω with $\partial\Omega$ piecewise C^1 and arbitrary $\mathcal{B}_\Omega$-measurable f, ≥ 0 or in $L^1(d\mu)$. From (3.8) and the fact that μ_Ω is globally Markov we have from (3.5) that (3.5) holds for $\Lambda = \Omega$, $C = \partial\Omega$, $f_+ = f$, $f_- = g$, which then together with (3.8) yields (3.9), with E the expectation with respect to μ. Hence we have proven the following

<u>Lemma 3.1:</u>

Any Gibbs measure has the local Markov property.

<u>Remark:</u> Results about the local Markov property of measures associated with Euclidean quantum fields are contained in [17],

[18].

We now define a pure Gibbs measure μ as a Gibbs measure such that if $\nu << \mu$ and ν is a Gibbs measure then $\nu = \mu$ (i.e. μ is such that there exists no other Gibbs measure absolutely continuous with respect to it). We observe also that if one can show that the family of all Gibbs measures to U consists of pure Gibbs measures then we have <u>uniqueness</u> i.e. there is only a Gibbs measure to U (and this measure is pure) (because if $\mu_1 \neq \mu_2$ were pure Gibbs measures then $\mu = \lambda\mu_1 + (1-\lambda)\mu_2$ ought to be also a pure Gibbs measure, for any $0 \leq \lambda \leq 1$, but this is impossible, because we would have $\mu_1 << \frac{1}{\lambda}\mu$, $\mu_2 \leq \frac{1}{1-\lambda}\,\mu$ and from the purity of μ then $\mu_1 = \frac{1}{\lambda}\,\mu, \mu_2 = \frac{1}{1-\lambda}\mu$, a contradiction against $\mu_1 \neq \mu_2$ for $\lambda = \frac{1}{2}$).

Let now μ be a Gibbs measure and let f be a $\mathcal{B}_\Lambda$-measurable function, for Λ a bounded Borel subset of $\mathbb{R}^2$. Suppose $f \geq 0$ or $f \in L^1(d\mu)$. From (3.10) we have that the family $\{E(f|\partial\Omega), \Omega \supset \Lambda\}$ is an inverse martingale with respect to the family of σ-algebras $\mathcal{B}_{\mathbb{R}^2-\Omega}$ (where subset of $\mathbb{R}^2$ are ordered by inclusion).

Let $\mathcal{B}_\infty \equiv \bigcap_\Omega \mathcal{B}_{\mathbb{R}^2-\Omega}$, then by the martingale convergence theorem

$$E(f|\Omega)(\eta) \rightarrow E(f|\infty)(\eta) \qquad (3.11)$$

for μ-almost all $\eta \in S'(\mathbb{R}^2)$, where

$$E(\cdot|\infty) \equiv E(\cdot\,|\mathcal{B}_\infty).$$

Let now h be a $\mathcal{B}_\infty$-measurable nonnegative function and such that $E(h) = 1$. It is easily seen that $hd\mu$ is again a Gibbs measure and any Gibbs measure absolutely continuous with respect to μ is of this form.

If $\mathcal{B}_\infty$ is trivial for μ in the sense that any $\mathcal{B}_\infty$-measurable function is constant μ-almost surely we say that μ is <u>trivial at infinity</u>. We can now summarize the above discussion around (3.11) in the following

<u>Lemma 3.2:</u>

μ is a pure Gibbs measure if and only if it is trivial at infinity. Moreover, if μ is trivial at infinity then $E(f|\partial\Omega) \to E(f)$ μ-almost surely as $\Omega \uparrow \mathbb{R}^2$, for all f which are $\mathcal{B}_{\Omega_0}$-measurable, where Ω_0 is any fixed bounded Borel subset of $\mathbb{R}^2$.

From this Lemma we see that in order to prove that a Gibbs measure μ is pure it suffices to prove that $E(f|\partial\Lambda) \to E(f)$ as $\Lambda \uparrow \mathbb{R}^2$ through the filter of bounded Borel sets Λ. From the basic equality (2.13) we can compute, with $C = \partial\Lambda$ and μ any Gibbs measure:

$$E(f \mid \partial\Lambda)(\eta) = E_0^{\partial\Lambda}(e^{-U_\Lambda(\cdot+\psi_\eta^{\partial\Lambda})})^{-1} \tag{3.12}$$
$$E_0^{\partial\Lambda}\;(f(\cdot+\psi_\eta^{\partial\Lambda})\; e^{-U_\Lambda(\cdot+\psi_\eta^{\partial\Lambda})}).$$

Let now f be a bounded function of the form $f(\xi) = e^{i<\xi,\phi>}$, for all $\xi \in S'(\mathbb{R}^2)$ and $\phi \in C_0^\infty(\mathbb{R}^2)$ (it is easy to see that it suffices to prove $E(f|\partial\Omega) \to E(f)$ for such f). For f of this form we have from (3.12)

$$E(e^{i<\cdot,\phi>}|\partial\Lambda)(\eta) = \tag{3.13}$$
$$= e^{i<\cdot,\psi_\eta^{\partial\Lambda}>}\; E_0^{\partial\Lambda}(e^{-U_\Lambda^\eta})^{-1}\;\; E_0^{\partial\Lambda}(e^{i<\cdot,\phi>}\; e^{-U_\Lambda^\eta}),$$

hence

$$E(e^{i<\cdot,\phi>}|\partial\Lambda)(\eta) = e^{i<\cdot,\psi_\eta^{\partial\Lambda}>}\; E_\eta^{\partial\Lambda}(e^{i<\cdot,\phi>}), \tag{3.14}$$

where $E_\eta^{\partial\Lambda}$ is the expectation with respect to the measure

$$d\mu_\eta^{\partial\Lambda} = E_0^{\partial\Lambda}(e^{-U_\Lambda^\eta})^{-1}\; e^{-U_\Lambda^\eta}\; d\mu_0^{\partial\Lambda}. \tag{3.15}$$

From (3.14) we see that to prove $E(e^{i<\cdot,\phi>}|\partial\Lambda)(\eta) \to E(e^{i<\cdot,\phi>})$ it suffices to prove that $e^{i<\cdot,\psi_\eta^{\partial\Lambda}>} \to 1$ for μ-almost all μ and $d\mu_\eta^{\partial\Lambda}$ converges weakly to μ, as $\Lambda \uparrow \mathbb{R}^2$. In order to control $\psi_\eta^{\partial\Lambda}$ as $\Lambda \uparrow \mathbb{R}^2$, for μ-almost all η, it is necessary to impose some regularity condition on μ. This is discussed in the next section. The procedure to prove uniqueness we have now sketched has some relations with those in [19], [20].

Let us now clarify the question of the global Markov property of a Gibbs measure μ, in a corresponding way as we clarified the uniqueness.

Let thus μ be a Gibbs measure. In order to show the global Markov property

$$E_\mu(f_+ f_- \mid C_o) = E_\mu(f_+ \mid C_o)\, E_\mu(f_- \mid C_o) \tag{3.16}$$

for any piecewise C^1 curve C_o such that

$$\mathbb{R}^2 - C_o = \Omega_+^{C_o} \cup \Omega_-^{C_o}, \quad \Omega_+^{C_o} \cap \Omega_-^{C_o} = \emptyset,$$

with $f_\pm$ $\mathcal{B}_{\Omega_\pm^{C_o}}$-measurable (and ≥ 0 or in $L^2(d\mu)$), let Λ be a bounded open set and let now $f_\pm$ be $\mathcal{B}_{\Lambda_o \cap \Omega_\pm^{C_o}}$-measurable, where $\Lambda_o \subset \Lambda$, Λ_o compact. By the global Markov property of μ_Λ we have

$$\begin{aligned} E_{\mu_\Lambda}(f_+ f_- \mid \partial\Lambda \cup C_o) &= E_{\mu_\Lambda}(f_+ f_- \mid \Lambda \vee C_o) = \\ &= E_{\mu_\Lambda}(f_+ \mid \Lambda \vee C_o)\, E_{\mu_\Lambda}(f_- \mid \Lambda \vee C_o) = E_{\mu_\Lambda}(f_+ \mid \partial\Lambda \cup C_o)\, E_{\mu_\Lambda}(f_- \mid \partial\Lambda \cup C_o), \end{aligned} \tag{3.17}$$

where $E(\cdot \mid \Lambda \vee C_o)$ is the conditional expectation with respect to the σ-algebra generated by $\mathcal{B}_\Lambda$ and $\mathcal{B}_{C_o}$. From (3.16) and (3.17) we see that to prove the global Markov property it is sufficient to prove

$$E_{\mu_\Lambda}(f \mid \partial\Lambda \cup C_o) \to E_\mu(f \mid C_o) \tag{3.18}$$

as $\Lambda \uparrow R^2$, for any bounded $\mathcal{B}_{\Lambda_o}$-measurable f. Using now as for the uniqueness proof discussed above the basic equality (2.13) and taking f of the special form $f(\xi) = e^{i<\xi,\phi>}$, $\phi \in C_o^\infty(\mathbb{R}^2)$ we see analogously as in (3.13) that the convergence (3.18) takes place, taking $\partial\Lambda = \partial B_n - C_o = C_n$, where B_n is a ball of radius n, when

$$e^{i<\cdot,\psi_\eta^{C_o \cup C_n}(x)>} \longrightarrow e^{i<\cdot,\psi_\eta^{C_o}(x)>} \quad \text{as } n \to \infty, \tag{3.19}$$

for μ - a.e.$\eta \in S'(\mathbb{R}^2)$ and all x in a bounded region and moreover the measure

$$d\mu_\eta^{C_o \cup C_n} = E_o^{C_o \cup C_n} (e^{-U_{C_o \cup C_n}^\eta})^{-1} e^{-U^\eta_{C_o \cup C_n}} d\mu_o^{C_o \cup C_n}$$

(3.20)

converges weakly to $\mu_\eta^{C_o}$, for μ-almost all $\eta \in S'(R^2)$, as $n \to \infty$, where $\mu_\eta^{C_o}$ is the conditional probability measure, which exists in a regular sense ($S'(R^2)$ being a Suslin space!), obtained by conditioning μ with respect to $\mathcal{B}_{C_o}$.

In the next section we shall get in particular the control on $\psi_\eta^{C_o \cup C_n}(x)$ required for (3.19).

4. Regular Random Fields

We shall call an $S'(\mathbb{R}^2)$-valued random field $\xi(x)$, $x \in \mathbb{R}^2$ regular if there exist positive constants C and m_o such that

$$E(|\langle\xi,\phi\rangle|^2) \leq C \iint G(x-y)\phi(x)\phi(y)dxdy, \tag{4.1}$$

where $G(x-y)$ is the kernel of $(-\Delta+m_o^2)^{-1}$ in $L^2(\mathbb{R}^2)$ and E is the espectation with respect to the distribution μ of ξ on $S'(\mathbb{R}^2)$.

Let now ξ be a regular random field. By (4.1) we have that for any $\rho \in H_{-1}(\mathbb{R}^2)$ $\xi \to \langle\xi,\rho\rangle \equiv \int \xi(x)\, d\rho(x)$ (in the sense of generalized function) is μ-measurable (and even in $L^2(d\mu)$). We are thus in a situation like for the free Euclidean field and, as in that case, we define the σ-algebra of events measurable within the Borel subset Λ of $\mathbb{R}^2$ as the σ-algebra $\mathcal{B}_\Lambda$ generated by all $\langle\xi,\rho\rangle$, supp $\rho \subset \Lambda$.

Let C be a piecewise C^1-curve and let $x \notin C$. We define $\psi_\xi^C(x)$ for all $\xi \in S'(\mathbb{R}^2)$ as in (2.6) i.e.

$$\psi_\xi^C(x) = \langle\xi, P_C(x,\cdot)\rangle ,$$

where $P_C(x,dy)$ is the Poisson kernel for the Dirichlet problem (2.3).

From (4.1) we see that $\xi \to \psi_\xi^C(x)$ is, for all $x \notin C$, in $L^2(d\mu)$ (and a fortiori μ-measurable), in fact we have

$$E(|\psi_\xi^C(x)|^2) \leq c\ K_C(x,x), \tag{4.2}$$

where $K_C(x,y)$ is defined by (2.5). Inserting now the inequality (2.8) i.e.

$$K_C(x,x) \leq a|\ln d(x,C)|e^{-m_o\ d(x,C)} \tag{4.3}$$

in (4.2) and integrating with respect to x over a compact subset $\tilde{\Lambda}$ of $\mathbb{R}^2 - C$ such that $d(\tilde{\Lambda},C) \geq 1$, we get

$$E(\int_{\tilde{\Lambda}} |\psi_\xi^C(x)|^2 dx) \leq \tilde{a}|\tilde{\Lambda}|e^{-\tilde{m}d(\tilde{\Lambda},C)} \tag{4.4}$$

where $\tilde{m} < m_o$ and $\tilde{a}$ depends only on m_o, $|\tilde{\Lambda}|$ being the volume of the Borel set $\tilde{\Lambda}$.

Let now $\tilde{\Lambda}_n$ and C_n be any sequence of Borel sets resp. piecewise C^1-curves such that

$$|\tilde{\Lambda}_n|\ e^{-\tilde{m}\ d(\tilde{\Lambda}_n,C_n)} \to 0 \tag{4.5}$$

as $n \to \infty$. Then we have from (4.4) that there exists a subsequence n' such that

$$|\tilde{\Lambda}_{n'}|^{-1}\ e^{md(\tilde{\Lambda}_{n'},C_{n'})} \int_{\tilde{\Lambda}_{n'}} |\psi_\xi^{C_{n'}}(x)|^2\ dx \to 0 \tag{4.6}$$

as $n' \to \infty$, for μ-almost all $\xi \in S'(\mathbb{R}^2)$, for all $m < \tilde{m}$. Take now $\Lambda_{n'} \supset \tilde{\Lambda}_{n'}$ such that $d(\tilde{\Lambda}_{n'}, \partial\Lambda_{n'}) \geq 2$ and $\Lambda_{n'}$ is contained in the interior of $C_{n'}$. Then we get, by using the properties of the Poisson kernel

$$|\psi_\xi^{C_{n'}}(x)|^2 \leq b \int_{\Lambda_{n'}} |\psi_\xi^{C_{n'}}(y)|^2 dy \tag{4.7}$$

for some constant b (see [13] for details).

Taking now $\Lambda_{n'}$ such that $d(\Lambda_{n'}, C_{n'})|\ d(\tilde{\Lambda}_{n'}, C_{n'}) \to 1$ we then get from (4.6) and (4.7)

$$|\tilde{\Lambda}_{n'}|^{-1/2}\ e^{\frac{m}{2}\ d(\tilde{\Lambda}_{n'},\ C_{n'})} \sup_{x\in\tilde{\Lambda}_{n'}} |\psi_\xi^{C_{n'}}(x)| \to 0. \tag{4.8}$$

In this way we get the

Theorem 4.1

Let $\xi(x)$ be a regular random field with distribution μ on $S'(\mathbb{R}^2)$. Let Λ_n, C_n be sequences of bounded Borel sets in $\mathbb{R}^2$ resp. piecewise C^1-curves in $\mathbb{R}^2$ such that $\Lambda_n \subset \mathbb{R}^2 - C_n$ and $d(\Lambda_n, C_n) \to \infty$ as $n \to \infty$, then for any $m < m_o$ there is a subsequence n' such that

$$e^{\frac{m}{2} d(\Lambda_{n'}, C_{n'})} |\Lambda_{n'}|^{-1/2} \sup_{x \in \Lambda_{n'}} |\psi_\xi^{C_{n'}}(x)| \to 0$$

for μ-almost all $\xi \in S'(\mathbb{R}^2)$.

For the proof of the global Markov property we saw in (3.19) that a control on the corresponding function $\psi_\xi^{C_o \cup C_n}$, C_o a fixed piecewise C^1-curve, $C_n = \partial B_n - C_o$, B_n ball of radius n, as $n \to \infty$, is required.

But if C is any piecewise C^1-curve we have that

$$\psi_g^{C, C_o}(x) = \psi_g^{C \cup C_o}(x) - \psi_g^{C_o}(x)$$

solves the Dirichlet problem

$$\left\{ \begin{array}{ll} (-\Delta + m_o^2)\psi_g^{C, C_o} = 0 & \text{in } R^2 - (C \cup C_o), \\ \psi_g^{C, C_o}(z) = 0 & z \in C_o, \ \psi_g^{C, C_o}(z) = g(z) - \psi_g^{C_o}(z), \ z \in C. \end{array} \right. \tag{4.9}$$

Thus $\psi_g^{C, C_o}(x) = \int P_{C \cup C_o}(x, dz)(g(z) - \psi_g^{C_o}(z))$ and hence defining for $x \notin C \cup C_o$

$$\begin{aligned} \psi_\xi^{C, C_o}(x) \equiv & \langle \xi, \ P_{C \cup C_o}(x, dz) \rangle - \\ & - \langle \psi_\xi^{C_o}, \ P_{C \cup C_o}(x, dz) \rangle \end{aligned} \tag{4.10}$$

we have

$$E(|\psi_\xi^{C, C_o}(x)|^2) \leq 2 c \int_C \int_C P_{C \cup C_o}(x, z) G(z - \tilde{z}) P_{C \cup C_o}(x, \tilde{z}) dz d\tilde{z}$$

$$+ 2c \int_C \int_C P_{C \cup C_o}(x,z)\, K_{C_o}(z,\tilde{z})\, P_{C \cup C_o}(x,\tilde{z})\, dz d\tilde{z}. \tag{4.11}$$

Using the properties of G, K_{C_o} and of the Poisson kernel this gives

$$E(|\psi_\xi^{C,C_o}(x)|^2) \leq a\, e^{-m'd(x,C)} \tag{4.12}$$

for some constants a and $m' < m_o$. Then in a similar way as for Theorem 4.1 we have

<u>Theorem 4.2</u>

Let ξ be a regular random field with distribution μ on $S'(\mathbb{R}^2)$ and let C_o be any fixed piecewise C^1-curve. Then with the same notations as in Theorem 4.1 we have

$$e^{\frac{m}{2} d(\Lambda_{n'},C_{n'})} |\Lambda_{n'}|^{-1/2} \sup_{x \in \Lambda_{n'}} |\psi_\xi^{C_o \cup C_{n'}}(x) - \psi_\xi^{C_o}(x)| \to 0$$

for μ-almost all $\xi \in S'(\mathbb{R}^2)$.

In the next sections we shall apply Theorems 4.1, 4.2 to obtain the uniqueness of Gibbs states and their global Markov property.

5. Uniqueness of the Gibbs Measure for the Trigonometric Interaction

We saw in section 3 that the idea for proving uniqueness of the Gibbs measures is to prove their purity, in turn this is reduced to an estimate of the functions $\psi_\xi^{\partial\Lambda}(x)$ for μ - a.e. $\xi \in S'(\mathbb{R}^2)$ as $\Lambda \uparrow \mathbb{R}^2$ (given by Theorem 4.1) and the weak convergence of the measures $\mu_\eta^{\partial\Lambda}$ of (3.15), for μ - a.e. $\eta \in S'(\mathbb{R}^2)$. The former control, as we saw in Theorem 4.1, is available if μ is the distribution of a regular random field. We shall now prove the weak convergence of $\mu_\eta^{\partial\Lambda}$ by assuming interactions U_Λ of a particular form, namely

$$U_\Lambda(\xi) \equiv \lambda \int_\Lambda :\cos(\alpha\xi(x) + \theta_o):\, dx \tag{5.1}$$

i.e. the so called "Sine-Gordon interactions". : : is the usual Wick ordering with respect to the free Euclidean field μ_o. It has been established by Fröhlich and Seiler [7] that (5.1) exists for all $\alpha^2 < 4\pi$, $0 \le \theta_o < 2\pi$ (and is the limit after removal of the ultraviolet cut-off of the interactions studied in [21]), moreover these authors have proven the existence of the limit μ_∞ for $\Lambda \uparrow \mathbb{R}^2$ of the corresponding measures μ_Λ, for all $|\lambda| \le \lambda_o\,(m_o,\alpha)$, and some λ_o. It follows from their results that μ_∞ is a Gibbs measure for U_Λ and is the distribution of a regular random field.

We shall now study the weak convergence of the measures $\mu_\eta^{\partial\Lambda}$ for μ- a.e. $\eta \in S'(\mathbb{R}^2)$, where by definition μ is any Gibbs measure to the interaction U given by (5.1), such that the corresponding random field is a regular one (i.e. μ could be μ_∞ but a priori any other regular Gibbs measure; we will show below that necessarily $\mu = \mu_\infty$). From the definitions in section 3 we have, for all $\eta,\xi \in S'(\mathbb{R}^2)$:

$$U_\Lambda^\eta(\xi) = U_\Lambda(\xi+\psi_\eta^{\partial\Lambda}) = \lambda\int :\cos\alpha\,\psi_\eta^{\partial\Lambda}(x):\,\cos(\alpha\xi(x)+\theta_o):dx$$

$$- \lambda\int :\sin\alpha\psi_\eta^{\partial\Lambda}(x):\,:\cos(\alpha\xi(x)+\theta_o):\,dx\,, \tag{5.2}$$

where we used the trigonometric formulae for Wick-ordered expressions (easily proven from the definition of the Wick ordering and the usual trigonometric formulae). Now we have

$$:\cos\,\alpha\psi_\eta^{\partial\Lambda}(x): \;=\; e^{\frac{\alpha^2}{2}K_{\partial\Lambda}(x,x)}\cos\,\alpha\psi_\eta^{\partial\Lambda}(x), \tag{5.3}$$

where we used the definition of the Wick-odering and (2.7). Similarly

$$:\sin\,\alpha\psi_\eta^{\partial\Lambda}(x): \;=\; e^{\frac{\alpha^2}{2}K_{\partial\Lambda}(x,x)}\sin\,\alpha\psi_\eta^{\partial\Lambda}(x). \tag{5.4}$$

Hence, from (5.1), (5.2), (5.3):

$$U_\Lambda^{\eta}(\xi) = \lambda\sum_{i=1,2}\int_\Lambda e^{\frac{\alpha^2}{2}K_{\partial\Lambda}(x,x)}\,g_{i,\eta}^{\partial\Lambda}(x):\cos(\alpha\xi(x)+\theta_o):dx \tag{5.5}$$

where $g_{i,\eta}^{\partial\Lambda}(x)$ are functions with $|g_{i,\eta}^{\partial\Lambda}(x)| \le 1$.

From the bound (2.8) on $K_{\partial\Lambda}(x,x)$ we see that $e^{\frac{\alpha^2}{2}K_{\partial\Lambda}(x,x)}$ is locally dx-integrable, for $\alpha^2 < 4\pi$, hence $e^{\frac{\alpha^2}{2}K_{\partial\Lambda}(x,x)} g_{i,\eta}^{\partial\Lambda}(x)$ is locally integrable, uniformly in $\eta \in S'(\mathbb{R}^2)$. By Theorem 4.1 we have

$$\cos \alpha\psi_\eta^{\partial\Lambda_{n'}}(x) \to 1 \quad \text{and} \quad \sin \alpha\psi_\eta^{\partial\Lambda_{n'}}(x) \to 0, \tag{5.6}$$

uniformly on $\Lambda'_{n'} \subset \Lambda_{n'}$, for some subsequence $n' \to \infty$, for μ-almost all $\eta \in S'(\mathbb{R}^2)$, whenever $d(\Lambda'_{n'}, \partial\Lambda_{n'}) \to \infty$. This then gives that the functions $g_{i,\eta}^{\partial\Lambda_{n'}}(x)$ converge to constants as $n' \to \infty$, for μ-a.e. $\eta \in S'(\mathbb{R}^2)$, locally uniformly in x.

From the definition (3.15) of $\mu_\eta^{\partial\Lambda_{n'}}$ and the formulae (3.15) and (5.4) we see that the weak convergence of $\mu_\eta^{\partial\Lambda_{n'}}$ as $n' \to \infty$ is "related" to that of $\mu_{\Lambda_{n'}}$ as $n' \to \infty$, in as much as it differs only by the presence of the functions $e^{\frac{\alpha^2}{2}K_{\partial\Lambda_{n'}}(x,x)} g_{i,\eta}^{\partial\Lambda_{n'}}(x)$ in (5.4), but these are locally integrable and tend to 1 respectively 0 as $n' \to \infty$. It looks at this point promising to adapt the method of the cluster expansion ([5]) to the present situation, along the lines used for μ_Λ in [7]. In fact one gets the weak convergence of $\mu_\eta^{\partial\Lambda}$ as $\Lambda \uparrow \mathbb{R}^2$ for μ-a.a. $\eta \in S'(\mathbb{R}^2)$, whenever $|\lambda|$ is sufficiently small. Moreover, the cluster expansion converges uniformly in the functions $g_{i,\eta}^{\partial\Lambda}$ (here $L^2_{loc}(dx)$-estimates are needed, involving $e^{\frac{\alpha^2}{2}K_{\partial\Lambda}(x,x)}$, and these work only for $\alpha^2 < 2\pi$) and from this fact we get that $\mu = \mu_\infty$. For details we refer to the sections 4, 5 of [13].

We shall formulate here the result

<u>Theorem 5.1</u>

For all $\alpha^2 < 2\pi$, $|\lambda| \le \lambda_o(m_o,\alpha)$, where λ_o is some positive number there exists one and only one Gibbs measure whose corresponding field is a regular random field, corresponding to the trigonometric interactions (5.1).

This theorem gives the uniqueness of the Gibbs measures μ, satisfying the condition $\int |<\xi,\phi>|^2 \, d\mu \leq c \int |<\xi,\phi>|^2 d\mu_o$, associated with the Sine-Gordon interactions, for small coupling constants and any α such that $\alpha^2 < 2\pi$. This measure is Euclidean invariant, with support on $S'(\mathbb{R}^2)$ and is known to give rise to a relativistic field theory satisfying all postulates. We shall see below that the field theory is in fact a canonical one, because of the global Markov property we shall discuss in the next section.

6. The Global Markov Property for the Trigonometric Interaction

Let μ be the unique Gibbs measure for the trigonometric interaction given in Theorem 5.1. Let C_o be a (bounded or unbounded) piecewise C^1-curve such that $\mathbb{R}^2 - C_o = \Omega_+ \cup \Omega_-$, $\Omega_+ \cap \Omega_- = \emptyset$. As discussed at the end of Section 3 to prove the global Markov property (3.16) it suffices to prove (3.19), which follows from Theorem 4.2, and the weak convergence of $\mu_\eta^{C_o \cap C_n}$ as $n \to \infty$ to $\mu_\eta^{C_o}$, for μ-a.e. $\eta \in S'(\mathbb{R}^2)$. In fact these imply (3.18), which is the purity of the Gibbs measure $\mu_\eta^{C_o}$ constructed similarly as μ was constructed out of μ_o and U_Λ (by μ_Λ), but now with μ_o replaced by $\mu_{o,\eta}^{C_o}$, the free Euclidean measure conditioned by the σ-Algebra $\mathcal{B}_{C_o}$, for μ-a.e. $\eta \in S'(\mathbb{R}^2)$. Using Theorem 4.2 and the cluster expansion, as for the uniqueness proof, but with the role of $\psi_\eta^{\partial\Lambda}$ now replaced by $\psi_\eta^{C_o \cup C_n, C_o}$, we obtain, similarly as for Theorem 5.1 (see Section 6 of [13] for details):

Theorem 6.1

Under the same assumptions as in Theorem 5.1 the unique Gibbs measure μ for the trigonometric interaction has the global Markov property i.e.

$$E(f_+ f_- | C_o) = E(f_+ | C_o) \, E(f_- | C_o)$$

for any (bounded or unbounded) piecewise C^1-curve C_o, such that

$\mathbb{R}^2 - C_o = \Omega_+ \cup \Omega_-$, $\Omega_+ \cap \Omega_- = \emptyset$, where E is expectation with respect to μ, and $f_\pm$ are $\mathcal{B}_{\Omega_\pm}$-measurable (and ≥ 0 or in $L^1(d\mu)$).

In the next section we shall mention some consequences of these results.

7. Relativistic Invariant Canonical Field Theories

A consequence of the global Markov property of a regular random field $\xi(y)$, $y \in \mathbb{R}^2$, is the fact that if we define $\xi_t(x) = \xi(t,x)$, $x \in \mathbb{R}$, $t \in \mathbb{R}$, then $t \to \xi_t$ is a Markov process with values in $S'(\mathbb{R})$ (such processes have been studied e.g. in [2], [3], [8] - [11], [22]. For a detailed study of the Gaussian case see also [23], [24]. Let ν be the invariant measure on $S'(\mathbb{R})$ for this process i.e. $\nu = \mu \restriction \mathcal{B}_o$, where $\mathcal{B}_o$ is the σ-algebra generated by the $\xi_o(x)$ (this restriction exists by the fact that μ is regular). Then the corresponding Markov semigroup, with invariant measure ν,

$$(e^{-tH} f)(\xi_o) = E(f(\xi_t) | \mathcal{B}_o) \tag{7.1}$$

is a strongly continuous symmetric semigroup in $L^2(S'(\mathbb{R}), d\nu)$. Its generator H is a positive self-adjoint operator which is the Hamiltonian for the quantum trigonometric interaction. $L^2(S'(\mathbb{R}), \nu)$ is the corresponding physical Hilbert space of the states of the quantum field. It is proven (see [10],[25]) that

$$(f,Hf)_2 = \int_R \int_{S'(\mathbb{R})} \left| \frac{\delta f(\xi_o)}{\delta \xi_o(x)} \right|^2 d\nu(\xi_o) dx \tag{7.2}$$

for all f which are C^1-cylinder functions on $S'(\mathbb{R})$, $(,)_2$ being the scalar product in $L^2(d\nu)$. Moreover, if Λ is the generator of the Lorentz boost we have

$$(f, \Lambda f)_2 = \int \int x \left| \frac{\delta f(\xi_o)}{\delta \xi_o(x)} \right|^2 d\nu(\xi_o) dx. \tag{7.3}$$

We also have proven that the Markov process $t \to \xi_t$ satisfies the stochastic differential equation

$$d\xi_t = \beta(\xi_t) dt + dw_t , \tag{7.4}$$

where β is a ν-measurable map of $S'(\mathbb{R})$ such that $<\beta(\xi),\phi> \in L^2(d\nu)$ for any $\phi \in S(\mathbb{R})$ and w_t is the Markov process with state space $S'(\mathbb{R})$ given by the free Euclidean field on $\mathbb{R}^2$, i.e. $w_t(x) = w(t,x)$ is the free Euclidean field.

Many other properties are known e.g. the closure of the quadratic form (7.2) is a Dirichlet form ([10]), in fact a regular one [11]. ν is $S(\mathbb{R})$-quasi invariant and $S(\mathbb{R})$ is represented unitarity, strongly continuously in $L^2(d\nu)$ by $\phi \to e^{i<\xi_o,\phi>}$ and $\phi \to \pi(\phi) \equiv (d\nu(\xi_o+\phi)/d\nu(\xi_o))^{1/2} f(\xi_o+\phi)$. $<\xi_o,\phi>$, $\pi(\phi)$ satisfy the Weyl commutation relations. $\pi(\phi)$ is essentially self-adjoint on $\{e^{i<\xi_o,\phi>}, \phi \in S(\mathbb{R})\}$, and is the physical momentum operator. For other properties of the Dirichlet form (7.2) and associated quantities see [10], [11], [25].

Λ has a Markov property, when suitably interpreted [25]. Let A be a bounded operator on $L^2(S'(\mathbb{R}),d\nu)$ and consider the vector state $A \to \omega(A) = (1,A1)_2$. ω is a cyclic and separating vector for the von Neumann algebra A_+ generated by the $<\xi_o,\phi>$ and $\pi(\phi)$, $\phi \in C_o^\infty(\mathbb{R}_+)$, and the commutant A_+' of A_+ is generated by $<\xi_o,\phi>$, $\pi(\phi)$, $\phi \in C_o^\infty(\mathbb{R}_-)$. The canonical map in Tomita's theory is induced by the reflection $x^1 \to -x^1$ and the Tomita automorphism is given by Λ. See [25].

8. The Global Markov Property for Lattice Systems

We shall here mention only shortly an application of the method for proving the global Markov property of Euclidean fields to the simpler case of lattice systems; for details we refer to our work with G. Olsen [15].

Consider a lattice $\mathbb{Z}^d$ in d dimensions. Let Φ be a finite range interaction on $\mathbb{Z}^d$ i.e. a real-valued function defined on the subsets of $\mathbb{Z}^d$, with support on finitely many finite subsets of $\mathbb{Z}^d$ and such that for $\Lambda \subset \mathbb{Z}^d$, $\Phi(\Lambda) = \Phi(-\Lambda)$. Define, for any subset $\Lambda \subset \mathbb{Z}^d$, $\|\Phi(\Lambda)\| = \Sigma(|\Lambda|-1)|\Phi(\Lambda)|$, where the sum is over all finite subsets of $\mathbb{Z}^d$. Let σ_i be the function $\sigma_i(\Lambda) = 1$ if i is a point of Λ and $\sigma_i(\Lambda) = 0$ if $i \notin \Lambda$. Let B_Λ be the σ-algebra generated by the σ_i, $i \in \Lambda$. Let $\Omega = \{0,1\}^{Z^d}$, let $d\alpha_i$ be a probability measure on the i-th component $\{0,1\}_i$. Define for $\sigma \in \Omega$

$$d\mu_\Lambda(\sigma) \equiv (e^{-U_\Lambda(\sigma)} \prod_{i\in\mathbb{Z}^d} d\alpha_i(\sigma_i))^{-1} e^{-U_\Lambda(\sigma)} \prod_{i\in\mathbb{Z}^d} d\alpha_i(\sigma_i) \qquad (8.1)$$

with

$$U_\Lambda(\sigma) = \sum_{i\in\Lambda} \sigma_i \sum_{\Lambda'} \Phi(\Lambda') \prod_{j-i\in\Lambda'} \sigma_j .$$

Note that $\sigma_i \sum_{\Lambda'} \Phi(\Lambda') \prod_{j-i\in\Lambda'} \sigma_j$ is the interaction between the lattice point i and the rest of the lattice. Clearly μ_Λ corresponds to the interaction (3.2) in the case of Euclidean random fields. μ is called a Gibbs measure for the interaction U_Λ if

$$E(f|\tilde{\Lambda}) = E_\Lambda(f|\tilde{\Lambda}) \qquad (8.2)$$

for any bounded measurable subset $\Lambda \subset \mathbb{Z}^d$, $\tilde{\Lambda} \equiv \mathbb{Z}^d - \Lambda$, and any bounded Borel function f on Ω, where E is the expectation with respect to μ. From Dobrushin's results about existence and uniqueness of Gibbs measures [14] it follows that if $\|\Phi\| < \frac{1}{2e}$ then μ is unique, $\mu \upharpoonright \mathcal{B}_\infty$ is trivial, where $\mathcal{B}_\infty \equiv \bigcap_\Lambda \mathcal{B}_{\tilde{\Lambda}}$.

We shall now look at the global Markov property. Consider the decomposition $\mathbb{Z}^d = \Omega_+ \cup C \cup \Omega_-$, where Ω_+, C, Ω_- are subsets of $\mathbb{Z}^d$ which are pairwise disjoint and C insulates Ω_+ and Ω_- in the sense that one cannot reach any finite subset of Ω_+ from Ω_- by translating with vectors out of a set in the support of Φ.

As for Euclidean fields one has easily (see [15]) the local Markov property

$$E(f_+f_-|C) = E(f_+|C)\; E(f_-|C) \qquad (8.3)$$

whenever Ω_+ is finite and $f_\pm$ are $\mathcal{B}_{\Omega_\pm}$-measurable.

Moreover, as for the Euclidean fields, (sects. 3,6) one gets

$$E_\eta^C(f_+f_-|\tilde{\Lambda}) = E_\eta^C(f_+|\tilde{\Lambda})\; E_\eta^C(f_-|\tilde{\Lambda}) \qquad (8.4)$$

for μ-a.e. $\eta \in \Omega$, for any bounded $\Lambda \subset \mathbb{Z}^d$. But the uniqueness result of Dobrushin holds for μ_η^C as well, and then yields from (8.4)

$$E_\eta^C(f_+f_-) = E_\eta{}^C(f_+)\ E_\eta{}^C(f_-)\ , \tag{8.5}$$

for μ-a.e. $\eta \in \Omega$. But $E_\eta{}^C(\cdot) = E(\cdot|C)(\eta)$, and (8.5) gives then (8.3), for arbitrary Ω_+, Ω_- insulated by C.

We have thus the

Theorem 8.1

Let Φ be any interaction of finite range on the lattice $\mathbb{Z}^d$, such that $\|\Phi\| < \frac{1}{2e}$ and consider $U_\Lambda(\sigma)$ as given by (8.1). Then there is one and only one Gibbs maeasure μ and for any disjoint partition $\mathbb{Z}^d = \Omega_+ \cup C \cup \Omega_-$ such that C insulates Ω_+ and Ω_- with respect to Φ one has the global Markov property

$$E(f_+f_-|C) = E(f_+|C)\ E(f_-|C),$$

for any $f_\pm$ $\mathcal{B}_{\Omega_\pm}$-measurable (≥ 0 or in $L^2(d\mu)$).

The global Markov Property can be used to deduce corresponding consequences as the one derived for Euclidean fields. In particular we have for any splitting $\mathbb{Z}^d = \mathbb{Z} \times \mathbb{Z}^{d-1}$, $i = (k,l)$ $k \in \mathbb{Z}$, $l \in \mathbb{Z}^{d-1}$ a symmetric Markov chain $k \to \sigma_k(l)$, with values in $\{0,1\}^{\mathbb{Z}^{d-1}}$ and invariant distribution $\nu = \mu \upharpoonright \mathcal{B}_{k=0}$. The corresponding symmetric transition operator e^{-tH}, $H \geq 0$, $t \geq 0$ on $L^2(d\nu)$ is the Onsagerian (transfermatrix) discussed in many papers of classical statistical mechanics (e.g.([26]).

Remark

(Added after the Colloquium) Recently J. Bellissard and P. Picco [27], have extended our results on the global Markov property to continuous valued lattice models, using their uniqueness results for such systems. H. Föllmer [28] has also remarked that the global Markov property for lattice systems follows whenever one has Dobrushin's uniqueness conditions. Moreover, he shows that if the interaction is attractive and admits a "high density state" and a "low density state" then each of the states has the global Markov property. These results have been recently extended by J. Bellissard and R. Høegh-Krohn.

Acknowledgements

We are grateful to Professor L. Streit and the Center of Interdisciplinary Research (ZIF) of the University of Bielefeld for the invitation to attend the Colloquium.

We also gratefully acknowledge stays at the Mathematics Institutes of the Universities of Bielefeld, Aix-Marseille (Luminy) and Oslo, as well as at the Physics Department of the University of Bielefeld and the Centre de Physique Théorique of the CNRS Marseille.

We thank Mrs. L. Jegerlehner heartily for her patience and excellent job with the typing.

References

[1] E. Nelson, The free Markov field, J. Funct. Anal. 12,221-227 (1973).

[2] E. Nelson, Construction of quantum fields from Markoff fields, J. Funct. Anal. 12, 97-112 (1973).

[3] B. Simon, The $P(\phi)_2$ Euclidean (quantum) field theory, Princeton University Press, Princeton (1974).

[4] E. Nelson, Probability theory and Euclidean field theory, in Constructive quantum field theory, Edts. G. Velo, A.A. Wightman, Lecture Notes in Physics 25, pp. 94-124, Springer, Berlin (1973).

[5] J. Glimm, A. Jaffe, T. Spencer, The particle structure of the weakly coupled $P(\phi)_2$ model and other applications of high temperature expansions, in Constructive quantum field theory, Edts. G. Velo, A.S. Wightman, Lecture Notes in Physics 25, pp. 132-242, Springer, Berlin (1973).

[6] S. Albeverio, R. Høegh-Krohn, The Wightman axioms and the mass gap for strong interactions of exponential type in two dimensional space-time, J. Funct. Anal. 16, 39-82 (1974).

[7] J. Fröhlich, E. Seiler, The massive Thirring-Schwinger model (QED_2): convergence of perturbation theory and particle structure, Helv. Phys. Acta 49, 889-924 (1976).

[8] S. Albeverio, R. Høegh-Krohn, Quasi invariant measures, symmetric diffusion processes and quantum fields, in Les méthodes mathématiques de la théorie quantique des champs, Coll. Int. CNRS 248, pp. 11-59, Publications du CNRS, Paris (1976).

[9] S. Albeverio, R. Høegh-Krohn, Topics in infinite dimensional analysis, in Mathematical problems in theoretical physics, Edts. G.F. Dell'Antonio, S. Doplicher, G. Jona-Lasinio, Lecture Notes in Physics, Vol. 80, pp. 279-302, Springer, Berlin (1978).

[10] S. Albeverio, R. Høegh-Krohn, Dirichlet forms and diffusion processes on rigged Hilbert spaces, Z. Wahrscheinlichkeitstheorie u. verw. Geb. 40, 1-57 (1977).

[11] S. Albeverio, R. Høegh-Krohn, Hunt processes and analytic

potential theory on rigged Hilbert spaces, Ann. Inst. H. Poincaré B13, 269-291 (1977).

[12] S. Albeverio, R. Høegh-Krohn, Local and global Markoff fields, Lectures given at Karpacz Winter School 1978.

[13] S. Albeverio, R. Høegh-Krohn, Uniqueness and the global Markov property for Euclidean fields. The case of trigonometric interactions, Commun. Math. Phys. 68, 95-128 (1979). The results were announced in: S. Albeverio, R. Høegh-Krohn, The global Markov property for Euclidean and lattice fields, Phys. Letters 84B, 89-90 (1979)

[14] R. Dobrushin, Prescribing a system of random variables by conditional distribution, Th. Prob. Appl. 15, 3, 485-486 (1970) (and references therein)

[15] S. Albeverio, R. Høegh-Krohn, G. Olsen, The global Markov property for Lattice systems, Bielefeld - Oslo Preprint, Nov. 1978. The results were announced in: S. Albeverio, R. Høegh-Krohn, The global Markov property of Euclidean and lattice fields, Phys. Letters 84B, 89-90 (1979)

[16] J.G.M. Molchan, Characterization of Gaussian fields with markovian property, Dokl. Akad. Nauk SSR 197 (1971) (transl. Soviet Math. Dokl. 12, 563-567 (1971)

[17] C. Newman, The construction of stationary two dimensional Markoff fields with an application to quantum field theory, J. Funct. Anal. 12, 211-227 (1973)

[18] J. Fröhlich, Schwinger functions and their generating functionals, I Helv. Phys. Acta 74, 265-306 (1974); II Adv. Math. 23, 119-180 (1977)

[19] R.L. Dobrushin, R. Minlos, Construction of a one dimensional quantum field via a continuous Markov field, Funct. Anal. and its Appl. 7, 81-82 (1973) (transl. 7, 324-325 (1973))

[20] R.L. Dobrushin, R. Minlos, The theory of Markov random fields and its applications to quantum field theory, in Functional and probabilistic methods in quantum field theory, I, pp. 23-49, B. Jancewitz Ed., Acta Univ. Wratisl. 368, Wroclaw 1976

[21] S. Albeverio, R. Høegh-Krohn, Uniqueness of the physical vacuum and the Wightman functions in the infinite volume

limit for some non-polynomial interactions, Commun. Math. Phys. 30, 171-200 (1973)

[22] F. Guerra, L. Rosen, B. Simon, The $P(\phi)_2$ Euclidean quantum field theory as classical statistical mechanics, Ann. Math. 101, 114-259 (1975)

[23] J.T. Cannon, Continuous sample paths in quantum field theory, Commun. Math. Phys. 36, 123-132 (1974)

[24] R. Carmona, Measurable norms and some Banach space valued Gaussian processes, Duke Math. J. 44, 109-127 (1977)

[25] S. Albeverio, R. Høegh-Krohn, Canonical relativistic quantum fields, Ann. Inst. H. Poincaré A (to appear)

[26] R. Minlos, Ya. Sinai, Spectra of stochastic operators arising in Lattice models of a gas, Theor. and Math. Phys. 2, 2, 230-243 (1976)

[27] J. Bellissard, P. Picco, Lattice quantum fields: Uniqueness and Markov property, CNRS-CPT Preprint, Dec. 1978 (and lecture given at this conference)

[28] H. Föllmer, On the global Markov property, ETH Preprint, Dec. 1978 (to appear in these Proceedings)

Martingale Convergence and the Exponential Interaction in $\mathbb{R}^n$

Sergio Albeverio and Raphael Høegh-Krohn

Fakultät für Mathematik, Universität Bielefeld
Federal Republic of Germany

Fakultät für Mathematik, Ruhr-Universität Bochum
Federal Republic of Germany

Matematisk Institutt, Universitetet i Oslo
Norway

Centre de Physique Théorique, CNRS Marseille
France

Université d'Aix-Marseille II, UER de Luminy
France

Abstract

In this lecture we discuss the exponential interaction in $\mathbb{R}^n$. We give a proof that the ultraviolet cut-off exponential interaction $m_k(\xi) = \int_\Lambda :e^{\alpha\xi_k(x)}: dx$, where Λ is a fixed bounded region in R^n and k the ultraviolet cut-off parameter, is a martingale in k with respect to the free Euclidean measure μ_o. Moreover $e^{-\lambda m_k(\xi)}$ is a positive bounded submartingale and we prove that $e^{-\lambda m_k(\xi)}$ converges pointwise μ_o-almost everywhere and strongly in $L_1(d\mu_o)$ as the ultraviolet cut-off k tends to infinity. Especially $e^{-\lambda m_k(\xi)} d\mu_o(\xi)$ converges weakly as $k \to \infty$ which implies that the ultraviolet cut-off Schwinger functions for the exponential interaction in $\mathbb{R}^n$ converge as $k \to \infty$. For $n \geq 4$ this limit is the free Euclidean field. Further results concerning the cases $n \leq 3$ are also mentioned, as well as applications to the study of the energy representations of groups of mappings from Riemann manifolds into compact Lie groups.

1. Introduction

The exponential interaction was introduced as a model of relativistic quantum fields in [1] and [2]. It was pointed out by us in [2] that in the case of space time dimension $n = 2$ the exponential interaction $m_\Lambda(\xi) = \int_\Lambda : e^{\alpha\xi} :(x)d\nu(\alpha)dx$ exists and is a positive function in $L_2(d\mu_o)$, if ν is a bounded positive measure with support in $(-2\sqrt{\pi}, 2\sqrt{\pi})$, μ_o being the free Euclidean field. We also proved for supp $\nu \subset (-4/\sqrt{\pi}, 4/\sqrt{\pi}))$ that the corresponding Schwinger functions are decreasing functions of Λ, and that the limit $\Lambda \uparrow R^2$ exists and defines Euclidean fields leading to relativistic quantum fields satisfying all the usual postulates. It follows also from the estimates in [2] that, for arbitrary n, one has a priori estimates for the Schwinger functions of a space-time and momentum cut off exponential interaction in terms of the free Schwinger functions. For this reason it is interesting to study the exponential interaction for arbitrary space-time dimension n.

Since our work in [2] closely related work has appeared [3] - [9] and in particular the range for existence of relativistic models with exponential interactions has been extended to the interval $(-2\sqrt{\pi}, 2\sqrt{\pi})$ [6]. Further discussion of the removal of the cut offs by subsequences is contained in [3],[5] - [10]. The first proof of the convergence of the ultraviolet cut-off measures when the cut-off is removed, in the case of arbitrary space-time dimension n,was given in [11] and will be given in this paper as well. There is a new proof of this fact in [9].

This lecture describes the results of [12], which show that for an arbitrary space time dimension n the measures defining the ultraviolet cut off exponential interaction converge weakly as the cut off is removed and the limit then defines the exponential interaction in a finite region of the Euclidean space of dimension n. Moreover, this limit is obtained as a limit of an a.e. convergent submartingale, and thus in particular the Schwinger functions are convergent. That is if $m_k(\xi) = \int_\Lambda :e^{\alpha\xi_k(x)}: dx$, where $\xi_k(x)$ is the ultraviolet cut-off free Euclidean field, then $e^{-\lambda m_k(\xi)}$ is an a.e. convergent submartingale in k with respect to μ_o, the probability measure for the free Euclidean field. In fact $m_k(\xi)$ is a positive martingale and we prove that, in any dimension $n, m_k(\xi)\, d\mu_o(\xi) \to \nu * \mu_o \equiv \int_\Lambda \mu_o(\xi - \alpha G_x) dx$, where $G_x(y) = G(x-y)$ is the kernel of $(-\Delta+m^2)^{-1}$, as $k \to \infty$, weakly as measures.

We then prove that, for $n \geq 4$, $\nu*\mu_o \perp \mu_o$ i.e. $\nu*\mu_o$ and μ_o have disjoint support. This together with the submartingale convergence permits us to prove that $e^{-\lambda m_k(\xi)} d\mu_o(\xi) \to d\mu_o(\xi)$ weakly as $k \to \infty$. This means that the exponential interaction is trivial for $n \geq 4$. As we recalled above the exponential interaction is non trivial for $n = 2$ and $\alpha^2 > 4\pi$. It is easily seen to be nontrivial for $n = 1$.

Remark 1

The situation for $n = 3$ for arbitrary α and for $n = 2$ and $\alpha^2 \geq \alpha_o^2$ for α_o^2 sufficiently big ($> 4\pi$) has been clarified by Gallavotti and ourselves in a paper [13] subsequent to the present lecture, by the coupling of a modification of the method of the present lecture together with tail estimates suggested by the analysis described in these Proceedings by G. Gallavotti [15].

Remark 2

The triviality of the exponential interaction is related to the irreducibility of the energy representation of the Sobolev-Lie groups [16]-[20] and our proof that the exponential interaction is trivial for dimension $n \geq 4$ is similar to the method used by Vershik, Gelfand and Graev [19] to prove irreducibility of the energy representation of Sobolev-Lie groups of mappings of $\mathbb{R}^n$ into a compact Lie group for $n \geq 4$. Irreducibility has been proven recently by D. Testard and ourselves [20] (subsequently to this lecture, by methods related to those of [13]) for $n \geq 3$ as well as for certain cases $n = 2$. [20] contains as well a partial analysis of the interesting situation $n = 1$, where one has reducibility.

2. The Exponential Interaction

Let $H_1(\mathbb{R}^n)$ be the Sobolev space of real-valued square integrable functions obtained by completing the real $C_0^\infty(\mathbb{R}^n)$ space in the Hilbert norm $(\phi,\phi)_1 \equiv \int_{\mathbb{R}^n} [(\nabla\phi)^2 + m^2 \phi^2]dx$, where m is a fixed constant, $m \geq 0$ for $n \geq 3$ and $m > 0$ for $n = 1,2$. Let $H_{-1}(\mathbb{R}^n)$ be the dual Hilbert space i.e. the Hilbert space of real distributions ϕ such that $\int\phi(x)\, G(x-y)\, \phi(y)\, dx\, dy < \infty$, where $G(x-y) \equiv (-\Delta+m^2)^{-1}(x-y)$ is the kernel of $(-\Delta+m^2)^{-1}$ in $L^2(\mathbb{R}^n)$ i.e.

$G(x-y) = (2\pi)^{-n} \int_{\mathbb{R}^n} \frac{e^{ik(x-y)}}{k^2+m^2} dk$. $H_{-1}(\mathbb{R}^n)$ is the completion of $C_0^\infty(\mathbb{R}^n)$ in the norm $\| (-\Delta+m^2)^{-1/2} f \|_2^2$. Consider the standard normal distribution μ_o with mean zero and covariance given by the scalar product in $H_{-1}(\mathbb{R}^n)$. It has support on a Hilbert space $\bar{H}(\mathbb{R}^n)$, such that the injection $H_1(\mathbb{R}^n) \hookrightarrow \bar{H}(\mathbb{R}^n)$ is Hilbert-Schmidt. We can consider the measure μ_o as a measure on $S'(\mathbb{R}^n)$. The probability space $(S'(\mathbb{R}^n), \mu_o)$ is the probability space for the free Euclidean field $\xi(x)$, $x \in \mathbb{R}^d$, of mass m, in $\mathbb{R}^n$. Since the continuous linear functions on $H_1(\mathbb{R}^n)$ are μ_o-measurable we have that, for any Borel measure ρ on $\mathbb{R}^n$, the linear function $\xi \to \langle\xi,\rho\rangle \equiv \int \xi(x)\, d\rho(x)$ on $S'(\mathbb{R}^n)$ is μ_o-measurable if and only if $\iint G(x-y)d\rho(x)d\rho(y)$ is finite i.e. iff ρ has finite energy i.e. $\rho \in H_{-1}(\mathbb{R}^n)$. Moreover we have

$$E(\xi(x)\; \xi(y)) = G(x-y)$$

and

$$E(|\int\xi(x)d\rho(x)|^2) = \iint G(x-y)d\rho(x)d\rho(y),$$

where E means expectation with respect to μ_o.

Let now Λ be any Borel subset of $\mathbb{R}^n$ and let B_Λ be the σ-algebra of μ_o-measurable subsets of $S'(\mathbb{R}^n)$ generated by the linear functions $\langle\xi,\rho\rangle$ with $\rho \in H_{-1}(\mathbb{R}^n)$ and supp $\rho \subset \Lambda$. By an interaction we shall understand a system (Λ, U_Λ) where U_Λ is, for any bounded regular Borel subset, such that both U_Λ and e^{-U_Λ} are B_Λ-measurable and in $L^1(d\mu_o)$ and for any two disjoint Λ_1, Λ_2 one has $U_{\Lambda_1 \cup \Lambda_2} = U_{\Lambda_1} + U_{\Lambda_2}$.
In other words U_Λ is an additive functional of the free field.
The limit $e^{-U_\Lambda} d\mu_o / \int e^{-U_\Lambda} d\mu_o$, as $\Lambda \uparrow \mathbb{R}^n$ if it exists, gives a Euclidean invariant measure. For n = 2 a large class of interactions

has been constructed (of polynomial [21], exponential [2] or trigonometric type [22]), and the limit $\Lambda \uparrow \mathbb{R}^2$ has yielded non Gaussian Euclidean invariant measures.

We shall now discuss for general n the possibility of defining interactions of exponential type, i.e. the object on $\mathbb{R}^n$ corresponding to the interaction $\int_\Lambda e^{\alpha\xi(x)}dx$ on $\mathbb{R}$. Let $G_x(\cdot) \equiv G(x-\cdot)$, then $x \to \alpha G_x$ is, for any fixed real α, a continuous map from $\mathbb{R}^n$ into $S'(\mathbb{R}^n)$, since $G_x(\cdot) \in L^1_{loc}(\mathbb{R}^n)$, for any fixed $x \in \mathbb{R}^n$. Note also that $G_x(\cdot) \geq 0$. We can thus consider $\alpha G_x(\cdot)$ as an element in $S'(\mathbb{R}^n)$ and consider now the translate of μ_o by the element $\alpha G_x \in S'(\mathbb{R}^n)$ i.e.

$$\mu_o(\xi-\alpha G_x).$$

This is a measure on $S'(\mathbb{R}^n)$ (singular with respect to μ_o, except for $\alpha = 0$ or $d = 1$, due to $\alpha G_x \notin H_1(\mathbb{R}^n)$), which depends continuously (in the weak topology) on x, in particular for $g \geq 0$, $g \in L^1_{loc}$ we have, for any Borel subset Λ of $\mathbb{R}^n$ such that $\int_\Lambda g(x)dx < \infty$, that

$$\mu_\Lambda^{\alpha,g}(\xi) \equiv \int_\Lambda \mu_o(\xi-\alpha G_x)\ g(x)dx$$

is a finite Borel measure on $S'(\mathbb{R}^n)$. One has for the total variation $\| \mu_\Lambda^{\alpha,g} \| \leq \int_\Lambda g(x)dx$.

In particular we can take g to be the characteristic function χ_Λ of a Borel set Λ of finite Lebesgue measure, obtaining $\mu_\Lambda^{\alpha,\chi_\Lambda}(\xi) \equiv \mu_\Lambda^\alpha(\xi)$. Let now ν_μ^α be the image in $S'(\mathbb{R}^n)$ of the restriction $d_\Lambda x$ of the Lebesgue measure to Λ under the mapping $x \to G_x$, where $x \in \mathbb{R}^n$. Then we have

$$\mu_\Lambda^\alpha(\xi) = \int_\Lambda \mu_o(\xi-\alpha G_x)dx = (\nu_\Lambda^\alpha * \mu_o)(\xi),$$

where * means convolution. The mapping $\Lambda \to \mu_\Lambda$ is a mapping from the Borel subsets of finite measure of $\mathbb{R}^n$ into $[0,\infty)$, which is positive and σ-additive in Λ i.e. $\Lambda \to \mu_\Lambda$ is a finite measure on the Borel subsets Λ of $\mathbb{R}^n$, bounded on compacts, and absolutely continuous with respect to dx. We shall call μ_Λ^α a local additive perturbation of μ_o. We can ask the question: is μ_Λ^α a measure on

$S'(\mathbb{R}^n)$ absolutely continuous with respect to μ_o?

If it is so i.e. if $\mu_\Lambda^\alpha << \mu_o$, then the Radon-Nikodym derivative $\frac{d\mu_\Lambda^\alpha}{d\mu_o}(\xi) \equiv U_\Lambda^\alpha(\xi)$ satisfies $U_\Lambda^\alpha \geq 0$, $U_\Lambda^\alpha \in L^1(d\mu_o)$ for any compact Λ, and $\Lambda \to U_\Lambda^\alpha$ is a function valued measure in Λ, absolutely continuous with respect to dx and $U_\Lambda^\alpha(\cdot)$ is a $\mathcal{B}_\Lambda$-measurable function. U_Λ^α is thus a positive additive functional of μ_o and $(\Lambda, U_\Lambda^\alpha)$ is an interaction. We now remark that for $n = 2$ and $|\alpha| < 2\sqrt{\pi}$ this is the exponential interaction constructed in [2]. In fact let χ_κ for any $\kappa > 0$ be the characteristic function of a ball of radius κ in Fourier transform space and consider for any $\xi \in S'(\mathbb{R}^2)$ the regularized distribution $\xi_\kappa(x) \equiv (\xi * \chi_\kappa)(x)$. Introduce the short notation: $:e^{\alpha\xi_\kappa(x)}:$ for $e^{-\frac{1}{2}\alpha^2 G_\kappa(0)}\, e^{\alpha\xi_\kappa(x)}$, where $G_\kappa \equiv \chi_\kappa * G * \chi_\kappa$ (one calls $:e^{\alpha\xi_\kappa(x)}:$ the Wick-ordered exponential). Clearly $:e^{\alpha\xi_\kappa(x)}:$ is positive and continuous in x. For any bounded Borel set $\Lambda \subset \mathbb{R}^2$, $U_{\Lambda,\kappa}^\alpha \equiv \int_\Lambda :e^{\alpha\xi_\kappa(x)}:dx$ exists, is in $L^1(d\mu_o)$ and one has, by a direct computation,

$U_{\Lambda,\kappa}^\alpha = \frac{d\mu_{\Lambda,\kappa}^\alpha}{d\mu_o}$, where $\mu_{\Lambda,\kappa}^\alpha(\xi) \equiv \int_\Lambda \mu_o(\xi_\kappa - \alpha G_x)\,dx_o$. We showed in [2] that for $|\alpha| < 2\sqrt{\pi}$ the $L^2(d\mu_o)$-limit of $U_{\Lambda,\kappa}^\alpha$ as $\kappa \to \infty$ exists and is equal to $U_\Lambda^\alpha = \frac{d\mu_\Lambda^\alpha}{d\mu_o}$. It is natural to write this also in the form

$U_\Lambda^\alpha = \int_\Lambda :e^{\alpha\xi(x)}:dx$, where the distribution $:e^{\alpha\xi(x)}:$ is defined for a.e. $\xi \in S'(\mathbb{R}^2)$ as the element $<:e^{\alpha\xi(\cdot)}:,g>$ in $L^2(d\mu_o)$, $g \in S(\mathbb{R}^2)$, such that $\int <:e^{\alpha\xi(\cdot)}:,g>\, e^{i<\xi,f>} d\mu_o =$

$e^{-\frac{1}{2}(f,Gf)} \int e^{\frac{1}{2} i\alpha(G_x,f)} g(x)dx$. It follows from the construction, see [2], that $(\Lambda, U_\Lambda^\alpha)$ is indeed an interaction, the "exponential interaction" (of index α). One has $U_\Lambda^\alpha \in L^2(d\mu_o)$, $U_\Lambda^\alpha \geq 0$, $e^{-U_\Lambda^\alpha} \in L^\infty(d\mu_o)$. Thus for $n = 2$, $|\alpha| < 2\sqrt{\pi}$ the answer to the above question is positive: μ_Λ^α is indeed absolutely continuous with respect to μ_o and the Radon-Nikodym derivative is the exponential interaction.

Remark:

We showed in [2] that, for $n = 2$ and $|\alpha| < \frac{4}{\sqrt{\pi}}$, $e^{-U^\alpha_\Lambda} d\mu_o / \int e^{-U^\alpha_\Lambda} d\mu_o$ converges weakly as $\Lambda \uparrow \mathbb{R}^2$ to a Euclidean invariant measure, the Euclidean measure for the exponential interaction. This was improved to all $|\alpha| < 2\sqrt{\pi}$ in [6] .

Remark:

The fact that for $n = 2$, $|\alpha| < 2\sqrt{\pi}$ one has $U^\alpha_\Lambda \in L^2(d\mu_o)$ is seen from the explicit formula

$$\int |U^\alpha_\Lambda|^2 \, d\mu_o = \int_\Lambda \int_\Lambda e^{\alpha^2 G(x-y)} dxdy,$$

observing that $G(x-y) \sim -\frac{1}{2\pi} \ln |x-y|$ as $x \to y$.

We shall now study the situation $n > 2$. We recall that for $n = 2$, $|\alpha| < 2\sqrt{\pi}$ we had $d\mu^\alpha_\Lambda(\xi) = U^\alpha_\Lambda(\xi) d\mu_o(\xi)$, with $U^\alpha_\Lambda(\xi)$ the $L^2(d\mu_o)$-limit as $\kappa \to \infty$ of the regularized exponential interaction $\int_\Lambda : e^{\alpha\xi_\kappa(x)} : dx$. It is thus natural to look also for $n > 2$ at the approximation $\mu^\alpha_{\Lambda,\kappa}(\xi)$ to $\mu^\alpha_\Lambda(\xi)$, where $\mu^\alpha_{\Lambda,\kappa}(\xi) \equiv \int_\Lambda \mu_o(\xi_\kappa - \alpha G_x) dx$, with $\xi_\kappa = \xi * \chi_\kappa$ and χ_κ the function on $\mathbb{R}^n$ whose Fouriertransform is the characteristic function of a ball of radius κ. We have $\mu^\alpha_{\Lambda,\kappa}(\xi) = \mu^\alpha_{\Lambda,\kappa}(P_\kappa \xi) = (\nu^\alpha_\Lambda * \mu_o)(P_\kappa \xi) =$

$$= \int d\nu^\alpha_\Lambda(\phi) \, d\mu_o(P_\kappa \xi - \phi) = \int d\nu^\alpha_\Lambda(\phi) \, e^{(P_\kappa \xi, G\phi)} = e^{-\frac{1}{2}(P_\kappa \phi, G P_\kappa \phi)} d\mu_o(P_\kappa \xi) =$$

$= \int_\Lambda : e^{\alpha\xi_\kappa(x)} : dx$, where $P_\kappa \phi \equiv \phi * \chi_\kappa$ for any $\phi \in S(\mathbb{R}^n)$, $(,)$ is the scalar product in $L^2(dx)$ and we used the definition $:e^{\alpha\xi_\kappa(x)}: \equiv e^{-\frac{\alpha^2}{2} G_\kappa(0)} e^{\alpha\xi_\kappa(x)}$ and the définition of ν^α_Λ.

Hence we have proven the following

Lemma 2.1:

Let Λ be a Borel subset of $\mathbb{R}^n$ of finite Lebesgue measure. Let χ_κ be the function on $\mathbb{R}^n$ whose Fourier transform is the charac-

teristic function of the ball of radius $\kappa > 0$.
Let $\mu^{\alpha}_{\Lambda,\kappa}(\xi) \equiv \int_{\Lambda} \mu_0(\xi_\kappa - \alpha G_x)dx = \nu^{\alpha}_{\Lambda} * \mu_0(\xi_\kappa)$, where $\xi_\kappa \equiv \xi * \chi_\kappa$, $\alpha \in \mathbb{R}_0$.

Then $\mu^{\alpha}_{\Lambda,\kappa}$ is absolutely continuous with respect to μ_0 and

$$U^{\alpha}_{\Lambda,\kappa}(\xi) \equiv d\mu^{\alpha}_{\Lambda,\kappa} / d\mu_0(\xi) = \int_{\Lambda} :e^{\alpha\xi_\kappa(x)}:dx, \text{ with}$$

$$:e^{\alpha\xi_\kappa(x)}: \quad \equiv \quad e^{-\frac{\alpha^2}{2} G_\kappa(0)} \, e^{\alpha\xi_\kappa(x)}.$$

$(\Lambda,\mu^{\alpha}_{\Lambda,\kappa})$ is called the regularized exponential interaction in $\mathbb{R}^n$.

We would like now to study the limit $\kappa \to \infty$ of $U^{\alpha}_{\Lambda,\kappa}$, as we did for $n = 2$, $|\alpha| < 2\sqrt{\pi}$. To do this it is useful to look at the corresponding question in a more abstract setting.

3. Martingale Convergence of Translates of Gaussian Measures

Let $H,\bar{H}$ be separable real Hilbert spaces, such that the injection of H in $\bar{H}$ is Hilbert-Schmidt. The standard normal distribution associated with H can be realized, by a well known theorem of Minlos-Sazonov, as a Borel probability measure μ on $\bar{H}$, with Fourier transform $e^{-\frac{1}{2}(y,y)}$, where $y \in H$ and $(,)$ is the scalar product in H. Let $\underline{H}$ be the dual of $\bar{H}$ with respect to the scalar product (x,y) in H i.e. $\underline{H}$ is the subspace of H consisting of vectors y such that $x \to (x,y)$ is continuous, for $x \in H$, in the $\bar{H}$ topology. If $y \in H$ we have that

$$d\mu\,(x+y)/d\mu\,(x) = e^{-\frac{1}{2}|y|^2} \, e^{-<x,y>}, \text{ for } x \in \bar{H}, \qquad (3.1)$$

where $<,>$ is the dualization between $\bar{H}$ and $\underline{H}$ and $|\cdot|^2 = (\cdot,\cdot)$, and if $y \notin H$ we have that $\mu\,(\cdot+y)$ is singular with respect to $\mu\,(\cdot)$.
Let ν be a probability measure on $\bar{H}$. Let $F\,C\,(\bar{H})$ be the space of bounded continuous cylinder functions on $\bar{H}$ i.e. $f \in F\,C\,(\bar{H})$ iff f is bounded continuous on $\bar{H}$ and there exists some finite dimensional orthogonal projection P on H (extended to $\bar{H}$) such that $P\,H \subset \underline{H}$ and $f(x) = f(Px)$ for all $x \in \bar{H}$. Then we have

$$\int f\, d(\nu * \mu\,) = \int_{\bar{H}} \int_{\bar{H}} f\,(x)\, d\nu(y)\, d\mu\,(x-y) =$$

$$= \int_{\bar H}\int_{\bar H} f(x+y)\, d\nu(y)\, d\mu\,(x) = \int_{\bar H}\int_{\bar H} f(Px+Py)\, d\nu(y)\, d\mu\,(x), \tag{3.2}$$

where we used $f(x) = f(Px)$. But by setting $z = x + Py$ we see that

$$\iint f(Px + Py)\, d\nu(y)\, d\mu\,(x) = \iint f(Px + z - x)\, d\mu(y)\, d\mu(z-py)$$

$$= \iint f\,(Px + Pz - Px)\, d\nu(y)\, d\mu\,(z - Py) = \tag{3.3}$$

$$= \iint f\,(Pz)\, d\nu(y)\, d\mu\,(z - Py),$$

where we used again $f(x) = f(Px)$.

Now by the properties of Gaussian measures we have

$$\int f(Pz)\, d\mu\,(z-z') =$$

$$= \int_{(\bar H \ominus PH)}\int_{PH} \int f(Pz)\, d\mu\,(P(z-z'))\, d\mu\,((1-P)(z-z')) = \tag{3.4}$$

$$= \int_{PH} f(Pz)\, d\mu\,(Pz-Pz'),\ \text{for all } z.z' \in \bar H.$$

But $\dfrac{d\mu\,(Pz-Pz')}{d\mu\,(Pz)} = e^{-\frac{1}{2}|Pz'|^2}\, e^{(z,Pz')}$.

Hence we see from (3.2) - (3.5) that

$$\int f\, d(\nu * \mu_H) = \int_{P\bar H} (f\circ P)(x) \int_{\bar H} e^{-\frac{1}{2}|Py|^2}\, e^{(Px,Py)}\, d\nu(y)\, d\mu\,(Px) \tag{3.6}$$

$$= \int_{P\bar H} (f\circ P)(x)\, m_P(x)\, d\mu\,(Px),$$

with (3.7)

$$m_P(x) \equiv \int_{\bar H} e^{-\frac{1}{2}|Py|^2}\, e^{(Px,Py)}\, d\nu(y),\ \text{for any } x \in \bar H.$$

Note that $m_P(x) = m_P(Px)$. Let B_P be the σ-algebra of $\bar H$ generated by the linear functions $x \to (x,y)$, $y \in P\bar H$. We have $B_{P'} \supseteq B_P$ for

$P' \geq P$ (in the sense that $P'\bar{H} \geq P\bar{H}$). Note that m_P is positive, B_P-measurable and in $L^1(d\mu)$ (since

$\int m_P(x)\, d\mu(x) = \int \frac{d\mu(x-Py)}{d\mu(x)}\, d\nu(y)\, d\mu(x) \leq 1$). Moreover we have denoting by $E(\cdot|B_P)$ the conditional expectation with respect to B_P:

$$E\Big(\int_{\bar{H}} e^{-\frac{1}{2}|P'y|^2} e^{(x,P'y)}\, d\nu(y)\,\Big|\,B_P\Big) = E\Big(\int_{\bar{H}} e^{-\frac{1}{2}|P'y|^2} e^{(Px,P'y)}\, d\nu(y)\,\Big|\,B_P\Big) = m_P.$$

μ a.s. Hence (m_P, B_P) is a positive martingale. By the known convergence theorem for martingales (see e.g. [23]) we have then that $m_P d\mu$ converges weakly as $P \uparrow 1$ along the filter of finite dimensional projections on $\bar{H}$ with range in $\underline{H}$ to the measure $\nu * \mu$. Hence we have[1)]

Theorem 3.1:

Let $H, \bar{H}$ be real separable Hilbert space such that the injection of H in $\bar{H}$ is a Hilbert-Schmidt operator. Let $\underline{H}$ be the dual of $\bar{H}$ with respect to the inner product in H. Let μ be the version on $\bar{H}$ of the standard normal distribution associated with H. Let ν be a Borel probability measure on $\bar{H}$. Let $\underline{P}$ be the filter of orthogonal projections P on H with PH a finite dimensional subspace of $\underline{H}$. Extend P naturally by continuity to $\bar{H}$. Then for $P \in \underline{P}$

$$m_P(x) \equiv \int_{\bar{H}} e^{-\frac{1}{2}|Py|^2} e^{(x,Py)}\, d\nu(y)$$

is a positive martingale with respect to the filter of σ-algebras B_P, $P \in \underline{P}$ generated by the linear functions $x \to (x,y)$, $y \in PH$ and the probability measure μ. One has in the sense of weak convergence of measures

$$\lim_{\underline{P}} m_P\, \mu = \nu * \mu . \qquad \square$$

---

1) This corresponds to Theorem 2.3 of [12].

Remark

From the definition of m_P one sees that

$$\int |m_P(x)|^2 \, d\mu_H(x) = \iint e^{(x,Py)} \, d\nu(x) \, d\nu(y). \tag{3.8}$$

From this one has the[2)]

Theorem 3.2

Let ν,μ be as in Theorem 3.1. Then $\nu*\mu$ is absolutely continuous with respect to μ with Radon-Nikodym derivative $m(x)$ in $L^2(d\mu\ )$ if and only if

$$\iint e^{(x,Py)} \, d\nu(x) \, d\nu(y)$$

is uniformly bounded for $P \in \underline{P}$. □

Recalling the result on $e^{-U^{\alpha}_{\Lambda,\kappa}}$ as $\kappa \to \infty$ discussed in Section 2, we shall now study $e^{-\lambda m_P(x)}$ as $P \uparrow 1$, for $\lambda \geq 0$. Since $m_P(x)$ is a positive martingale and $\beta \to e^{-\lambda\beta}$ is for all $\lambda \geq 0$ a convex function on the real line, we have that $e^{-\lambda m_P(x)}$ is a bounded submartingale. In fact by Jensen's inequality for $P' \geq P$:

$E(e^{-\lambda m_{P'}} | B_P)(x) \geq e^{-\lambda E(m_{P'}|B_P)}(x) \geq e^{-\lambda m_P}(x)$. By the submartingale convergence theorem (see e.g. [23]) we then have that $e^{-\lambda m_P(x)}$ converges pointwise and strongly in $L^1(d\mu_H)$ to a limit which we denote by $e^{-\lambda m(x)}$. Hence we have the following theorem[3)]:

Theorem 3.3

Let $\mu\ ,\nu,m_P$ be as in Theorem 3.1. Then for all $\lambda \geq 0$

$$\lim_{\underline{P}} e^{-\lambda m_P(x)} = e^{-\lambda m(x)}$$

μ -almost surely and strongly in $L^1(d\mu\ )$. In particular we have

---

2) This is Theorem 2.4 of [12]

3) This is Theorem 2.6 of [12]

$$e^{-\lambda m_P(x)} d\mu(x) \to e^{-\lambda m(x)} d\mu(x)$$

weakly as measures on $\bar{H}$. □

Remark

The results in Theorem 3.1, 3.2, 3.3 hold also, as one sees easily looking at the definition of m_P, when one replaces the filter $\underline{P}$ of orthogonal projection with finite dimensional range in $\underline{H}$ by any filter of orthogonal projections P, such that the norm $|Py|$ is ν-measurable, for any $y \in \bar{H}$.

We shall see in Section 4 how we can apply these results to control the limit of $e^{-U^{\alpha}_{\Lambda,\kappa}} d\mu_o \,|\int e^{-U^{\alpha}_{\Lambda,\kappa}} d\mu_o$ as $\kappa \to \infty$ for the exponential interaction.

It is of course interesting, in general and in view of this application, to have criteria that enable to identify the above limit function m(x), in particular to decide when $\nu*\mu$ is singular with respect to μ and m(x) is μ-almost surely zero.

Theorem 3.4[4)]

Let $\underline{H} \subset H \subset \bar{H}$ be a Hilbert-Schmidt rigging of the real separable Hilbert space H, as in Theorem 3.1. Let $A \geq 0$ be a Hilbert-Schmidt operator on H with eigenvalues λ_i and normalized eigenvectors $e_i \in \underline{H}$. Let ν be a Borel probability measure on $\bar{H}$. Then if $(x,Ax) = \Sigma\, \lambda_i (x,e_i)^2 = \infty$ for ν-almost all $x \in \bar{H}$, then one has $\nu*\mu \perp \mu$.

Proof:

Consider $\sum_{i=1}^{\infty} [\lambda_i (x,e_i)^2 - \lambda_i]$. It is easily seen that the sum is $L^2(d\mu)$-convergent and thus defines a function in $L^2(d\mu)$, which we denote by $:(x,Ax):$. Then $e^{-\frac{1}{2}:(x,Ax):}$ is a positive function on $\bar{H}$ which is strictly positive μ-almost everywhere. On the other hand we have

---

4) This is Theorem 2.5 of [12].

$$\int_{\bar{H}} e^{-\frac{1}{2}:(x,Ax):} d(\nu*\mu)(x) = \int_{\bar{H}}\int_{\bar{H}} e^{-\frac{1}{2}:(x+y,A(x+y)):} d\mu(x)$$

$$d\nu(y) = \int_{\bar{H}}\int_{\bar{H}} e^{-\frac{1}{2}:(x,Ax):} e^{(x,Ay)} d\mu(x)\, e^{-\frac{1}{2}(y,Ay)} d\nu(y). \tag{3.9}$$

Let H_{1+A} be the Hilbert space obtained by completing H in the norm $|x|^2_{1+A} = |x|^2 + (x,Ax)$. It is easy to see that the normal distributions associated with H_{1+A} and H resp., $\mu_{H_{1+A}}$ and $\mu_H \equiv \mu$, are equivalent measures on $\bar{H}$ and in fact

$$d\mu_{H_{1+A}} / d\mu(x) = c\, e^{-\frac{1}{2}:(x,Ax):}, \tag{3.10}$$

where c is a finite constant. Using this we have

$$\int_{\bar{H}} e^{-\frac{1}{2}:(x,Ax):} e^{(x,Ay)} d\mu(x) = $$
$$= c^{-1}\int e^{(x,Ay)} d\mu_{H_{1+A}}(x) = c^{-1} \exp \frac{1}{2}(y, \frac{A^2}{1+A} y). \tag{3.11}$$

Now using

$$A^2(1+A)^{-1} = -A(1+A)^{-1} \tag{3.12}$$

we get from (3.9) and (3.11) that (3.9) is equal to

$$c^{-1}\int e^{-\frac{1}{2}(y,\frac{A}{1+A}y)} d\nu(y). \tag{3.13}$$

Now

$$(y, \frac{A}{1+A} y) \geq \|1+A\|^{-1} (y,Ay). \tag{3.12}$$

By the assumption we have $(y,Ay) = \infty$ for ν-a.e. y, thus the integrand $e^{-\frac{1}{2}(y,\frac{A}{1+A}y)}$ in (3.13) vanishes ν-a.e., so that (3.13) is zero, which then implies that (3.9) is zero i.e.

$$e^{-\frac{1}{2}:(x,Ax):} d(\nu*\mu) = 0. \tag{3.15}$$

This implies $e^{-\frac{1}{2} :(x,Ax):} = 0$ $\nu * \mu$ -a.e. We remarked at the beginning, however, that $e^{-\frac{1}{2} :(x,Ax):} > 0$ μ -a.e. This then proves the theorem.

Remark

Recently this result has been sharpened considerably by W.D. Wick [24]. He also gives other necessary conditions and, in the case where ν is Gaussian, necessary and sufficient conditions for $\nu * \mu \ll \mu$.

Theorem 3.5 [5)]

Under the same assumption as in Theorem 3.4 we have

$$\lim_{P \in \underline{P}_\nu} e^{-\lambda m_P(x)} = 1$$

μ -almost surely and strongly in $L^1(d\mu)$, where the limit is taken along the filter $\underline{P}_\nu$ of orthogonal projections P on $\bar{H}$ which are such that the norm $|Py|$, $y \in \bar{H}$, is ν-measurable. In particular one has

$$e^{-\lambda m_P(x)} d\mu (x) \rightarrow d\mu (x)$$

weakly as measures on $\bar{H}$.

Proof:

Let P_n be the orthogonal projection in H onto the subspace spanned by the eigenvectors $e_1, \dots, e_n$ of A. By assumption $e_i \in \underline{H}$, hence $P_n \in \underline{P}_\nu$. We compute

$$\int_{\bar{H}} e^{-\frac{1}{2} :(x.Ax):} m_{P_n}(x) \, d\mu_H(x) =$$

$$= \int e^{-\frac{1}{2} :(x,P_n Ax):} m_{P_n}(x) \, d\mu_{P_n H}(x) \int e^{-\frac{1}{2} :(x,(1-P_n)Ax):} d\mu_{(1-P_n)H}(x) \tag{3.16}$$

5) This is Theorem 2.7 of [12].

By (3.6) and the definition (3.7) of m_P we have

$$\int e^{-\frac{1}{2}:(x,P_nAx):} m_{P_n}(x)\, d\mu_{P_nH}(x) =$$

$$= \int e^{-\frac{1}{2}:(x,P_nAx):} d(\nu*\mu_{P_nH})(x) \tag{3.17}$$

and since $:(P_nx,\ P_nAP_nx): = :(x,P_nAx):$, we have by (3.6) that

$$\int e^{-\frac{1}{2}:(x,P_nAx):} d(\nu*\mu_{P_nH}) = \int e^{-\frac{1}{2}:(x,P_nAx):} d(\nu*\mu_H)(x). \tag{3.18}$$

Now let $a_n \equiv \int e^{-\frac{1}{2}:(x,(1-P_n)Ax):} d\mu_{(1-P_n)H}(x)$. (3.19)

We compute, using the definition of $:(x,Ax):$

$$a_n = \prod_{i=n+1}^{\infty} e^{\frac{1}{2}\lambda_i}(1+\lambda_i)^{-1/2} = \tag{3.20}$$

$$= \exp\left(\frac{1}{2}\sum_{i=n+1}^{\infty}(\lambda_i - \ln(1+\lambda_i))\right). \tag{3.21}$$

Since the series $\sum_i \lambda_i - \ln(1-\lambda_i)$ is convergent (due to $\Sigma\ \lambda_i^2 < \infty$), we see that $a_n \to 1$ as $n \to \infty$. From (3.16) - (3.19) we have

$$\int_{\bar{H}} e^{-\frac{1}{2}:(x,Ax):} m_{P_n}(x)\, d\mu\ (x) = a_n \int e^{-\frac{1}{2}:(x,P_n\ Ax):} \tag{3.22}$$

$$d(\nu*\mu\ )(x).$$

On the other hand from the equality of (3.9) and (3.13), with A replaced by P_nA, we get

$$\int e^{-\frac{1}{2}:(x,P_nAx):} d(\nu*\mu\ ) = c_n^{-1} \int e^{-\frac{1}{2}\left(y,\ \frac{P_nA}{1+P_nA}\, y\right)} d\nu(y) \tag{3.23}$$

with c_n given by

$$c_n^{-1} = \int e^{-\frac{1}{2}:(x,P_nAx):} d\mu\ (x) = \prod_{i=1}^{n} e^{\frac{1}{2}\lambda_i}(1+\lambda_i)^{-1/2}. \tag{3.24}$$

By (3.22) - (3.24) we then have

$$\int e^{-\frac{1}{2}:(x,Ax):} m_{P_n}(x)\, d\mu\,(x) = a \int e^{-\frac{1}{2}\left(y,\, \frac{P_nA}{1+P_nA}\, y\right)} d\nu(y), \tag{3.25}$$

where $a \equiv \prod_{i=1}^{\infty} e^{\frac{1}{2}\lambda_i}(1+\lambda_i)^{-1/2}$ is the inverse square root of the modified Fredholm determinant of A. Now

$$e^{-\frac{1}{2}\left(y,\, \frac{P_nA}{1+P_nA}\, y\right)} \to e^{-\frac{1}{2}\left(y,\, \frac{A}{1+A}\, y\right)} \tag{3.26}$$

for all $y \in \bar{H}$, as $n \to \infty$. By the assumption, however, $(y,Ay) = \infty$ for ν-a.e. $y \in \bar{H}$, and this implies that $(y,\frac{A}{1+A}\, y) = \infty$ for ν-a.e. $y \in \bar{H}$. Thus (3.26) can be replaced by

$$e^{-\frac{1}{2}\left(y,\, \frac{P_nA}{1+P_nA}\, y\right)} \to 0 \text{ for } \nu\text{-a.e. } y \in \bar{H}.$$

By dominated convergence the right hand side of (3.25) then converges to zero and we have thus proven

$$\int e^{-\frac{1}{2}:(x,Ax):} m_{P_n}(x)\, d\mu\,(x) \to 0 \tag{3.27}$$

as $n \to \infty$. On the other hand by Theorem 3.3

$$e^{-\frac{1}{2}:(x,Ax):} m_{P_n}(x) \to e^{-\frac{1}{2}:(x,Ax):} m(x) \tag{3.28}$$

μ -a.e. By Fatou Lemma we then have

$$\int e^{-\frac{1}{2}:(x,Ax):} m_{P_n}(x)\, d\mu\,(x) \to \int e^{-\frac{1}{2}:(x,Ax):} m(x)\, d\mu \tag{3.29}$$

By (3.29) and (3.27) we see that

$$\int e^{-\frac{1}{2}:(x,Ax):} m(x)\, d\mu\,(x) = 0. \tag{3.30}$$

Since $e^{-\frac{1}{2}:(x,Ax):} > 0$ for μ -almost all $x \in \bar{H}$ we get then

$m(x) = 0$ μ -a.e. This proves the theorem. □

In the next section we apply these results to the study of the exponential interaction in $\mathbb{R}^n$.

4. Application of the Results to the Exponential Interaction

Let us now consider the free Euclidean field measure μ_o on $S'(\mathbb{R}^n)$, described in Section 2. In the framework of Section 3 we have $H = \mathcal{H}_1(\mathbb{R}^n)$ and $\bar{H}$ is taken to be a real separable Hilbert space of tempered distributions such that supp $\mu_o = \bar{H}$. We can as in Section 2 consider μ_o as a probability measure on $S'(\mathbb{R}^n) \supset \bar{H}$. We have $\underline{H} \subset \mathcal{H}_{-1}(\mathbb{R}^n)$.

Let χ_κ be as in Section 2 i.e. the Fourier transform of the characteristic function of the ball of radius κ in $\mathbb{R}^n$.
$P_\kappa: \phi \to P_\kappa\phi \equiv \chi_\kappa * \phi$ is then an orthogonal projection on H.
Let Λ be a Borel subset of $\mathbb{R}^n$ with finite Lebesgue measure and let ν_Λ^α be as in Section 2 i.e. the image of the restriction of the Borel measure on $\mathbb{R}^n$ to Λ by the mapping $x \to \alpha G_x$ of $\mathbb{R}^n$ into $S'(\mathbb{R}^n)$, where α is real and $G_x(y) \equiv G(x-y)$, $G(x-y)$ being the kernel of $(-\Delta+m^2)^{-1}$ in $L^2(\mathbb{R}^n)$. It can be shown easily that one can apply to the pair $(P_\kappa, \nu_\Lambda^\alpha)$ the consideration of the pair (P_ν, ν) of Section 3, in particular $\phi \to |P_\kappa\phi|$ being ν_Λ^α-measurable. The quantity m_P defined in Section 3 becomes in this case the quantity

$$U_{\Lambda,\kappa}^\alpha(\xi) = \frac{\int d\mu_o(\xi_\kappa - \alpha G_x)\,dx}{d\mu_o(\xi_\kappa)} = \frac{d(\nu_\Lambda^\alpha * \mu_o)(\xi_\kappa)}{d\mu_o(\xi_\kappa)} =$$

$$= e^{-\frac{1}{2}\alpha^2 G_\kappa(0)} \int_\Lambda e^{\alpha\xi_\kappa(x)}\,dx = \int_\Lambda :e^{\alpha\xi_\kappa(x)}:\,dx$$

already discussed in Section 2, where $\xi \in S'(\mathbb{R}^n)$, $\xi_\kappa \equiv \xi * \chi_\kappa$, $G_\kappa(\cdot) = \chi_\kappa * G * \chi_\kappa(\cdot)$. We then have, from Theor. 3.4, 3.5:

Theorem 4.1[6)]

Let n be an arbitrary natural number and let μ_o be the probability

6) This corresponds to Theorem 3.2 of [12].

measure on $S'(\mathbb{R}^n)$ given by the free Euclidean field ξ of mass m. Let Λ be any Borel measurable subset of $\mathbb{R}^n$ with finite Lebesgue measure. Then the regularized exponential interaction

$$U^{\alpha}_{\Lambda,\kappa}(\xi) = \int_{\Lambda} : e^{\alpha\xi_{\kappa}(x)} : dx$$

has the property that for any $\lambda > 0$

$$\lim_{\kappa\to\infty} e^{-\lambda U^{\alpha}_{\Lambda,\kappa}(\xi)} \quad \text{exists,}$$

for μ_o-a.e. $\xi \in S'(\mathbb{R}^n)$ and strongly in $L^1(d\mu_o)$. Call $e^{-\lambda U^{\alpha}_{\Lambda}(\xi)}$ the limit. Clearly $U^{\alpha}_{\Lambda} \geq 0$. We have in particular

$$e^{-\lambda U^{\alpha}_{\Lambda,\kappa}} d\mu_o \to e^{-\lambda U^{\alpha}_{\Lambda}} \; d\mu_o$$

weakly as measures on $S'(\mathbb{R}^n)$. □

This theorem defines the exponential interaction $(\Lambda, U^{\alpha}_{\Lambda})$ in $\mathbb{R}^n$ having the properties required in Section 2. For $n = 2$, $|\alpha| < 2\sqrt{\pi}$ we already recalled in Section 2 that $U^{\alpha}_{\Lambda} \geq 0$, $U^{\alpha}_{\Lambda} \in L^2(d\mu_o)$ [2].

We shall now see that for $n \geq 4$ we have $U^{\alpha}_{\Lambda} = 0$ μ_o-a.e. To prove this we follow the lines of Theorem 3.4 and 3.5.

Let Λ be a compact subset of R^n with a non empty interior and set

$$(\xi, S_{\Lambda}\xi) = \int_{\Lambda}\int_{\Lambda} S(x-y)\; \xi(x)\; \xi(y)\; dx\; dy \tag{4.1}$$

with (,) the inner product in H and

$$S(x-y) = (2\pi)^{-n} \int e^{\,i(x-y)p}\, d\sigma(p)\; dp \tag{4.2}$$

where $\sigma(p) = \sigma(-p)$ is a positive and tempered (in the sense of tempered distributions) (not necessarily bounded) measure on $\mathbb{R}^n$. The Hilbert-Schmidt norm of the operator S_{Λ} in $H_1(R^n)$ is then given by

$$\int_{\Lambda}\!..\!\int_{\Lambda} S(x-y)\; G(y-z)\; S(z-w)\; G(w-x)\; dx\; dy\; dz\; dw. \tag{4.3}$$

Since S(x-y) is tempered and G(y-z) tends to zero exponentially as $|y-z| \to \infty$ we have that (4.3) is finite if and only if

$$\int_{\Lambda} \int_{R^n} \int_{R^n} \int_{R^n} S(x-y_1)\ G(y_1-y_2)\ S(y_2-y_3)\ G(y_3-x)\ dy_1\ dy_2\ dy_3\ dx \tag{4.4}$$

is finite. From translation invariance however (4.4) is equal to

$$|\Lambda| \int_{R^n} \int_{R^n} \int_{R^n} S(-y_1)\ G(y_1-y_2)\ S(y_2-y_3)\ G(y_3)\ dy_1\ dy_2\ dy_3. \tag{4.5}$$

By Parseval identity we have that

$$\int_{R^n} \int_{R^n} \int_{R^n} S(-y_1)\ G(y_1-y_2)\ S(y_2-y_3)\ G(y_3)\ dy_1\ dy_2\ dy_3$$

$$= \int_{R^n} \sigma(p)^2\ (p^2+m^2)^{-2}\ dp. \tag{4.6}$$

Take now

$$\sigma(p) = (p^2+m^2)^{-\frac{n}{4}+1} (\ln\ (p^2+m^2+1))^{-1}. \tag{4.7}$$

Then (4.6) is finite and therefore S_Λ is a Hilbert-Schmidt operator in $H_1(R^n)$. Let us compute with $\xi(y) = G_x(y) = G(x-y)$, from (4.1) we have

$$(G_x, S_\Lambda\ G_x) = \int_{\Lambda} \int_{\Lambda} G(x-y)\ S(y-z)\ G(z-x)\ dy\ dz. \tag{4.8}$$

Again we have that (4.8) is finite if and only if

$$\int_{R^n} \int_{R^n} G(x-y)\ S(y-z)\ G(z-x)\ dy\ dz$$

$$= \int_{R^n} \sigma(p)\ (p^2+m^2)^{-2}\ dp \tag{4.9}$$

is finite. (4.9) is by (4.7) equal to

$$\int (p^2+m^2)^{-\frac{n}{4}-1} (\ln\ (p^2+m^2+1))^{-1}\ dp \tag{4.10}$$

which is finite if $n \leq 3$ and infinite if $n \geq 4$. If $n \geq 4$ we therefore have that $(\xi, S_\Lambda \xi) = \infty$ ν_Λ^α-almost everywhere, where S_Λ given by (4.1) is a Hilbert-Schmidt operator on $H_1(\mathbb{R}^n)$. By theorem 3.5 we then have $U_\Lambda^\alpha(\xi) = 0$ μ_o-almost everywhere on $S'(\mathbb{R}^n)$ and this implies that the measures $\int_\Lambda d\mu_o(\cdot - \alpha G_x)dx$ and $d\mu_o(\cdot)$ have disjoint support. Hence we have[7)]

Theorem 4.2

Let the assumptions be as in theorem 4.1. If the dimension n satisfies $n \geq 4$ then $U_\Lambda^\alpha(\xi) = 0$ μ_o-almost everywhere. In particular

$$e^{-\lambda m_{\Lambda,\kappa}^\alpha(\xi)} d\mu_o \to d\mu_o$$

weakly as measures on $S'(R^n)$. The measures $\int_\Lambda d\mu_o(\cdot - \alpha G_x)dx$ and $d\mu_o(\cdot)$ have disjoint support.

Remark

In connection with the lecture, G. Gallavotti pointed out to us that methods of [15] could be used together with a modification of the above method to control in a similar way the cases of dimension $n = 3$ and $n = 2$ for $|\alpha|$ sufficiently big. In fact it was then proven in [13] that for $n = 2$ there exists $\alpha_o > 0$ such that for all $|\alpha| > \alpha_o$ and all $\lambda \geq 0$ the exponential interaction is trivial (in the sense that the moments i.e. the Schwinger functions of the regularized interaction converge to the Schwinger functions of the free Euclidean field, as the regularization is removed); moreover it is shown in [13] that the same conclusion holds for all α if $n \geq 3$. For an application of the results for $n = 2$ see [24]. Recently the value $\sqrt{8\pi}$ for α_o has been found by Osipov [25], who has also a very nice method to derive the triviality of the Schwinger functions for $n \geq 3$ or $n = 2$, $|\alpha| > \sqrt{8\pi}$.

As mentioned in the introduction, the question of the triviality of the exponential interaction in $\mathbb{R}^n$ is very related to the question of the irreducibility of the energy representation of the group of mappings of a Riemannian manifold of dimension n

---

7) This corresponds to Theorem 3.3 of [12].

[14] S. Albeverio, G. Gallavotti, R. Høegh-Krohn, The exponential interaction in $\mathbb{R}^n$, Phys. Letters 83B, 177-179 (1979)

[15a] G. Gallavotti, these Proceedings

[15b] G. Benfatto, G. Gallavotti, F. Nicolò, Elliptic equations and Gaussian processes, IHES Preprint, Jan. 1979

[16] R. Ismagilov, On unitary representations of the group $C_o^\infty(X,G)$, $G = SU_2$, Math Sbornik 100 (2), 117-131 (1976) (transl. Math. USSRSb. 29, 105-117 (1976))

[17] S. Albeverio, R. Høegh-Krohn, The energy representation of Sobolev-Lie groups, Compos. Mathem. 36, 37-52 (1978)

[18] A.M. Vershik, I.M. Gelfand, M.I. Graev, Representation of the group of smooth mappings of a manifold into a compact Lie group, Compositio Mathematica 35, 299-334 (1977)

[19] A.M. Vershik, I.M. Gelfand, M.I. Graev, Remarks on the representation of the group of functions with values in a compact Lie group, Preprint N° 17, Inst. Prikl. Mat., Ak. Nauk (Moscow 1979) (russ.) (to appear in Compos. Mathem.)

[20] S. Albeverio, R. Høegh-Krohn, D. Testard, Irreducibility and reducibility for the energy representation of Sobolev-Lie groups, Bochum Preprint 1980

[21] J. Glimm, A. Jaffe, T. Spencer, The particle structure of the weakly coupled $P(\phi)_2$ model and other applications of high temperature expansions, in Constructive quantum field theory, Edts. G. Velo, A.S. Wightman, Lecture Notes in Physics 25, pp. 132-242, Springer, Berlin (1973)

[22] J. Fröhlich, E. Seiler, The massive Thirring-Schwinger model $(QED)_2$: convergence of perturbation theory and particle structure, Helv. Phys. Acta 49, 889-924 (1976)

[23] P. A. Méyer, Probabilités et potentiel, Hermann, Paris (1966)

[24] W.D. Wick, On the absolute continuity of a convolution with an infinite dimensional Gaussian measure, University of Washington, Seattle, Sept. 1979, Preprint

[25] E.P. Osipov, On the triviality of the $:\exp\lambda\phi:_4$ quantum field theory in a finite volume, Preprint TPh-103, USSR Acad. Sci. Siberian Div. January 1979.

[23] [illegible] Meyer, Probabilités et potentiel, Hermann, Paris (1966)

[24] W.D. Wick, On the absolute continuity of a convolution with an infinite dimensional Gaussian measure, University of Washington, Seattle, Sept. 1979, Preprint

[25] E.P. Osipov, On the triviality of the :exp φ:$_2$ quantum field theory in a finite volume, Preprint TPh-10?, USSR Acad. Sci. Siberian Div. January 1979.

On Dia- and Paramagnetic Properties of Yang-Mills Potentials

H. Hogreve

R. Schrader

R. Seiler

Institut für Theoretische Physik

Freie Universität Berlin

Abstract

Schrödinger Hamilton operators describing particles moving in external Yang-Mills potentials are considered. The dia- and paramagnetic properties of scalar and spin 1/2 particles respectively are discussed.

The increasing interest in gauge field theories naturally leads one to consider Hamilton operators in external Yang-Mills potentials A. These are operators obtained from ordinary quantum mechanical Hamilton operators by the substitution $p \to p+eA$. In particular one would like to know properties of the Laplacean $\Delta_A = -(p+eA)^2$ and the spinor Laplacean $\not\Delta_A = -(\not p+e\not A)^2$. They act on the Hilbert spaces of Lebesgue square integrable functions on $\mathbb{R}^n$ with values in the internal symmetry space L and the space $L \otimes S$ respectively, where S is the finite dim.spin space. Also $\not p = \sum_{\nu=1}^{n} p_\nu \gamma^\nu$ with γ^ν being hermitean matrices on S satisfying $\gamma^\nu\gamma^\mu + \gamma^\mu\gamma^\nu = 2\delta^{\mu\nu}$. A is a function on $\mathbb{R}^n$ with values in the hermitean operators on L. Only local regularity conditions are needed to make Δ_A and $\not\Delta_A$ selfadjoint operators [1]. An important tool to control Δ_A is Kato's inequality [1], [2], by which Δ_A is compared with the ordinary Laplacean Δ on $\mathbb{R}^n$ in the following way: Consider the following absolute value map

$$| \ | : L^2(\mathbb{R}^n, L) \to L^2(\mathbb{R}^n)$$

given as

$$|f|(x) = \langle f(x), f(x)\rangle^{1/2}$$

where $\langle \ \rangle$ is the scalar product in L. Also for an arbitrary unit vector e in L let

$$\operatorname{sign}_e f(x) = \begin{cases} \frac{1}{|f|(x)} f(x) & \text{if } f(x) \neq 0 \\ e & \text{otherwise} \end{cases} .$$

Then in the sense of distributions

$$\Delta|f| \geq \operatorname{Re} \langle \Delta_A f, \operatorname{sign}_e f\rangle .$$

This inequality has an equivalent formulation in terms of the corresponding contraction semigroups [3], [2]. To cover the more general case, let V be an ordinary potential, i.e. a multiplication operator on $L^2(\mathbb{R}^n)$ and set

$$H = -\Delta + V$$

$$H_A = -\Delta_A + V \otimes \mathbb{1}_S$$

$$H_A^S = -\not{\Delta}_A + V \otimes \mathbb{1}_L \otimes \mathbb{1}_S .$$

We will assume $V \geq 0$ and sufficiently regular to make these Hamiltonians selfadjoint. Let $\exp\text{-}t\, H_A(x,y) \varepsilon$ End. (L) and $\exp\text{-}t\, H(x,y)$ $(x,y \in \mathbb{R}^n)$ be the kernels of the corresponding contraction semigroups. Then

$$\| \exp - t\, H_A(x,y) \| \leq \exp - t\, H\, (x,y) \tag{1}$$

where $\| \cdot \|$ is the norm of operators in L. If V grows sufficiently strongly at infinity (in fact $\int e^{-tV} d^n x < \infty$ for all $t > 0$ is sufficient), then the semigroups are of trace class for $t > 0$ and we have the a priori estimate

$$\text{Trace } \exp\text{-}tH_A \leq (\dim L) \text{ Trace } \exp\text{-}tH \quad (t > 0) \tag{2}$$

which says that on the average the spectrum of H_A is pushed upward relative to that of H. More precisely, the free energy is increased by A at all temperatures. In particular

$$\inf \text{spec } H_A \geq \inf \text{spec } H . \tag{3}$$

Thus Kato's inequality reflects the diamagnetic properties of the Yang-Mills potential A.

Next consider a (complex) Gaussian random process with mean zero and covariance $(-\Delta_A + m^2)^{-1}$. Then the estimate

$$\|(-\Delta_A + m^2)^{-1}(x,y)\| \leq (-\Delta + m^2)^{-1}(x,y) \; (\leq C \exp\text{-}m|x-y| \text{ for } |x-y| \geq 1)$$

resulting from the estimate (1) indicates that the minimal coupling of a Yang-Mills field to a scalar massive field ϕ does not tend to decrease the mass of ϕ. This seems to persist in 2 space-time dimension when this scalar field is polynomially self coupled [4]. Also, as a consequence of Kato's inequality, the effective Lagrangian $L(A)$ for A obtained in this way by

$$\exp - L(A) = \frac{\int d\phi\, d\phi^* \, \exp - \phi^* (-\Delta_A + m^2)\phi}{\int d\phi\, d\phi^* \, \exp - \phi^* (-\Delta + m^2)\phi}$$

$$= \left[\frac{\det (-\Delta + m^2)}{\det (-\Delta_A + m^2)} \right]^{\frac{1}{2}}$$

satisfies the stability property $L(A) \geq 0$. This also persits in the case of curved space, since there Bochner Laplacians are also related to the Laplace Beltrami operator via a Kato inequality [5]. For the spinor Laplacean we expect a quite different behaviour, in fact the spin should have a paramagnetic influence. More precisely we conjecture [6]

$$\dim (L \otimes S) \text{ Trace } \exp - tH \leq \text{Trace } \exp - tH_A^S \tag{4}$$

such that in particular

$$\inf \text{spec } H \geq \inf \text{spec } H_A^S \, . \tag{5}$$

Also the effective Lagrangean for A now defined by

$$\exp - L(A) = \frac{\int d\psi\, d\psi^* \, \exp - \psi^* (\not{p} + e\not{A})\psi}{\int d\psi\, d\psi^* \, \exp - \psi^* \not{p}\, \psi}$$

$$= \left[\frac{\det (\not{p} + \not{A})^2}{\det \not{p}^2} \right]^{\frac{1}{2}}$$

with ψ being a spinor field, should again satisfy the stability property

$$L(A) \geq 0. \tag{6}$$

Estimate (6) has been verified to order h^2 for electromagnetic A's and to order e^2 for the general case [6].

Estimate (5) has been verified for special A's and V's [7].

Estimate (6), if true, sheds a new light on the dynamical importance of the theorem of spin and statistics as stressed e.g. by Feynman [8]. This estimate is known for the two-dimensional case with electromagnetic A (see e.g. [9]) and $L(A)$ may be calculated explicitly in the 4-dimensional case for homogeneous A [6]. The corresponding relativistic case has first been treated by Euler and Heisenberg [10]. Finally relation (6) has been verified on the lattice [11].

References

[1] T. Kato, Israel J. Math. 13, 135-148 (1972)
B. Simon, Math. Z. 131, 361-370 (1973)
M. Schechter, J. Funct. Anal. 20, 93-104 (1975)

[2] H. Hess, R. Schrader, D.A. Uhlenbrock, Duke Math. J. 44, 893-903 (1977)

[3] B. Simon, Indiana Math. J. 26, 1067-1073 (1977)
B. Simon, J. Funct. Anal. 32, 97-101 (1979)

[4] R. Schrader, Comm. Math. Phys. 58, 299-312 (1978)

[5] H. Hess, R. Schrader, D.A. Uhlenbrock, J. Diff. Geom. (to appear)

[6] H. Hogreve, R. Schrader, R. Seiler, Nucl. Phys. B142, 525-534 (1978)

[7] E. Lieb, priv. comm.
Y. Avron, R. Seiler, Phys.Rev.Lett. 42, 931-934 (1979)

[8] R.P. Feynman, Phys. Rev. 76, 749-759 (1949)

[9] N.K. Nielsen, B. Schroer, Nucl. Phys. B120, 62-76 (1977)

[10] W. Heisenberg, H. Euler, Z. Phys. 714-732 (1936)
V. Weisskopf, Kgl. Dansk Vid. Selsk. Math. Phys. Medd. XIV No 6 (1936)
V. Fock, Physik. Z. Sowjetunion 12, 404-425 (1937)
J. Schwinger, Phys. Rev. 82, 664-679 (1951)

[11] D. Brydges, J. Fröhlich, E. Seiler, Ann. Phys. 121, 227-284 (1979); Commun. Math. Phys. 71, 159-205 (1980)

A New Look at Generalized, Non-Linear σ-Models and Yang-Mills Theory

Jürg Fröhlich

Institut des Hautes Etudes Scientifiques

Bures-sur-Yvette

France

Abstract

First a list of recent papers on (lattice) gauge theories and non-linear σ-models is presented which serves as an introduction to the subject.

Subsequently, a new, quantum mechanical interpretation of the formalism used by Atiyah et al. and Corrigan et al. for the construction of self-dual Yang-Mills fields is attempted and criticized: Yang-Mills theory turns out to be a natural generalization of non-linear σ-models which has many conserved (Noether) currents. Confinement is linked to the presence of an "intrinsic (or parton) flavour" symmetry, at least in the case of the σ-models.

0. A Short Guide to the Literature

Before we come to the main part of this contribution, describing some new ideas and results concerning non-linear σ-models and Yang-Mills theories, we present a short list of papers which is a tiny selection out of a huge number of publications on non-linear σ-models and gauge theory. It replaces an introduction to the subject. Our selection does not represent a value judgment. Many (in certain respects perhaps most) important papers are missing in our list. The author has been guided largely by his lack of comprehension and time, ignorance and taste.

First, we quote some papers on classical, non-linear σ-models:

- Refs. 9 and 13 of the bibliography.
- K. Pohlmeyer, Commun. math. Phys. 46, 207, (1976).
- K. Pohlmeyer, in "New Developments in Quantum Field Theory and Statistical Mechanics", M. Lévy and P. Mitter (eds.), Plenum Press, New York 1977.
- A. D'Adda, P. Di Vecchia and M. Lüscher, in ref. 6 of the bibliography, and Preprint, Niels Bohr Institute, 1978.
- V.L. Golo and A.M. Perelomov, Phys. Lett. 79 B, 112, (1978).
- E. Brézin, C. Itzykson, J. Zinn-Justin and J.-B. Zuber, "Remarks about the Existence of Non-Local Charges in Two-Dimensional Models", to appear in Phys. Lett. B, (1979). (This paper contains an explicit construction of infinitely many conserved currents for all two-dimensional σ-models considered to be interesting).

Questions of complete integrability of two-dimensional, non-linear σ-models are discussed by the Russian Inverse-Scattering school.

Some important, recent papers on the quantum field theory of two-dimensional, non-linear σ-models are:

- Refs. 4, 14, 16, 17, 18 of the bibliography.
- A.M. Polyakov, Preprint ICTP 77/122.

- E. Witten, Instantons, the Quark Model, and the 1/N Expansion, Harvard Preprint HUTP-78/A042.
- K. Symanzik, in ref. 6 and these proceedings, and refs. given there.

Further very interesting, recent papers have been written by the Leningrad -, Berlin - and Sato (Japan) groups.

Next, we collect some convenient references concerning classical Yang-Mills theory:

- Refs. 1, 2, 3, 5, 6 of the bibliography.

A readable survey of classical Yang-Mills theory, describing the developments and including all important references prior to fall 1977 is:

- R. Stora, in "Invariant Wave Equations", G. Velo and A.S. Wightman (eds.), Lecture Notes in Physics 73, Springer-Verlag, Berlin-Heidelberg-New York, 1978.

Moreover,

- M.F. Atiyah, in "Mathematical Problems in Theoretical Physics", G.F. Dell'Antonio, S. Doplicher and G. Jona-Lasinio (eds.), Lecture Notes in Physics 80, Springer-Verlag, Berlin-Heidelberg-New York, 1978.
  More recent reviews, by R. Stora and E. Corrigan, may be found in ref. 6.

Additional, recent papers (among numerous others) are:

M.F. Atiyah, N. Hitchin and I.M. Singer, Proc. Nat. Acad. Sci. 74, 2662, (1977), and Proc. Royal Soc., to appear.

I.M. Singer, Commun. math. Phys. 60, 7, (1978).

M.F. Atiyah and J.D.S. Jones, Commun. math. Phys. 61, 97, (1978).

An extensive review, "Gauge Theories and Differential Geometry", by T. Eguchi, P.B. Gilkey and A.J. Hanson is to appear in Physics Reports (1979).

A list of references to work on lattice gauge theories, with emphasis on recent papers, follows: (among) the classic papers are:

- Ref. 19, bibliography.
- A.M. Polyakov, Phys. Letts. 59 B, 79, 82, (1975).

- J. Kogut and L. Susskind, Phys. Rev. D11, 395, (1975).

Reviews are e.g.:
- K. Osterwalder
- K. Wilson
in: "New Developments in Quantum Field Theory and Statistical Mechanics", loc.cit.

- L. Kadanoff, Rev. Mod. Phys. 49, 267, (1977), and refs. given there.
- A. Jaffe
- F. Guerra
- E. Seiler
in: "Mathematical Problems in Theoretical Physics", loc.cit.

More recent, useful papers are:
- K. Osterwalder and E. Seiler, Ann. Phys. (N.Y.) 110, 440, (1978).
- E. Seiler, Phys. Rev. D18, 482, (1978).
- M. Lüscher, Commun. math. Phys. 54, 283, (1977).
- J. Glimm and A. Jaffe, Phys. Letts. 66 B, 67, (1977).
- K. Gawedçki, Commun. math. Phys. 63, 31, (1978).
- J. Challifour and E. Weingarten, University of Indiana, Preprint 1978.
- G. t'Hooft, Nucl. Phys. B138, 1, (1978).

Very recent papers are:
- D. Brydges, J. Fröhlich and E. Seiler, "On the Construction of Quantized Gauge Fields, I: General Results", to appear in Ann. Phys. (N.Y.) (1979), and preprint in preparation.
- R. Israel and C. Nappi, "Quark Confinement in the Two-Dimensional Lattice Higgs-Villain Model", to appear in Commun. math. Phys. (1979).

(For earlier, related results see also: J. Fröhlich, in "Math. Problems of Theor. Phys." loc. cit.)

- D. Brydges, J. Fröhlich and E. Seiler, "Diamagnetic and Critical Properties of Higgs Lattice Gauge Theories", to appear in Nucl. Phys. B, (1979).
- G. Mack, "Confinement of Static Quarks in Two Dimensional Lattice Gauge Theories", to appear in Commun. math. Phys. (1979).
- T. Yoneya, "Topological Excitations in Yang-Mills Theories: Duality and Confinement", Preprint, City College, 1978.
- J. Glimm and A. Jaffe, "Charges, Vortices and Confinement", Harvard Preprint, 1978.
- G. Mack and V.B. Petkova, "Comparison of Lattice Gauge

Theories with Gauge Groups $\mathbb{Z}_2$ and SU(2)", Preprint, Hamburg, Dec. 1978.
- G. Mack and V.B. Petkova, "Sufficient Condition for Confinement of Static Quarks by a Vortex Condensation Mechanism", Preprint, Hamburg, Dec. 1978.
- A.M. Polyakov, "String Representation and Hidden Symmetries for the Gauge Fields", Preprint 1978.

(In this paper the existence of infinitely many, non-local conserved charges for three dimensional Yang-Mills theory is suggested).

- J. Fröhlich, "Confinement in $\mathbb{Z}_n$-Lattice Gauge Theories Implies Confinement in SU(n) Lattice Higgs Theories", Preprint, IHES, Feb. 1979.

(Some extensions of this paper, due to B. Durhuus and J. Fröhlich, are in preparation).

For a recent survey of quantized Yang-Mills theory in the continuum limit see:
- S. Coleman, "The Uses of Instantons",Erice Lectures 1977, to be published, A. Zichichi (ed.), and refs. to the original papers given there.

Work by D. Brydges, E. Seiler and the author concerning the quantized Higgs theory in two space-time dimensions is in preparation. (In that work, some earlier results by R. Schrader and R. Seiler, "A Uniform Lower Bound on the Renormalized Functional Determinant", to appear in Commun. math. Phys. (1979), and by B. Simon, "Kato's Inequality and the Comparison of Semi-Groups",to appear in J. Funct. Anal. (1979), and refs. given there, were very useful).

We omit references to "infinitely many" papers on perturbative renormalization of Yang-Mills theory, applications of the renormalization group (e.g. asymptotic freedom), model building or axiomatic investigations of gauge theories, although much of the present faith in gauge theories is certainly founded on the results of those earlier papers. Many of the out-standing references are by now so well known that we need not give them here again.

We now proceed to the main part of this contribution, where we suggest a somewhat novel way of looking at σ-models and Yang-Mills theory. (It is quite doubtful, though whether it will provide a useful point of view for the quantization of Yang-Mills theories.)

1. Introduction

In the following I propose and discuss a quantum mechanical interpretation of the construction - due to Atiyah, Drinfeld, Hitchin and Manin [1,2] - of all self-dual, Euclidean Yang-Mills fields as described and elaborated on by Corrigan et al. [3]. My program received stimulation from the work of D'Adda et al. [4] on the $\mathbb{C}P^{N-1}$ non-linear σ-models*). After the main results described below had been found a preprint of Dubois-Violette and Georgelin [5] appeared in which a program related to mine is announced. However, the main ideas presented below and the way the emphasis is placed differ much from theirs. Those ideas may briefly be summarized as follows: It is known from [1,2,3] that every self-dual U(n) Yang-Mills connection, A, can be written as

$$A_\mu(x) = \Phi(x)^*(\partial\Phi/\partial x^\mu)(x) \equiv \Phi(x)^* \Phi_\mu(x), \tag{1}$$

where Φ: $S^4 \ni x \mapsto \Phi(x)$ is a mapping from S^4 into orthonormal n-frames in $\mathbb{C}^N$,

$$\Phi(x) = (\phi_1(x),\ldots,\phi_n(x)), \quad \phi_j(x) \in \mathbb{C}^N,$$

$$\text{for some } N > n, \text{ and } (\phi_i(x), \phi_j(x)) = \delta_{ij}, \tag{2}$$

for all $i,j = 1,\ldots,n$.

It is natural to try to abstract from self-duality, in such a way that one views Φ, resp. the gauge-invariant, projection-valued field $P = \Phi\Phi^*$ of ref. [3], as the fundamental fields of the theory, and $A = A_\mu(\cdot)$ as derived, via equ. (1). For this purpose, one reexpresses the Yang-Mills action, A_{YM}, as a func-

*) This program started taking shape after talks of E. Corrigan and M. Lüscher on refs. [3,4], resp., at the Les Houches work shop on gauge theories, in spring 1978 [6].

tional of P (or Φ),

$$A_{YM} = \int d^4x \, \mathrm{tr}[(P_{[\mu}(1-P)P_{\nu]})^2 \, (x)], \tag{3}$$

where $f_\mu(x) \equiv (\partial f/\partial x^\mu)(x)$, for any f; see [3]. With this action one now tries to associate a Euclidean field theory [7] for P satisfying a suitable form of Osterwalder-Schrader axioms [8].

The natural analogue of the Yang-Mills action, A_{YM}, in a two dimensional space-time is

$$A_\sigma = \int d^2x \, \mathrm{tr}[(P_\mu(1-P)P_\mu)(x)] \tag{4}$$

which defines a generalized, non-linear σ-model. For n = 1, it coincides with the $\mathbb{C}P^{N-1}$-model of [9,4]. This observation suggests that there are close, mathematical connections between four dimensional Yang-Mills theories and two dimensional, generalized, non-linear σ-models. One of the main purposes of this contribution is to exhibit such connections. The main results can be found in §§ 3 and 5. Details of these and other results and proofs will be given elsewhere.

I thank P. Collet, H. Epstein, M. Lüscher and K. Osterwalder for valuable discussions and M. Dubois-Violette and Y. Georgelin for informing me of their independent results [5] prior to publication.

2. Mathematical Preliminaries

Euclidean space-time is denoted $\mathbb{E}^\nu$, (related to the sphere S^ν by stereographic projection). Let $\mathbb{D}$ be the real ($\mathbb{R}$), complex ($\mathbb{C}$) or quaternionic ($\mathbb{H}$) numbers, and $\mathbb{D}^N$ the linear space of N-tuples of elements in $\mathbb{D}$. If A is an $i\times j$ matrix with entries in $\mathbb{D}$ then A^* denotes the $j\times i$ matrix defined by $(A^*)_{ij} = \overline{A_{ji}}$, with $\bar{a}$ the conjugate of a in $\mathbb{D}$. If B is a $j\times k$ matrix, AB is the matrix product of A and B which is an $i\times k$ matrix.

An orthonormal n-frame in $\mathbb{D}^N$ is given by an $N\times n$ matrix, $\Phi = (\phi_1,\ldots,\phi_n)$, with $\phi_j \in \mathbb{D}^N$ and

$$\Phi^*\Phi = 1_n, \text{ i.e. } (\phi_i,\phi_j) = \delta_{ij}, \text{ for all } i,j = 1,\ldots,n. \tag{5}$$

The manifold of all n-frames satisfying (5) is denoted $S_{N,n}(\mathbb{D})$. Clearly

$$P \equiv \Phi\Phi^* \tag{6}$$

is the orthogonal projection onto the hyperplane spanned by $\phi_1,\ldots,\phi_n$ [3]. The manifold of all such hyperplanes is denoted $G_{N,n}(\mathbb{D})$, the "Grassmannian of n-planes", and can be written as a coset space [10].

Let H be the group of all linear transformations, h, of $\mathbb{D}^N$ with $h^*h = 1_N$, and $G_P \cong G$ the group of linear transformations, g, of an n-plane, $P\,\mathbb{D}^N$, into itself, with $g^*g = 1_n$; H is called "intrinsic flavour group", G gauge group.

$$\begin{aligned} &\text{For } \mathbb{D} = \mathbb{R} : H = O(N), \quad G = O(n), \\ &\text{for } \mathbb{D} = \mathbb{C} : H = U(N), \quad G = U(n), \\ &\text{for } \mathbb{D} = \mathbb{H} : H = Sp(N), \ G = Sp(n). \end{aligned} \tag{7}$$

Given Φ, one sets

$$^h\Phi^g = h\Phi g, \ h \in H, \ g \in G_P. \tag{8}$$

Next, consider arbitrary mappings

$$\begin{aligned} &\Phi : \mathbb{E}^\nu \ni x \mapsto \Phi(x) \in S_{N,n}(\mathbb{D}) \text{ , resp.} \\ &P : \mathbb{E}^\nu \ni x \mapsto P(x) = \Phi(x)\ \Phi(x)^* \in G_{n,N}(\mathbb{D}) \text{ ,} \end{aligned} \tag{9}$$

and define a field $A = A_\mu(\cdot)$ by equ. (1). By (1) and (6),

$$A_\mu(x) = \Phi(x)^* \ \Phi_\mu(x) = \tfrac{1}{2}(\Phi(x)^* \ \Phi_\mu(x) - \Phi_\mu(x)^* \ \Phi(x)), \tag{10}$$

is in $\mathfrak{g}$, the Lie algebra of G. Let $g(\cdot)$ be an arbitrary G-valued function on $\mathbb{E}^\nu$ (a gauge transformation). Then

$$A^g_\mu(x) = {}^1\Phi^g(x)^* \ {}^1\Phi^g_\mu(x) = g(x)^* A_\mu(x) g(x) + g(x)^* g_\mu(x). \tag{11}$$

Thus $A = A_\mu(\cdot)$ is the connection form of a principal G-bundle with base space $\mathbb{E}^\nu$ and fiber G.

Remark In an interesting preprint [5], Dubois-Violette and Georgelin quote an important theorem (see ref. [2] of [5]) saying that for $\mathbb{D} = \mathbb{C}$, there exists a finite integer, N, depending on ν and n $(N \geq \frac{\nu+1}{2} n)$ such that, to each given A, there exists a Φ such that $A_\mu = \Phi^* \cdot \Phi_\mu$.

Next, let $h(\cdot)$ be a U(N)-valued function on $\mathbb{E}^\nu$. Then

$$ {}^hA_\mu \equiv ({}^h\Phi^1)^* \, {}^h\Phi^1 = A_\mu \text{ if and only if} \qquad (12) $$
$$ \Phi^* h^* h_\mu \Phi = 0 \text{ , i.e. } P h^* h_\mu P = 0. $$

An h satisfying (12) is called an "intrinsic flavour transformation". Clearly a constant $h \equiv h_o \in U(N)$ obeys (12).

The curvature of A is given in terms of Φ by

$$ F_{\mu\nu} = \Phi^*{}_{[\mu}\Phi_{\nu]} + \Phi^*\Phi_{[\mu}\Phi^*\Phi_{\nu]} \text{ ,} \qquad (13) $$

and the gauge-invariant form of $F_{\mu\nu}$ by

$$ F^1_{\mu\nu} \equiv \Phi F_{\mu\nu}\Phi^* = P_{[\mu}(1-P)P_{\nu]} = P[P_\mu,P_\nu] = [P_\mu,P_\nu]P \text{ ,} \qquad (14) $$

where the parentheses denote anti-symmetrization, and we have used $(1-P)P = P(1-P) = 0$, i.e.

$$ P_\mu P = (1-P)P_\mu \quad \text{and} \quad PP_\mu = P_\mu(1-P). \qquad (15) $$

With the help of (14) the Yang-Mills equations and the action (see (3)) can be rewritten in terms of P [3]. Let γ be a closed curve in $\mathbb{E}^\nu$, and $\{x^m_j\}^{m+1}_{j=1}$, $x^m_{m+1} = x^m_1 = x_o$ a family of points on γ, with $\sup_j |x^m_{j+1} - x^m_j| \to 0$, as $m \to \infty$. Let

$$ P_{x_o}(\gamma) = \lim_{m\to\infty} \prod_{j=1}^{m} P(x^m_j). \qquad (16) $$

Then $W(\gamma) = \text{tr}[P_{x_o}(\gamma)]$ is the Wilson loop [3].

3. Actions, Conserved Currents and Infrared Critical Dimension

Let $D_\mu^\Phi = \partial/\partial x^\mu - A_\mu^L$, where A_μ^L is left multiplication by $A_\mu = \Phi^*\Phi_\mu$. Then

$$(D_\mu^\Phi \Phi)\Phi^* = (1-P)P_\mu = P_\mu P. \tag{17}$$

Using (5) and (15) one sees that

$$\begin{aligned} A_\sigma = \int d^2x \, tr[|D_\mu^\Phi \Phi|^2(x)] &= \int d^2x \, tr[|(D_\mu^\Phi \Phi)\Phi^*|^2(x)] \\ &= \frac{1}{2}\int d^2x \, tr[P_\mu(x)^2]. \end{aligned} \tag{18}$$

Thus, the models with action A_σ are natural generalizations of the $\mathbb{C}P^{N-1}$ models [9,4] with which they coincide when $\mathbb{D} = \mathbb{C}$, $n = 1$. The following further <u>isomorphisms</u> are noteworthy: The $\mathbb{R}P^1$ model is isomorphic to the S^1 (XY - or rotator) model, the $\mathbb{C}P^1$ model to the S^2- (classical Heisenberg) model, and the $\mathbb{H}P^1$ (or $G_{2,1}(\mathbb{H})$ -) model to the S^4- (5 vector) model. The last two isomorphisms are obtained by expressing P in terms of Pauli-, resp. γ-matrices. The last one can also be reduced to the simpler isomorphism between the S^3- (4 vector) model and the SU(2) model with action

$$A_\sigma = \int d^2x \, tr[g_\mu(x)^* g_\mu(x)], \; g(x) \in SU(2) \cong S^3.$$

For the lattice σ-models defined in (27), §5, the proofs of these isomorphisms are simple.

Finally, we note that equations (17) and (18) remain meaningful for $\mathbb{D} = \mathbb{O}$, the octonions or Cayley numbers [11], with $G_{N,n}(\mathbb{D}) = G_{3,1}(\mathbb{O})$. In this case the projections P(x) label the points of the Moufang projective plane [11,12], and one obtains an octonionic non-linear σ-model whose symmetry group is the exceptional Lie group F_4. These models and Yang-Mills versions thereof may be of interest to strong interaction physics [12].

On a classical level, the discussion of the $G_{N,n}(\mathbb{D})$- σ-models proceeds as in [4,13]*). In two dimensional, Euclidean space-

*) After completion of this paper, a preprint by E. Brézin, C. Itzykson, J. Zinn-Justin and J.-B. Zuber appeared, where, in addition to one local and one non-local, infinitely many other non-local, conserved currents are constructed explicitly: see Section 0.

time the $\mathbb{C}P^{N-1}$ models are presumably the most interesting ones, since, for $n > 1$, the $G_{N,n}(\mathbb{D})$ models do generally not admit new classes of instanton solutions with non-trivial homotopy [10]. (E.g. the $\mathbb{H}P^1$ model does not have such solutions). We conjecture, however, that the results of [14] extend to the quantized version of these models. Finally, we note that the infrared critical dimensions of the σ-models is $\nu_{crit.} = 2$. This is related to the fact that the naive dimension of the conserved currents is unity; see (25) and §5.

The Yang-Mills action is obtained from $F_{\mu\nu} = [D^{\Phi}_{\mu}, D^{\Phi}_{\nu}]$, by setting

$$A_{YM} = \int d^4x \, \mathrm{tr}[F_{\mu\nu}(x)^2] = \int d^4x \, \mathrm{tr}[F^I_{\mu\nu}(x)^2], \tag{19}$$

which by (14) coincides with (3).

Next, one tries to form higher tensors, $T_{\mu_1 \ldots \mu_d}$, such that $A_d = \int d^{2d}x \, \mathrm{tr}[T_{\mu_1 \ldots \mu_d}(x)^2]$ can serve as an action for a conformally invariant theory in $\nu = 2d$ dimensions. If one insists on formal Osterwalder-Schrader positivity [8] and conformal invariance, T is necessarily totally anti-symmetric and of rank d. (Unfortunately, the tensors $T_{\mu_1 \ldots \mu_d} = P_{[\mu_1}(1-P)P_{\mu_2}(1-P)\ldots P_{\mu_d]}$, a natural generalization of $F^I_{\mu\nu}$, vanish identically, for $d > 2$, because of the identity:

$$\ldots(1-P)P_{\mu_j}(1-P)\ldots = \ldots P_{\mu_j}P(1-P)\ldots = 0 \text{ ; see (15)).}$$

In general, the correct expression for $\mathrm{tr}[T_{\mu_1 \ldots \mu_d}(x)^2]$ can be obtained from the following limiting procedure: Choose a hypercube, Δ_ε, in the $(\mu_1, \ldots, \mu_d)$-hyperplane with sides of length ε, centered at x. Let $\partial\Delta_\varepsilon$ be its boundary. Then

$$- \mathrm{tr}[T_{\mu_1 \ldots \mu_d}(x)^2] = \lim_{\varepsilon \downarrow 0} \varepsilon^{-d}\{\mathrm{tr}[P(\partial\Delta_\varepsilon)] - c(n,\mathbb{D})\}, \tag{20}$$

where $P(\partial\Delta_\varepsilon) = \prod_{x \in \partial\Delta_\varepsilon} P(x)$, see (16), and $c(n,\mathbb{D})$ is a constant. These expressions make sense and are the desired ones if and only if $d = 1$ (σ-model) or $d = 2$ (Yang-Mills theory). When $d > 2$, $P(\partial\Delta_\varepsilon)$ is ill defined because of ordering problems. This rules out the existence of admissible actions, A_d, for $d > 2$, (i.e. $\nu > 4$).

(If one does not insist on Osterwalder-Schrader positivity then, of course, there are plenty of actions for $d > 2$.)

Clearly, the actions A_σ and A_{YM} have a global H-symmetry. In addition, A_{YM} has a <u>local symmetry:</u> By (12) A_{YM} is invariant under

$$P(x) \mapsto P^h(x) \equiv h(x)P(x)h(x)^*, \text{ with} \qquad (21)$$

$$Ph^*h_\mu P = 0.$$

This equation has non-trivial solutions. Associated with these symmetries are Noether currents, J_μ^δ. They are given by

$$J_\mu^\delta(x) = (\partial L_{YM}(x)/\partial P_\mu(x)_{ij})(\delta P)_{ij}(x), \qquad (22)$$

where L is the Lagrange density corresponding to A. (In Euclidean field theory, L is replaced by the action density [15].) Furthermore, δ is the derivation associated with a symmetry transformation, i.e.

$$\delta P = [B,P], \text{ with } PB_\mu P = 0, \qquad (23)$$

where $B(x) \in H$, the Lie algebra of H, for all x. Apart from $B(x) = B = \text{const.}$ there will be other solutions of (23). They form a linear space. By solving (23) in terms of P one gets a linear space of conserved currents. (After quantization they ought to determine a Lie algebra of conserved charges properly containing H.)

Heuristically, quantization consists in associating with an action A the formal measure on the space of $G_{N,n}(\mathbb{D})$- valued distributions

$$d\mu(P) = e^{-g^{-2}A} \prod_{x\in\mathbb{E}^\nu} \mathcal{D}P(x), \qquad (24)$$

where $\mathcal{D}P$ is, heuristically, the uniform measure on $G_{N,n}(\mathbb{D})$, and g is a coupling constant (dimensionless for $\nu = 2$, $A = A_\sigma$ and $\nu = 4$, $A = A_{YM}$). In the process of renormalization, $P^2 = P$ will have to be replaced by $P^2 = fP$, f divergent [4], g is renormalized, and the conserved currents yield many Ward identities

[16]. Formally, (i.e. disregarding from the existence problem), $d\mu$ satisfies Osterwalder-Schrader positivity for the fields P, when $A = A_\sigma$, resp. $W(\gamma)$, see (16), when $A = A_{YM}$. This is shown by approximating A by an action A_ε constructed in terms of $P(\partial\Delta_\varepsilon)$; see (20), (16). Note that the naive dimension of P is zero. In four dimensions there are no dimensionless fields satisfying positivity. Hence, for $A = A_{YM}$ ($\nu = 4$), presumably only loop observables, $W(\gamma)$, and functionals thereof, survive the $\varepsilon \downarrow 0$ limit*). Thus, whereas for $A = A_\sigma$, $\nu = 2$, spin wave theory about $P = P_o$ = const. makes sense, providing an expansion in g - at least after adding a term $m^2 \int d^2x\, tr[P(x)P_o]$ to A_σ - this is not clear, at all, for $A = A_{YM}$. It is further complicated by the symmetries (21), (23). (Moreover, it is doubtful whether one can add a term $m^4 \int d^4x [tr\, P(x)P_o]$ to A_{YM}, destroying those symmetries, to eliminate infrared divergences). A natural question is whether the symmetries of A can be broken spontaneously. The arguments of [15] suggest that the symmetry associated with a current J_μ cannot be broken spontaneously in dimension

$$\nu \leqq \nu_{crit.} \equiv [J_\mu] + 1, \tag{25}$$

where $[J_\mu]$ is the infrared dimension of J . For $A = A_{YM}$, we obtain from (22), by naive dimensional analysis, $[J_\mu] \geqq 3$, i.e. $\nu_{crit.} \geqq 4$! (For $A = A_\sigma$ one obtains of course $\nu_{crit.} = 2$.) In order for these arguments to be convincing they should be reformulated in terms of the loop observables, $W(\gamma)$.

4. Couplings to Quark Fields

Let $\bar\psi_\alpha$, ψ_α be Dirac spinors transforming under the same representation of G as Φ, Φ^*, resp.; (α is the G-(colour) index). In addition $\bar\psi$ and ψ may transform under some flavour group, i.e. carry a flavour index j = 1,...,F. The components $\psi_{\alpha j}$ are the matrix elements of an n×F matrix, denoted ψ, and $\bar\psi$ is the corresponding, conjugate F×n matrix. One defines the gauge-invariant fields [3] $\bar\psi^1 = \bar\psi\Phi^*$, $\psi^1 = \Phi\psi$. When $\mathbb{D} = \mathbb{H}$, there are further

*) If, for $A = A_{YM}$, $\nu = 4$, P exists as a quantized field its ultraviolet dimension must be ≥ 1, (Källen-Lehmann representation). Since its naive dimension is 0, P may, in fact, not survive quantization. See also Section 5.

invariants, $\psi^2 = \psi^T \varepsilon\ \Phi^*$ and $\bar{\psi}^2 = \Phi\ \varepsilon\ \bar{\psi}^T$, where

$$\varepsilon = \begin{pmatrix} \varepsilon_1 & & 0 \\ & \ddots & \\ 0 & & \varepsilon_n \end{pmatrix}, \quad \varepsilon_j = \begin{pmatrix} 0 & 1 \\ -1 & 0 \end{pmatrix}, \quad j = 1,\ldots,n.$$

One observes that N and F play symmetric roles. Thus one may speculate that N = F, i.e. "intrinsic flavour" = flavour. The result quoted in [5] then suggests that $F = N \geq (\nu+1)n/2$, i.e. $F \geq 8$, for $\nu = 4$, $n = 3$.

The minimal coupling matter actions are

$$A_j^M = \int d^\nu x\ \mathrm{tr}[\bar{\psi}^j(x)(\not{D}\psi^j)(x)] \ , \tag{26}$$

$j = 1,2$. For N = F, there are the following flavour invariants: $\psi_I^j = \mathrm{tr}[\psi^j]$, $\bar{\psi}_I^j = \mathrm{tr}[\bar{\psi}_j]$. One can then form the action $A_{j,I}^M = \int d^\nu x\ \bar{\psi}_I^j(x)(\not{\partial}\psi_I^j)(x)$. However, they give rise to trivial interactions, and the Green function for $\bar{\psi}$ and ψ can be calculated in an arbitrary Φ-field.

5. Lattice Models

The $G_{N,n}(\mathbb{D})$ non-linear σ-models can be put onto the lattice $\mathbb{Z}^\nu$, in two different ways:

$$A_\sigma' = \sum_{n.n.} \mathrm{tr}[P_x P_y] \ ,$$
$$A_\sigma'' = \sum_{n.n.} \mathrm{tr}[\Phi_x g_{xy} \Phi_y^*] \ , \tag{27}$$

where $g_{xy} \in G$, for all nearest neighbors (n.n.)x,y. The gauge field, $g = \{g_{xy}\}$, plays the role of a Lagrange multiplier field [4].

As a priori distributions one chooses the uniform measure on $G_{N,n}(\mathbb{D})$ for P_x, resp. the Haar measure on G for g_{xy}, and for Φ and Φ^* the measure

$$d\lambda(\Phi) = \prod_{\alpha\leq\beta} \delta([\Phi^*\Phi]_{\alpha\beta} - \delta_{\alpha\beta}) \prod_{i\alpha} d\Phi_{i\alpha}\, d\overline{\Phi_{i\alpha}} \ .$$

This yields two different lattice models with identical, formal continuum limit, (18). The Mermin-Wagner argument [15] excludes spontaneous breaking of the intrinsic flavour symmetry group H, when $\nu = 2$. For the discussion of the breaking of H in $\nu \geqq 3$ dimensions, the methods of [17] can be applied to the model with action A'_σ. Good results concerning the symmetric phase are achieved by applying the methods of [18] to the model with action A''_σ. Those methods vaguely correspond to partially resummed 1/N expansions and are rigorous. It would be of interest to develop a double expansion in 1/N and 1/n, or, for $A = A_{YM}$, one in $\frac{N-n}{N}$ (about pure gauges). The following result may be useful.

<u>Theorem:</u> <u>If one replaces</u> $d\lambda$ <u>by</u>

$$d\lambda^{(N)}(\Phi) = \prod_{\alpha \leqq \beta} \delta([\Phi\ \Phi]_{\alpha\beta} - N^{3/4}\ \delta_{\alpha\beta}) \prod_{i\alpha} d\Phi_{i\alpha}\ d\overline{\Phi_{i\alpha}} ,$$

<u>and defines</u>

$$d\mu^{(N)}(g) / \prod_{n.n.} dg_{xy} \equiv Z^{(N)^{-1}} \int e^{-g^{-2} A''_\sigma} \prod_{x \in \mathbb{Z}^\nu} d\lambda^{(N)}(\Phi(x))$$

<u>then</u> $w^*\text{-}\lim_{N\to\infty} d\mu^{(N)}(g) \equiv d\mu^W(g)$ <u>exists and is the measure of Wilson's pure Yang-Mills lattice theory [19] with group G.</u>

Analogous limit theorems hold for Gaussian $d\lambda^{(N)}$ and $|\Phi|^4$-type $d\lambda^{(N)}$. In the last case, one obtains a formal $|\Phi|^4$-theory which is conformal invariant, for $\nu = 4$, and whose $N\to\infty$ limit is pure Yang-Mills. One may speculate that that theory is asymptotically free.

On the lattice one can try to mimick A_{YM} by the action

$$A_S = -\sum_p (\mathrm{tr}[P_x P_y P_z P_u] + \mathrm{tr}[P_x P_u P_z P_y]) ,$$

where p is the plaquette 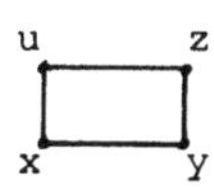

. We note that $\mathrm{tr}[P_x P_y P_z P_u]$ is a lattice approximation to $W(\partial p)$; see (16), (20). In spite of this formal relationship with A_{YM}, the naive continuum limit of A_S is not A_{YM}. (The model with action A_S is related to an Ising type model of Slawny [20].) A better approximation to A_{YM}

is

$$A'_{YM} = - \sum_{\Delta} \left(\mathrm{tr}\left[\prod_{x \in \partial\Delta}^{\circlearrowleft} P_x \right] + \mathrm{tr}\left[\prod_{x \in \partial\Delta}^{\circlearrowright} P_x \right] \right), \qquad (29)$$

where Δ is an arbitrary lattice square parallel to two axes of $\mathbb{Z}^{\nu}$ each side of which contains <u>three</u> sites. Both actions, A_S and A'_{YM} , admit a transfer matrix formalism with selfadjoint, generalized transfer matrix [17]. This guarantees Osterwalder-Schrader positivity. Wilson's lattice gauge theory [19] is recovered by choosing $A_W = - \sum_p W(\partial p)$, where $W(\partial p)$ is given by (16), and as a priori distribution " $\prod_{x \in \partial p} \mathcal{D}P(x)$" the product of the Haar measures for G on the sides of ∂p.

If one discusses confinement for the models with actions A'_{σ}, or A'_{YM} in terms of a lattice version of the Wilson loop, $W(\gamma)$, one arrives at the heuristic picture that <u>confinement breaks down if the intrinsic flavour symmetry</u>, H, is spontaneously broken. Such breaking is expected for $\nu > \nu_{crit.}$ (i.e. $\nu \geqq 5$ for A'_{YM}) and small g.

One can show that if one couples quark fields to P with a large number, F, of flavours and a mass $\propto F^{1/2}$, $F >> 1$, by means of a lattice version of A^M_j, this enhances the spontaneous breaking of H which appears to become possible in $\nu \geqq 3$ dimensions.

Rigorous proofs are so far restricted to the σ-models, because neither the techniques of [17] nor the ones of [18] , in their present form, apply to Yang-Mills theory.

References

[1] M.F. Atiyah, N.J. Hitchin, V.G. Drinfeld and Yu.I. Manin, Phys. Lett. 65 A, 185, (1978).

[2] V.G. Drinfeld and Yu.I. Manin, Commun. Math. Phys. 63, 177, (1978).

[3] E.F. Corrigan, D.B. Fairlie, S. Templeton and P. Goddard, Nucl. Phys. B 140, 31, (1978).

[4] A. D'Adda. P. Di Vecchia and M. Lüscher, Preprint, Niels Bohr Institute, 1978; and in ref. [6].

[5] M. Dubois-Violette, and Y. Georgelin, Preprint, Orsay 1978.

[6] E. Brézin and J.-L. Gervais (eds.), Physics Reports, to appear.

[7] "Constructive Quantum Field Theory", G. Velo and A.S. Wightman (eds.), Lecture Notes in Physics 25, Springer-Verlang, Berlin-Heidelberg-New York, 1973.

[8] K. Osterwalder and R. Schrader, Commun. Math. Phys. 42, 281, (1975).

[9] H. Eichenherr, Ph. D. thesis, Heidelberg, 1978.

[10] N. Steenrod, "The Topology of Fibre Bundles", Princeton University Press, Princeton, 1951.

[11] N. Jacobson, "Lie Algebras", Wiley, New York-London, 1962.

[12] M. Gunaydin, C. Piron and H. Ruegg, Commun. Math. Phys. 61, 69, (1978); and refs. given there.

[13] M. Lüscher and K. Pohlmeyer, Nucl. Phys. B 137, 46, (1978).

[14] M. Lüscher, Nucl. Phys. B 135, 1, (1978); see also A.B. Zamolodchikov and A.B. Zamolodchikov, Nucl. Phys. B 133, 525, (1978).

[15] Predictions that $\nu_{crit.} \geq 4$, for Yang-Mills, were previously made by A. Migdal and G. Parisi. Our argument extends that in: K. Symanzik, Commun. Math. Phys. 6, 288, (1967); also J. Fröhlich and T. Spencer, in "New Developments in QFT and Stat. Mech.", M. Lévy and P. Mitter (eds.), Plenum, New York, 1977.

[16] E. Brézin, J. Zinn-Justin and J.C. Le Guillou, Phys. Rev. D 14, 2615, (1976).

[17] J. Fröhlich, B. Simon and T. Spencer, Commun. Math. Phys. 50, 79, (1976); J. Fröhlich, R. Israel, E. Lieb and B. Simon, Commun. Math. Phys. 62, 1, (1978).

[18] K. Symanzik, in "Local Quantum Theory", R. Jost (ed.), Academic Press, New York, 1969; D. Brydges and P. Federbush, Commun. Math. Phys. 62, 79, (1978); J. Fröhlich and T. Spencer, unpublished.

[19] K. Wilson, Phys. Rev. D 10, 2445, (1974); R. Balian, J.M. Drouffe and C. Itzykson, Phys. Rev. D 10, 3376, (1974), D 11, 2098, 2104, (1975).

[20] J. Slawny, Commun. Math. Phys. 46, 75, (1976).

1/N Expansions and the O(N) Nonlinear σ-Model in Two Dimensions

K. Symanzik

DESY, Hamburg

Federal Republic of Germany

1/N Expansions and the O(N) Nonlinear σ-Model in Two Dimensions

For four-dimensional realistic QFT models, renormalized perturbation theory is so far the only systematic method of (formal) solution. It is known, however, that ordinary renormalized perturbation theory is inapplicable to two classes [1] of physically interesting models: Nonrenormalizable theories [2] (e.g. the nonlinear σ (NLσ) model in three dimensions) and massless superrenormalizable theories [3] (e.g. $(\vec{\phi}^2)_2$ and $(\vec{\phi}^2)_3$ theory). In these cases with O(N) symmetry, an entirely satisfactory method of formal solution (for N≥3 at least) is the 1/N expansion [4]. If one further expands the 1/N expansion terms as asymptotic expansions in the ordinary renormalized coupling constant g, one obtains a triple power series expansion in g. lng, and 1/N, whereby upon summing again the 1/N expansion a double power series in g and lng would result [1], [2], [3]. However, for a suitable convention concerning the definition [2], [3] of g, the parametric functions β(g), γ(g) etc. of the renormalization group equation remain power series in g only.

An interesting special case hereby is the NLO_2 model [5], [6], which is UV-renormalizable and asymptotically free. Ordinary perturbation expansion is possible, however, only in the presence of a constant external (magnetic) field that breaks the O(N) invariance explicitly and thus gives to the N-1 would be Goldstone bosons of nonlinearly realized O(N)/O(N-1) symmetry a mass. Switching off the external field may be expected to lead to infrared divergences in this expansion, related to the fact that in two dimensions, spontaneous breaking of a continuous symmetry is impossible [7].

It has been suggested by A. Jevicki [8] that the infrared divergences in perturbation theory, based on formally broken O(N) invariance, of the model cancel in the vacuum self energy, and S. Elitzur [9] has offered a proof of this cancellation to all orders in the expansion of O(N)-symmetric combinations. This proof, however, is inconclusive.

We show how to pass from the manifestly O(N)-invariant 1/N expansion of the NLO_2 model systematically to an "ordinary" perturbation expansion in powers of g and lng, for O(N)-symmetric combinations; hereby, exponentially small terms of order exp[-const/g] with const > 0 are dropped resulting in consistency of the "massive" 1/N expansion with the formally "massless" perturbation expansion since the mass is of the neglected order. To order N^o and N^{-1}, the logarithms cancel in conformity with Elitzur's explicit calculations [9]. A check on higher orders requires to extend the present calculation to order N^{-2}. We hope to present the result hereof elsewhere. - These results should be relevant [10] also for the CP_2^{n-1} model, which has in perturbation theory infrared divergences analogous to the ones of the NLO_2 model.

References

[1] K. Symanzik, Acta Phys. Austr., Suppl. XVI, 177 (1976)

[2] K. Symanzik, Comm. Math. Phys. 45, 79 (1975)
K. Symanzik, lectures (DESY 75/24, unpublished) at GIFT '75 seminar

[3] K. Symanzik, Lett. Nuovo Cimento 8, 771 (1973)
K. Symanzik, 1973 Cargèse lectures in physics (ed. E. Brézin, unpublished) (DESY 73/58)

[4] K. Symanzik, DESY 77/05 and unpublished

[5] E. Brézin, J. Zinn-Justin, Phys. Rev. B14, 3110 (1976)
E. Brézin, J. Zinn-Justin, J.C. Le Guillou, Phys. Rev. D14, 2615 (1976)

[6] W.A. Bardeen, B.W. Lee, R.E. Schrock, Phys. Rev. D14, 985 (1976)

[7] N.D. Mermin, H. Wagner, Phys. Rev. Lett. 17, 1133 (1960)
S. Coleman, Comm. Math. Phys. 31, 259 (1973)

[8] A. Jevicki, Phys. Lett. 71B, 327 (1977)
A. Jevicki, IAS preprint (Aug. 1978)

[9] S. Elitzur, Princeton University preprint (Nov. 78)

[10] M. Lüscher, private communication

On the $\mathbb{Z}_2$ Lattice Higgs System

S. Miracle-Sole

Centre de Physique Théorique CNRS

Marseille, France

Abstract

Some aspects of the phase structure of the $\mathbb{Z}_2$-lattice Higgs system are studied by means of convergent expansions.

1. Introduction

In these notes we study one of the Wegner's generalized Ising systems [1], a simple model for a gauge theory on a lattice according to Wilson's ideas [2]. Mainly, we discuss some results of joint works with G. Gallavotti and F. Guerra [3] and with R. Marra [4] concerning this system. More general lattice gauge theories are treated in other lectures at this conference and we refer to [5,6] for a general outline and results related to the present study.

The system can be described as follows. Let there be given a cubical lattice L. To each bond (i,j) of neighbouring sites a variable $A_{ij} = A_{ji}$ taking values in $\mathbb{Z}_2 = \{-1, +1\}$ is assigned. A configuration A of the system is specified by giving the values of A_{ij} at each bond. An elementary square of four adjacent bonds is a plaquette p = (i,j,k,l) and we write $A(p) = A_{ij}A_{jk}A_{kl}A_{li}$. Let B be a given configuration on $L\backslash\Lambda$ where Λ is finite. The energy (or the euclidean action in the field theory language) of the system restricted to Λ with boundary condition B outside Λ is

$$H_\Lambda^B(A) = -\{\beta_\ell \sum_{\{i,j\}\cap\Lambda\neq\emptyset} A_{ij} + \beta_p \sum_{p\cap\Lambda\neq\emptyset} A(p)\} \tag{1}$$

where the configuration A coincides with B outside Λ. The coefficients β_ℓ and β_p are the coupling constants.

A gauge invariant version of this system can be presented by introducing extra $\mathbb{Z}_2$-valued variables K_i at each site and replacing $\beta_\ell \Sigma A_{ij}$ by $\beta_\ell \Sigma K_i A_{ij} K_j$. In this form the system can be viewed as a gauge field theory on the lattice, the Higgs and the Yang-Mills fields being respectively described by the variables K_i and A_{ij}. However, the gauge transformation $A_{ij} \to K_i A_{ij} K_j$ eliminates the K's while leaving the plaquette term invariant, and reduces this gauge field system to the model (1).

To describe the infinite system one introduces the thermodynamic free energy $f(\beta_p, \beta_\ell)$ as the limit when $\Lambda \to \infty$ of the free energy per unit volume, and the Gibbs states or equilibrium measures. A probability measure on the set of configurations on L (equipped with the product topology and the corresponding Borel σ-algebra) is a Gibbs state if for all Λ its conditional probability given the configuration B outside Λ is equal to the Gibbs measure of the finite system on Λ defined by the hamiltonian (1).

We shall be concerned with the problem of the occurrence and properties of the phase transitions when β_p and β_ℓ vary. In the study of this problem we use the convergent expansions of the free energy and the correlation functions, a very familiar technique in statistical mechanics. For the sake of concreteness we shall consider the case of a three dimensional lattice.

2. Preliminary Remarks

First, we introduce some notations. Let $b(\Lambda)$ be the set of bonds and $p(\Lambda)$ the set of plaquettes contained in Λ, and let $|b(\Lambda)|$ and $|p(\Lambda)|$ be the number of elements of these sets. A set S of plaquettes on L can be viewed as a surface of area $|S|$. The boundary ∂S of this surface, i.e. the set of bonds at which an odd number of plaquettes of S occurs, is a set of closed circuits on the lattice. Similarly, if C is a set of bonds, of total length $|C|$, its boundary ∂C is the set of sites at which an odd number of bonds of C occurs. Finally we introduce the dual lattice L^* as a cubic lattice, the bonds of which cut perpendicularly the plaquettes of L in such a way that the centres of the plaquettes coincide with the middle points of the bonds. We call Λ^* the sublattice of L^* consisting of the bonds which intersect the plaquettes contained in Λ.

We next derive two expressions for the partition function from which two different convergent expansions can be obtained.

The first expression follows in a standard way. Consider the partition function of the system in a parallelepipedic box Λ with free boundary conditions

$$Z_\Lambda = \sum_{\{A\}} \exp\{\beta_\ell \sum_{\{i,j\}\subset\Lambda} A_{ij} + \beta_p \sum_{p\subset\Lambda} A(p)\} .$$

By expanding the exponential we have

$$Z_\Lambda = R_\Lambda \sum_{\{A\}} \prod_{\{i,j\}\subset\Lambda} (1+\text{th}\beta_\ell A_{ij}) \prod_{p\subset\Lambda} (1+\text{th}\beta_p A(p))$$

where $R_\Lambda = (\text{ch}\beta_\ell)^{|b(\Lambda)|}(\text{ch}\beta_p)^{|p(\Lambda)|}$. By developping the products and noticing that the contributions of terms containing some A_{ij} to an odd power are zero, we get

$$Z_\Lambda = R_\Lambda \sum_{S\subset p(\Lambda)} (\text{th}\beta_p)^{|S|}(\text{th}\beta_\ell)^{|\partial S|} \tag{2}$$

where the sum runs over the family of all subsets S of the set of plaquettes contained in Λ.

The second expression, which leads to a new expansion, is obtained as follows. Consider the system in a parallelepipedic box $\Lambda \subset L$ with the boundary condition for which $A_{ij} = 1$ if $\{i,j\}$ is on the boundary of Λ. To each configuration A of this system we associate the set C' of bonds on the dual lattice L^* that intersect the plaquettes of L at which $A(p) = -1$. The fact that $\prod_{p\in K} A(p) = 1$ for any elementary cube K of Λ implies $\partial C' = \emptyset$, i.e., C' is a set of closed circuits of bonds on Λ^*. Let $Q_\Lambda = e^{\beta_p|p(\Lambda)|}(\text{ch}\beta_\ell)^{|b(\Lambda)|}$. Then the partition function becomes

$$Z_\Lambda = Q_\Lambda \sum_{\substack{C'\subset b(\Lambda^*)\\ \partial C'=\emptyset}} e^{-2\beta_p|C'|} \sum_{\{A\}}{}' \prod_{\{i,j\}\subset\Lambda} (1+\text{th}\beta_\ell A_{ij})$$

where the sum Σ' is restricted to the configurations $\{A\}$ such that $A(p) = -1$ if p intersects C' and $A(p) = 1$ if p does not intersect C'.

Develop now the product $\Pi\ (1+\text{th}\beta_\ell A_{ij})$ and observe that only the terms of the form $\prod_{\{i,j\}\in C} (\text{th}\beta_\ell)A_{ij}$ where $\partial C = \emptyset$ give a contribution different from zero. Now, the product $\prod_{\{i,j\}\in C} A_{ij}$ is equal to $\prod_{p\in S} A(p)$, for any set S of plaquettes on Λ such that $\partial S = C$, and hence its value +1 or -1 is determined. Denote by n(C,C') this number. Then

$$Z_\Lambda = Q_\Lambda \sum_{\substack{C\subset b(\Lambda)\\ \partial C=\emptyset}} \sum_{\substack{C'\subset b(\Lambda^*)\\ \partial C'=\emptyset}} e^{-2\beta_p|C'|} (\text{th}\beta_\ell)^{|C|}\, n(C,C'). \qquad (3)$$

Expressions (2) and (3) show two different geometrical aspects of the interaction defining the system.

They imply that in the three dimensional case the model is self-dual, namely, that (up to an unimportant additive term) the thermodynamic free energy is invariant under the transformation $(\beta_p,\beta_\ell) \to (\beta_p^*,\beta_\ell^*)$ where

$$\text{th}\beta_p^* = \exp - 2\beta_\ell \ , \ \text{th}\beta_\ell^* = \exp - 2\beta_p \ . \qquad (4)$$

This property is evident from (3). As usual, the dual system is introduced by assigning new $\mathbb{Z}_2$-valued variables to the bonds of L^*. If $S\subset p(\Lambda)$ let $A'_{ij} = -1$ if the bond $(i,j) \subset L^*$ intersects a plaquette of S and let $A'_{ij} = 1$ otherwise. Then $|S| = \sum_{\{i,j\}\subset b(\Lambda^*)} (1/2)(1-A'_{ij})$ and $|\partial S| = \sum_{p\subset p(\Lambda^*)} (1/2)(1-A'(p))$, and (2) implies again the selfduality property of the model. This symmetry property extends in an appropriate way to the correlation functions with particular boundary conditions.

If $\beta_\ell = 0$ the system (also called the pure gauge system) with some coupling β_p is related by duality to the usual three dimensional Ising model with coupling

$$\beta^* = - (1/2)\log \text{th}\, \beta_p \qquad (5)$$

and zero magnetic field. This can easily be seen from (2) or (3) which, when $\beta_\ell = 0$, take the form

$$Z_\Lambda = R_\Lambda \sum_{\substack{S\subset p(\Lambda)\\ \partial S=\emptyset}} (\text{th}\beta_p)^{|S|} \ , \quad Z_\Lambda = Q_\Lambda \sum_{\substack{C\subset b(\Lambda^*)\\ \partial C=\emptyset}} e^{-\beta_p|C|} \ .$$

The selfdual property of the complete system implies an analogous result for $\beta_\ell = + \infty$.

3. Analyticity Properties

Consider now the expression (2) and suppose that $|\beta_p|$ is small enough. Because of the factor $(\text{th}\beta_p)^{|S|}$ the terms with S large will contribute very little to the partition function, independently of the value of β_ℓ. Decompose the surface S into its connected components S_i, i = 1,2,... Since most of the important configurations are then associated to small S's, the system can be viewed as a low density gas, the particles of which are the small surfaces S_i made of plaquettes, with an activity

$$z_i = (\text{th}\beta_p)^{|S_i|} (\text{th}\beta_\ell)^{|\partial S_i|} .$$

The same can be said for the dual system if $|\beta_p^*|$ is sufficiently small, that is, if $\beta_p \geq 0$ and $|\beta_\ell|$ is sufficiently large.

On the other hand, consider the sets C,C' of closed circuits of bonds appearing in formula (3). By smoothing if necessary some corners of the circuits, the set CUC' can be decomposed into a set of self-avoiding closed curves of bonds on L and L^*. Two closed curves γ and γ' respectively on L and L^* will be said to be connected if $n(\gamma,\gamma') = -1$. From the definition of $n(\gamma,\gamma')$ we see that this happens whenever γ gives an odd number of tours around γ'. We decompose the set of curves CUC' into its components Γ_i, i = 1,2,... which are connected in the sense just defined. These sets $\Gamma_i = C_i \cup C_i'$, where C_i is a set of closed curves on L and C_i' a set of closed curves on L^* , will be called contours. If we associate to each contour Γ_i an activity

$z_i' = e^{-2\beta_p|C_i'|} (\text{th}\beta_\ell)^{|C_i|} n(C_i,C_i')$ the system can then be viewed as a gas of contours. If $\exp -2\beta_p$ and $\text{th}\beta_\ell$ are small enough most of the important configurations of this gas consist on a low density set of small contours.

Two different expansions for the thermodynamic free energy and the correlation functions with appropriate boundary conditions follow as the low activity expansions for the two "gases" just described. As the above discussion suggests, it can be proved that these expansions converge, respectively,

I) <u>if</u> $|\beta_p|$ <u>is sufficiently small of if</u> $\beta_p \geq 0$ <u>and</u> $|\beta_\ell|$ <u>is suf-</u>

<u>ficiently large</u>

II) <u>If</u> β_p <u>is positive and sufficiently large and</u> $|\beta_\ell|$ <u>is sufficiently small</u>.

When these expansions converge, they define analytic functions of β_p and β_ℓ. The unicity of the translation invariant equilibrium state also follows in the convergence regions, and we have

<u>Proposition 1</u>

<u>In the regions I and II the thermodynamic free energy of the system</u> $f(\beta_p,\beta_\ell)$ <u>is an analytic function of its arguments. Moreover, in these regions the translation invariant Gibbs state is unique and analytic with respect to</u> β_p <u>and</u> β_ℓ.

Details on the proof of this proposition will be given in [4]. In fact, the first expansion corresponds to the well known weak coupling (or high temperature) expansion for the system. The second is treated in [4] by a method inspired from the work by Minlos and Sinai [7] on the Ising model at low temperature. Lebowitz's arguments in [8] provide the easiest way of proving the unicity of the state. The regions I and II are invariant under the transformation (4) and the obvious symmetry $\beta_\ell \to -\beta_\ell$.

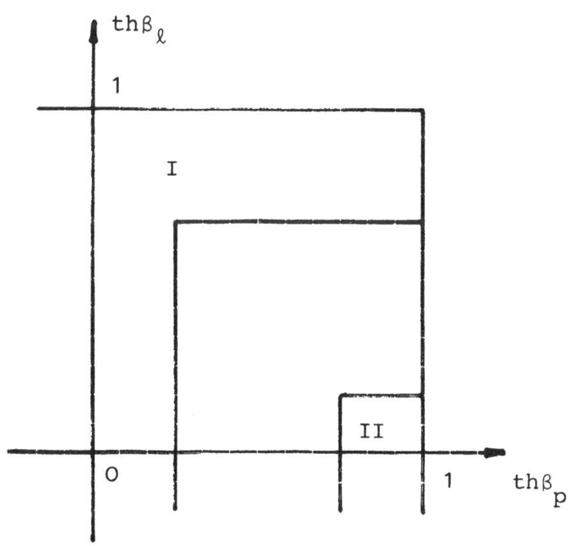

Fig. 1: The analyticity regions

Proposition 1 agrees with the conjectured phase diagram for the system (see fig. 1). According to this conjecture [1,9], a critical line, at which a second order phase transition would take place, joins two critical points respectively situated on the lines $\beta_\ell = 0$ and $\beta_p = +\infty$ (where the system is related by duality to the threedimensional Ising model) and divides the plane (β_p, β_ℓ) into two regions, corresponding to the pure phase domains. The existence of this critical line is supported in part by mean field theory calculations [1].

It has been suggested [9] that a qualitative different behaviour marks the difference between the two analyticity regions, which would correspond to a phase of particle confinement and a phase where separated charge excitations are allowed. Although the convergent expansions may also be a useful tool for studying the structure of the pure phases we are unfortunately not able to comment on this conjecture. A related problem in the case $\beta_\ell = 0$ will be discussed in the next section.

4. On the Pure Gauge System

For characterizing the phase transition in the pure gauge system $(\beta_\ell = 0)$ a non local order parameter was introduced in [1] which corresponds for this model to the Wilson loop integral appearing in the confinement problem. In this section we apply the convergent expansions to the study of this parameter.

Let γ_o be a rectangular loop in a lattice plane with sides of length T and L parallel to the axis. One defines the order parameter as the expectation value

$$\phi(\gamma_o) = \langle \prod_{\{i,j\} \in \gamma_o} A_{ij} \rangle \tag{6}$$

for the translation invariant Gibbs state.

Proposition 2

Let $\beta_\ell = 0$ and let L and T tend to infinity. Then, $\phi(\gamma_o)$ behaves as $\exp - \tau(L.T)$ for β_p small enough and as $\exp - \lambda(L+T)$ if β_p is sufficiently large, where τ and λ are positive functions of β_p. If $\beta_p > 0$, $\phi(\gamma_o)$ behaves always as $\exp - \lambda'(L+T)$.

Let $\phi_\Lambda(\gamma_o)$ be the expectation $< \prod_{\{i,j\}\in\gamma_o} A_{ij} >$ for the system on a finite box Λ. The same calculations that led to formula (2) imply

$$\phi_\Lambda(\gamma_o) = \{ \sum_{\substack{S\subset p(\Lambda)\\ \partial S=\emptyset}} (\mathrm{th}\beta_p)^{|S|} \}^{-1} \sum_{\substack{S\subset p(\Lambda)\\ \partial S=\gamma_o}} (\mathrm{th}\beta_p)^{|S|} . \tag{7}$$

Let R be, for instance, the set of plaquettes which form the rectangle of boundary γ_o. Because of

$$\sum_{\substack{S\subset p(\Lambda)\\ \partial S=\gamma_o}} (\mathrm{th}\beta_p)^{|S|} = \sum_{\substack{S\subset p(\Lambda)\\ \partial S=\emptyset}} (\mathrm{th}\beta_p)^{|(S\cup R)\setminus(S\cap R)|} \tag{8}$$

and the duality property (5) we have

$$\phi_\Lambda(\gamma_o) = Z^I_\Lambda(\beta^*)^{-1} \; Z^I_\Lambda(\beta^*_1) \tag{9}$$

where $Z^I(\beta^*)$ is the partition function for the Ising model with coupling β^* and $Z^I(\beta^*_1)$ is the partition function for the same model with a coupling β^* at all bonds Λ^* except the bonds that intersect R where the coupling is $-\beta^*$. Since this antiferromagnetic coupling tends to produce opposite phases in both sides of the surface R one can then interpret, following the Fisher's definition [10],

$$\tau = \lim_{T,L\to\infty} \lim_{\Lambda\to\infty} -\frac{1}{L.T} \log \{Z^I_\Lambda(\beta^*)^{-1} \; Z^I_\Lambda(\beta^*_1)\} \tag{10}$$

as the surface tension of the Ising model. Proposition 2 implies then

Corollary

The surface tension of the three dimensional Ising model, defined by (10), is strictly positive at sufficiently low temperature and it is zero at high temperature.

We now sketch the proof of proposition 2.

First, if β_p is small enough, from (7) one derives the bound

$$\phi_\Lambda(\gamma_o) \leq \sum_{\substack{S'\subset p'(\Lambda)\\ \partial S'=\gamma_o}} (\mathrm{th}\beta_p)^{|S'|} \tag{11}$$

where the sum runs over all connected and non selfintersecting surfaces S' such that $\partial S' = \gamma_o$. This estimate is an easy generalization to our system of the Fisher's bound [11] for the pair correlation function of the Ising model using selfavoiding walks. The same arguments used in that work, essentially an adaptation of the Peierls' argument, show that if β_p is sufficiently small, (11) can be bounded by $\exp-\tau(T.L)$, where $\tau = \tau(\beta_p) > 0$ and $\tau(\beta_p) \to \infty$ when $\beta_p \to 0$.

Next we consider the case of a large β_p. Arguments analogous to those leading to (3) show that

$$\phi_\Lambda(\gamma_o) = \{\sum_{\substack{C\subset b(\Lambda^*)\\ \partial C=\emptyset}} e^{-\beta_p|C|}\}^{-1} \sum_{\substack{C\subset b(\Lambda^*)\\ \partial C=\emptyset}} e^{-\beta_p|C|}\, n(C,\gamma_o). \tag{12}$$

Decomposed the set C of bonds into a set of disjoint selfavoiding closed curves $\{\gamma_1,\ldots,\gamma_n\}$ on the lattice as in the discussion of expression (3). The sums appearing in the numerator and in the denominator of (12) can then be considered as the partition functions Z_Λ^N and Z_Λ^D for a gas of these closed curves, the activity of which may be negative for the system in the numerator. By expanding $\log Z_\Lambda^N$ and $\log Z_\Lambda^D$ into a power series of $e^{-\beta_p}$, we find

$$\phi_\Lambda(\gamma_o) = \frac{\exp\{A_o + e^{-4\beta_p}\sum_{\gamma'} \mu^N(\gamma') + O(e^{-6\beta_p})\}}{\exp\{A_o + e^{-4\beta_p}\sum_{\gamma'} \mu^D(\gamma') + O(e^{-6\beta_p})\}} \tag{13}$$

where the lowest order term comes from the contribution of the configurations containing one of the curves γ' formed by the sides of an elementary square of the lattice. The remaining terms are of order at least $e^{-6\beta_p}$. Since $\mu^N(\gamma')n(\gamma',\gamma_o) = \mu^D(\gamma') = \mu > 0$ the first difference between the two series appears at the lowest order term due to the curves γ' such that $n(\gamma',\gamma_o) = -1$.

Fig. 2: The behaviour of $\phi(\gamma_o)$

Their number is 2(T+L) (see fig. 2) and therefore

$\phi_\Lambda(\gamma_o) \sim \exp - 2\mu e^{-4\beta_p}(T+L)$ which proves the second part of proposition 2, provided that the series converge. Further details on the proof are given in [3].

Finally, suppose β_ℓ strictly positive. The Griffiths-Kelly-Sherman inequalities [12] show that $\phi_\Lambda(\gamma_o)$ at some positive β_p and β_ℓ is larger then $\phi_\Lambda(\gamma_o)$ at $\beta_p = 0$ and the same β_ℓ. But at this point one finds $\phi_\Lambda(\gamma_o) = (\mathrm{th}\beta_\ell)^{|\gamma_o|}$ showing the perimeter behaviour of $\phi_\Lambda(\gamma_o)$ whenever $\beta_\ell > 0$.

5. Concluding Remarks

In relation with proposition 2 it is of interest to note some recent results due to Fröhlich [13] (see also the previous works mentioned in that paper) allowing to compare the lattice Higgs theories with compact group G with the pure gauge lattice theories the gauge group of which is the centre Z(G) of G. In particular, the behaviour of the Wilson parameter (corresponding to $\phi(\gamma_o)$) as exp-τ(L.T) is ensured for the SU(2) Higgs system, if the coupling constant is small enough, since it holds for the $\mathbb{Z}_2$ pure gauge system. This fact is also true for the U(1) theory. In the case of the U(1) or the SU(2) pure gauge systems this result can be obtained directly by applying the Ginibre inequalities [12] to these models in which the group matrices have been parametrized by the angle variables.

In order to simplify the exposition we made explicit use of

particular properties of the system in three dimensions. However, propositions 1 and 2 can be proved in any dimensions by arguments similar to those described in these notes.

The extension of the same results to the Higgs lattice system with the discrete gauge group $\mathbb{Z}_n$ should also be possible along the same lines.

References

[1] F. Wegner, J.Math.Phys. 12, 2259, (1971)

[2] K.G. Wilson, Phys.Rev., D10, 2445, (1974)

[3] G. Gallavotti, F. Guerra, S. Miracle-Sole in: "Mathematical Problems in Theoretical Physics"; G.F. Dell'Antonio, S. Doplicher, G. Jona-Lasinio, eds., Lecture Notes in Physics, 80, Springer, Berlin, (1978)

[4] R. Marra, S. Miracle-Sole, "On the Statistical Mechanics of the Gauge Invariant Ising Model", Preprint CNRS, Marseille (1979)

[5] R. Balian, J.M. Drouffe, C. Itzykson, Phys.Rev. D10, 3376, (1974); D11 , 2098, (1975) and D11, 2104, (1975)

[6] K. Osterwalder, E. Seiler, Ann. of Phys. 110, 440, (1978)

[7] R.A. Minlos, J.G. Sinai, Trans. Moscow Math.Soc. 17, 237, (1967) and 19, 121, (1968)

[8] J.L. Lebowitz, J. Statistical Phys. 16, 463, (1977)

[9] G.F. de Angelis, D. De Falco, F. Guerra, R. Marra, Acta Phys. Austriaca, supp. 19, 205, (1978)

[10] W.J. Camp, M.E. Fisher, Phys.Rev. B6, 946, (1972)

[11] M.E. Fisher, Phys.Rev. 162, 475, (1967)

[12] R.B. Griffiths, in: "Phase Transitions and Critical Phenomena", C. Domb, M.S. Green, eds., Academic Press, London, (1972)

[13] J. Fröhlich, "Confinement in $\mathbb{Z}_n$ Lattice Gauge Theories Implies Confinement in SU(n) Lattice Higgs Theories", Preprint I.H.E.S., Bures-sur-Yvette, (1979)

Fluctuation of the Interface of the Two-Dimensional Ising Model

Yasunari Higuchi

Institut für angewandte Mathematik

Universität Heidelberg

Federal Republic of Germany

Abstract

We show that the interface of the two-dimensional Ising model, when it is suitably normalized, converges to a one-dimensional Brownian bridge at sufficiently low temeperatures.

Let V_N be a N×N square lattice. We consider an interacting system on V_N; the energy of the system is given for each $\eta \in \Omega_N \equiv \{+1,-1\}^{V_N}$, by

$$U_N(\eta) \equiv - \sum_{\substack{<x,y> \\ x,y \in V_N}} \eta(x)\eta(y) \quad - \sum_{\substack{<x,y> \\ x \in V_N,\ y \in \partial V_N}} \eta(x)\omega(y)$$

where, $\omega(x) = +1$ if x is in the upper half boundary of V_N, and $\omega(x) = -1$ if x is in the lower half boundary of V_N.

The Gibbs state of this system is then of the form:

$$P_N(\eta) \equiv [\sum_{\sigma \in \Omega_N} e^{-\beta U_N(\sigma)}]^{-1} \quad e^{-\beta U_N(\eta)} \qquad \eta \in \Omega_N .$$

Now, we are given a probability space (Ω_N, P_N). What we are looking at is the following: Take $\eta \in \Omega_N$ arbitrarily. To every bond connecting two neighbouring points, say x and y, which satisfy $\eta(x)\eta(y) = -1$ (or $\eta(x)\omega(y) = -1$), we draw a segment of length 1 perpendicularly to this bond. Then we obtain some connected curves. There is one curve which starts from one side of ∂V_N and ends at the other side of ∂V_N. This is called the interface because it separates the sea of + sign from the sea of - sign. We denote it by $\lambda_N(\eta)$ (see Fig. 1,2). The problem is to understand the limiting behavior of $\lambda_N(\cdot)$ as $N \to \infty$.

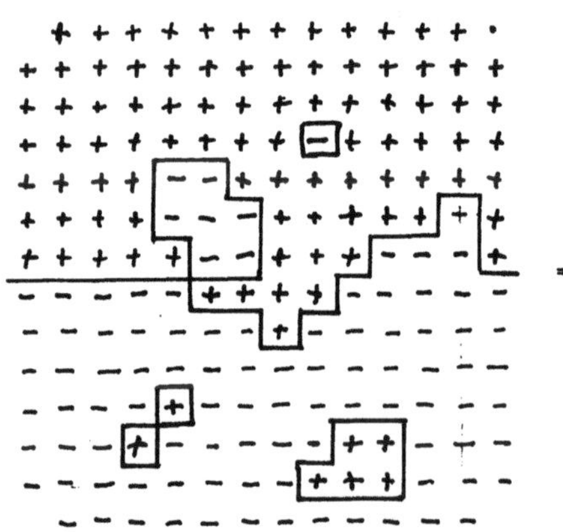

Fig. 1

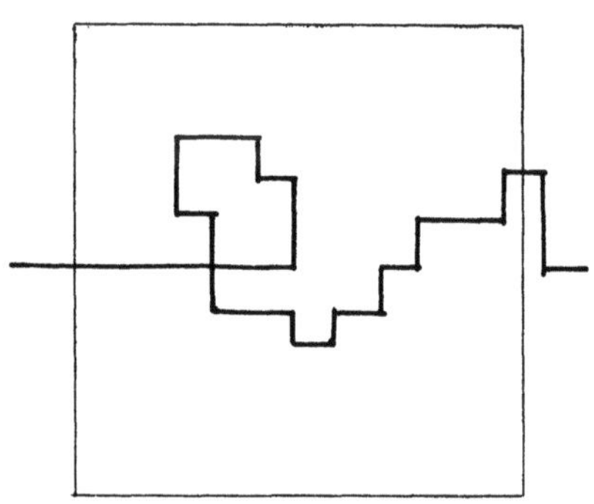

Fig. 2 $\lambda_N(\eta)$

The following fact is a consequence of Gallavotti's result [2].

Theorem O: If $\beta > 0$ is sufficiently large, then

(i) $P_N \{\mathrm{diam}(\lambda_N \cap q_k) \geq (\log N)^2 \text{ for some } k \in [0,N]\}$

$= O(N^{-\ell})$ as $N \to \infty$ for any $\ell \in \mathbb{N}$,

where q_k is the vertical line passing through $(k,0)$.

(ii) The law of $\max \{\frac{\ell}{\sqrt{N}} : (\frac{N}{2}, \ell) \in \lambda_N\}$ converges to a Gaussian distribution. (Of course we get the same limiting distribution when we take minimum instead of maximum).

Theorem O implies that we will obtain some one-dimensional Gaussian process when we normalize λ_N by $\sqrt{N}$. Since the both ends of λ_N are fixed, the limiting process may be a one-dimensional Brownian bridge. Our aim is to show this convergence in the strong sense.

Let $A_N: \mathbb{R}^2 \to \mathbb{R}$ be $A_N(x_1,x_2) = (\frac{x_1}{N}, \frac{x_2}{\sqrt{N}})$, and $C \equiv \{C\text{: connected compact set in } [0,1]\times\mathbb{R}^1,\ C \cap q_0 \neq \emptyset,\ C \cap q_1 \neq \emptyset\}$. Then $A_N(\lambda_N) \in C$. $\bar{\rho}$ is a metric in C defined by

$$\bar{\rho}(C,C') \equiv \frac{1}{2}\left[\sup_{x\in C} \inf_{y\in C'} |x-y| + \sup_{y\in C'} \inf_{x\in C} |x-y|\right].$$

Then $(C,\bar{\rho})$ is a complete separable metric space.

Theorem 1: If $\beta > 0$ is sufficiently large, then there exist a probability space $(\hat{\Omega},\hat{B},\hat{P})$ and C-valued random variables $\hat{\lambda}_1,\hat{\lambda}_2,\ldots, B_0$ such that (i) the law of $\hat{\lambda}_N$ is the same as $(A_N(\lambda_N),P_N)$. (ii) B_0 is a one-dimensional Brownian bridge, and

$$\bar{\rho}(\hat{\lambda}_N, \sigma B_0) \to 0 \quad \text{as} \quad N \to \infty \qquad \hat{P} - \text{a.e.},$$

where $\sigma = \sigma(\beta) > 0$ is a constant depending only on $\beta > 0$.

Idea of the Proof

Let $\tilde{P}_N$ be the distribution of $A_N(\lambda_N)$ induced by P_N. $\tilde{P}_N$ is a probability measure on C. We can easily check that C[0,1] is measurable in $(C,\bar{P})$. We can define the distribution $P_{0,0}^{1,0}$ of a one-dimensional Brownian bridge as a probability measure on C. The theorem of Skorokhod (see [4], p. 10,11) tells us that the statement of Theorem 1 is true if we can prove that

$$\int_C f(c)\tilde{P}_N(dc) \underset{N\to\infty}{\longrightarrow} \int_C f(c)P_{0,0}^{1,0}(dc) \tag{2}$$

for any bounded uniformly continuous function f.

We don't know much about the convergence of measures on C as on C[0,1]. So we reduce (2) to the convergence of measures on C[0,1]. Define a continuous process $\{Y_N(t)\}_{0\le t\le 1}$; $\Omega_N \to C[0,1]$ by

$$\begin{aligned} &Y_N(0) \equiv Y_N(1) \equiv 0, \\ &Y_N(\tfrac{k}{N}) \equiv \max\{U \in \mathbb{R};\ (\tfrac{R}{N}, U) \in A_N\lambda_N\}, \end{aligned} \tag{3}$$

and connect them linearly (Fig. 3).

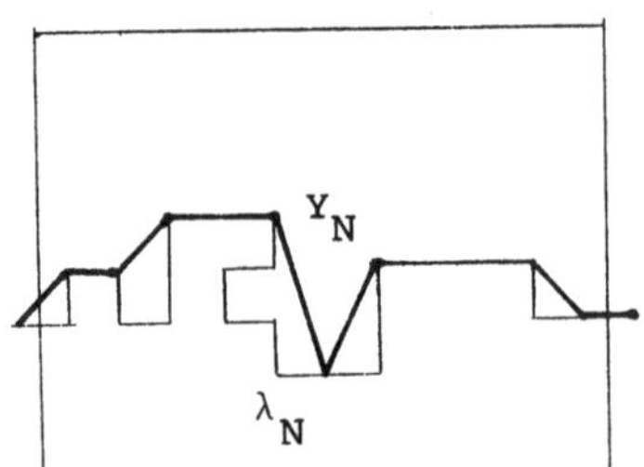

Fig. 3 λ_N and Y_N

Let us denote by μ_N the distribution of Y_N induced by P_N. For any bounded uniformly continuous f on C, we have

$$\begin{aligned} &\left| \int_C f(c)\tilde{P}_N(dc) - \int_C f(c)\mu_N(dc) \right| \\ &= \left| \int_{\Omega_N} [f(A_N\lambda_N(\eta)) - f(Y_N(\eta))]\, P_N(d\eta) \right| \\ &\le \left| \int_{\{\rho(A_N\lambda_N, Y_N)\ge\varepsilon\}} \right| + \left| \int_{\{\rho(A_N\lambda_N, Y_N)<\varepsilon\}} \right|. \end{aligned} \tag{4}$$

The first term of the right hand side vanishes as $N \to \infty$ from (i) of Theorem 0 and the boundedness of f and the second term vanishes as $\varepsilon \downarrow 0$ by the uniform continuity of f. Since $f|_{C[0,1]}$ with respect to the supremum norm topology in C[0,1], we can reduce the problem to the convergence of μ_N to $P^{1,0}_{0,0}$ on C[1,0].

For the convergence of μ_N to $P^{1,0}_{0,0}$, it is enough to show the following two conditions.

I. Convergence of the Finite Dimensional Distribution

For any $0 = t_0 < t_1 < t_2 < \ldots < t_k < t_{k+1} = 1$ and $-\infty < a_i < b_i < \infty$ $i = 1,2,\ldots,k$.

$$\mu_N(\{C \in ([0,1] \; ; \; C(t_i) \in [a_i, b_i] \quad i = 1,2,\ldots,k\})$$

$$\longrightarrow P^{1,0}_{0,0}(\{C \in C[0,1]; \quad C(t_i) \in [a_i\sigma, b_i\sigma] \quad i = 1,2,\ldots k\})$$

$$= \mathrm{Const} \times \int_{a_1\sigma}^{b_1\sigma} P(t_1,x_1)dx_1 \int_{a_2\sigma}^{b_2\sigma} P(t_2-t_1,x_2-x_1)dx_2 \ldots$$

$$\int_{a_k\sigma}^{b_k\sigma} P(t_k-t_{k-1},x_k-x_{k-1}) \; P(1-t_k, -x_k)dx_k$$

where $p(t,x) = (2\pi t)^{-1/2} \exp(-x^2/2t)$-

II. Compactness of Measures $\{\mu_N\}_{N=1}^{\infty}$

(a) There exist constants $M > 0$, $Y > 0$ such that

$$\int_{C[0,1]} |C(0)|^{\gamma} \mu_N(dc) < M \qquad N = 1,2,\ldots .$$

(b) There exist constants $L > 0$, $\alpha > 0$, $\beta > 0$ such that

$$\int_{C[0,1]} |C(t) - C(S)|^{\alpha} \mu_N(dc) \le L \; |t-S|^{1+\beta}$$

$$N = 1,2,\ldots, \quad t,S \in [0,1].$$

(a) is trivial by the definition of $Y_N(0)$. Moreover, since we have the asymptotic formula for the finite dimensional distribution of μ_N ($\mu_N \to P^{1,0}_{0,0}$), and $P^{1,0}_{0,0}$ satisfies the condition (b), we can guess that μ_N satisfies (b).

So, the only remaining problem is to show the condition (I). Theorem 0, (ii) is the case when $k = 1$. For general $k \geq 2$, C. Cammarota essentially proved (I) in 1977, however he could not publish it before he was taken in military service. The proof of (I) given in [3] is done independently from his work, although the idea of the proof is just the same as Cammarota's.

Let us illustrate the idea. First we consider the following system.

$$\hat{\Lambda}_N \equiv \{\lambda;\ \text{subgraph of } [0,N]\times\mathbb{Z}^1,\quad \lambda \ni (0,0),\ \lambda \cap q_N \neq \emptyset$$

(see Fig. 4).

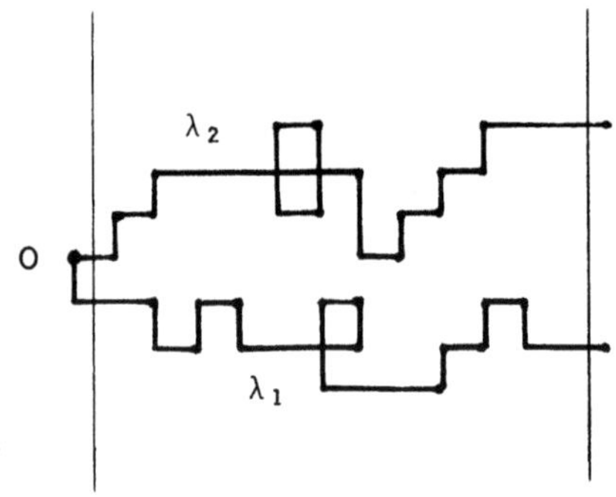

Fig. 4
Elements in $\hat{\Lambda}_N$

We define $\hat{P}_N$ on $\hat{\Lambda}_N$ as

$$\hat{P}_N(\lambda) \propto \exp\{-\,2\beta|\lambda| - 2\beta \sum_{\substack{\Gamma\cap\lambda\neq\emptyset \\ \Gamma\subset[0,N]\times\mathbb{Z}^1 \\ \Gamma:\ \text{closed}}} |\Gamma|\}^{\sim}$$

Then,

$$P_N(\cdot) \sim \frac{P_N(\cdot\cap\{\lambda\ni(N,0)\})}{\hat{P}_N(\lambda\ni(N,0))} \quad \text{as } N\to\infty.$$

The difference between these two values is of order $o(N^{-\ell})$ for any $\ell \in \mathbb{N}$.

Let us decompose λ into cluster parts i.e.

$$\lambda = (\xi_1, \xi_2, \dots \xi_s\ ;\ S_{\xi_s}, \dots S_{\xi_s}) \equiv S_x,$$

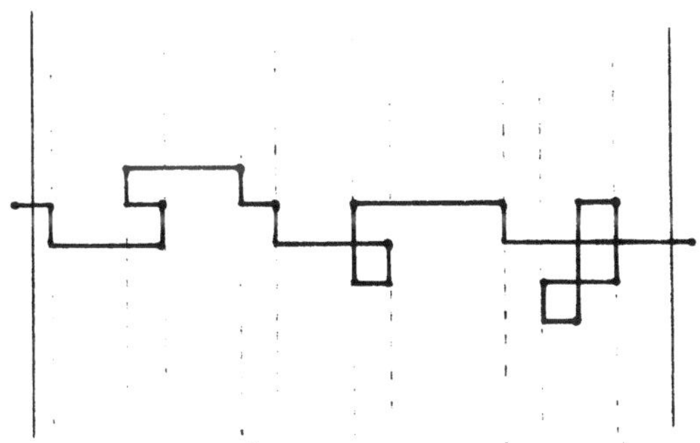

ξ_1 ξ_2 ξ_3 ξ_4 ξ_5 ξ_6 ξ_7

Fig. 5

where ξ_i indicates the horizontal location of the i-th cluster and S_{ξ_i} indicates the shape of λ over ξ_i (Fig. 5). Let $\delta S_{\xi_i} \equiv$ (the height of the first point of S_{ξ_i}) - (the height of the last point of S_{ξ_i}). In order to prove (I), it is sufficient to show the asymptotic independence of each jumps; $\sum_{i=1}^{s} \chi_j(\xi_1)\cdot\delta S_{\xi_i}$, $j = 1,2,\ldots, k$, where

$$\chi_j(\xi) = \begin{cases} 1 & \text{if } \xi\subset[t_{j-1}N,\ t\,N] \\ 0 & \text{otherwise} \end{cases} \qquad j = 1,2,\ldots, k+1.$$

For simplicity we consider the case k = 1, but the idea applies to any k. Let $J_j(S\ ) \equiv \sum_{i=1}^{s} \chi_j(\xi_i)\delta S_{\xi_i}$ $j = 1,2$. Then from [1], [2], for any real y_1, y_2 we have

$$\int_{\hat{\Delta}_N} e^{iy_1J_1(S_x)+iy_2J_2(S_x)}\ \hat{P}_N(dS_x)$$
$$= \exp\{\sum_{x\subset[0,N]}\ \sum_{S_x} \Phi_N(S_x)(e^{iy_1J_1(S_x)+iy_2J_2(S_x)}-1\} \tag{5}$$

for some function Φ_N, which satisfies

$$\sup_{P\in[0,N]}\ \sum_{P\in X\subset[0,N]} |\Phi_N(S_x)| < +\infty. \tag{6}$$

and Φ_N is asymptotically translatinary invariant.

The summation in the right hand side of (5) splits into

$$\sum_{X\subset[0,\frac{N}{2}]} + \sum_{X\subset[\frac{N}{2},0]} + \sum_{\substack{X\subset[0,N] \\ X\cap[0,\frac{N}{2}]\neq\emptyset \\ X\cap[\frac{N}{2},N]\neq\emptyset}} .$$

For the independence of J_1 and J_2 as $N \to \infty$, it is sufficient to show that the third term converges to zero as $N \to \infty$. If Φ_N is translationally invariant, then it follows from (6). Since Φ_N is asymptotically translationally invariant, we can show (I). The complete proof is given in [3].

References

[1] Del Grosso, G., On the Local Central Limit Theorem for Gibbs Processes, Comm. Math. Phys. 37, 141-160, 1974.

[2] Gallavotti, G., The Phase Separation Line in the Two-Dimensional Ising Model, Comm. Math. Phys. 27, 103-136.

[3] Higuchi, Y., On Some Limit Theorems Related to the Phase Separation Line in the Two-Dimensional Ising Model, Z. Wahrsch. verw. Gebiete 50, 287-315 (1979).

[4] Skorokhod, A.V., Studies in the Theory of Random Processes, Addison-Wesley, 1961.

The Stability Problem in ϕ^4 Scalar Field Theories

Giovanni Gallavotti

Istituto Matematico

Università di Roma

Italy

Abstract

In these notes the renormalization theory for the ground state energy in ϕ^4 field theory in 2 and 3 dimensions is examined for the hierarchical fields as well as for the Euclidean fields and the similarity between the two cases is stressed.

In these notes I collect, together with the talk that I gave at the meeting (represented by §6,7) some other introductory topics like the theory of the Bernoulli fields [1],[2] in a form which takes into account the results in [3], [4], [5].

It seems to me that the Bernoulli field theory already contains the problems typical of the euclidean field theory (i.e., the renormalization of the ultraviolet divergences) in a form essentially identical to the one in which they appear in the more interesting euclidean fields. However, such models are simpler as they do not present some additional difficulties which do not seem to me to have much to do with the ultraviolet stability. See concluding remarks for references and historical comments.

1. The Hierarchical Model

The role that the hierarchical model will play in these lectures is that of providing a guide to the understanding of the mechanism which allows the cancellation of the various divergent contributions to the ground state energy to actually occur, in a controllable way. The hierarchical model is a model of field theory in which the free field is replaced by some new free field of simpler structure but which leads to essentially the same ultraviolet problems.

Let $(Q_n)_{n=k}^{\infty}$ be a sequence of pavements of R^d with cubic tesserae: the tesserae of Q_k have side size 2^{-k} and the pavements will be supposed compatible.

To each tessera Δ we associate a gaussian random variable z_Δ normalized so that $E(z_\Delta^2) \equiv 1, \forall \Delta \in \cup Q$; we shall also suppose that z and $z_{\Delta'}$ are independent unless $\Delta, \Delta' \in Q_k$ for some $k = 0,1,\ldots$

The variables associated with the same pavement Q_k of R^d will be supposed distributed with the following formal density

$$P\Big(\prod_{\Delta\in Q_k} dz_\Delta\Big) \propto \exp\left(-\frac{\alpha^2}{2} \sum_{\Delta,\Delta' n.n} (z_\Delta - z_{\Delta'})^2 - \frac{\beta^2}{2} \sum_\Delta z_\Delta^2\right) \prod_\Delta dz_\Delta \tag{1.1}$$

i.e. z_Δ, $\Delta \in Q_k$, is the gaussian process on the lattice of the centers of the tesserae with invariance operator

$$-\alpha^2 D + \beta^2 \tag{1.2}$$

when D is the second difference.

We define the "free field":

$$\phi_\xi = \sum_{\substack{\Delta\ni\xi \\ \Delta\in Q_k}} \sqrt{2^{(d-2)k}}\, z_\Delta \qquad \xi \in R^d \tag{1.3}$$

and the cut-off field

$$\phi_\xi^{[\leq N]} = \sum_{k=0}^{N} \sqrt{2^{(d-2)k}}\, z_{\Delta_k} \qquad \xi \in R^d \tag{1.4}$$

where $\Delta_o, \Delta_1, \ldots, \Delta_N$ are in $Q_o, \ldots, Q_N$ and contain ξ. To resolve the ambiguities at the boundaries let us suppose, for instance, that the Δ's are open. The field $\phi^{[\leq N]}$ is normalizable, in contrast to the field ϕ. Its normalized copy

$$x_\xi^{(N)} = \sum_{k=0}^{N} \frac{\sqrt{2^{(d-2)k}}\, z_\Delta}{\sqrt{\sum_{k=0}^{N} 2^{(d-2)k}}} \qquad \xi \in R^d \tag{1.5}$$

verifies a recursion relation which stems from its definition:

$$x_\Delta^{(N)} = \frac{z_\Delta^{(N)} + \sqrt{\Gamma_N}\, x_{\Delta'}^{(N-1)}}{\sqrt{1 + \Gamma_N}} \tag{1.6}$$

where $\Delta \subset Q_N$, $\Delta' \subset Q_{N-1}$, $\Delta' \supset \Delta$ and $x_\Delta^{(N)}$ denotes the constant value of $x_\xi^{(N)}$ on A, finally

$$\Gamma_N = \sum_{k=0}^{N-1} \sqrt{2^{(d-2)(k-N)}} = \begin{cases} N & \text{if } d = 2 \\ \dfrac{1-2^{-N(d-2)}}{2^{d-2}-1} & \text{if } d > 2 \end{cases} \tag{1.7}$$

so if d = 3, $\Gamma_N \tilde{\equiv} 1$, if d = 4, $\Gamma_N \tilde{\equiv} \frac{1}{3}$ etc..

The "interaction" that we wish to consider is defined by

$$V_{O,I}^{(N)}(\phi) = -\lambda \int_I \phi_\xi^{[\leq N]^4} d\xi \tag{1.8}$$

where I is a cube paved by Q_O.

The above quantity is the "nonrenormalized" interaction.

We define for an arbitrary gaussian random variable x the random variables $:x^n:$ by

$$:x^n: = \sqrt{2c}^n H_n\left(\frac{x}{\sqrt{2c}}\right), \quad c = E(x^2) \tag{1.9}$$

where H_n = n-th hermite polynomial (e.g. $H_2(y) = y^2 - 1/2$, $H_4(y) = y^4 - 3y^2 + 3/2$ etc.). The "renormalized interaction" to first order in λ" is

$$V_{1,I}^{(N)}(\phi) = -\lambda \int_I :(\phi_\xi^{[\leq N]})^4: d\xi. \tag{1.10}$$

We shall later define the higher order renormalized interactions $V_{t,I}^{(N)}(\phi)$, $t \geq 2$. The ultraviolet problem to order t is to show the existence (or nonexistence) of a function $E_t(\lambda)$ such that

$$\exp - E_t(\lambda)(I) \leq \int \exp V_{t,I}^{(N)}(\phi)\ P(\phi) \leq \exp E_t(\lambda)(I)$$

$$\lim_{\lambda \to O} \lambda^{-t} E_t(\lambda) = O.$$

We shall consider only d = 2 or d = 3. The formulation of the ultraviolet stability in d ≥ 4 would be different and, anyway, it is still an open problem.

The d = 2, d = 3 cases are special because the models are "superrenormalizable" in such dimensions. The mathematical aspect of superrenormalizability can be seen by writing $V_{1,I}^{(N)}$, say, as a sum using that $X_\xi^{(N)}$ is piecewise constant: since

$$:(\phi_\xi^{[\leq N]})^4: \equiv 2^{2(d-2)N}(1+\Gamma_N)^2\ H_4(X_\Delta^{(N)})$$

we find

$$V_{1,I}^{(N)} = - \lambda \sum_{\Delta' \in Q_N} 2^{(d-4)N} (1+\Gamma_N) H_4(X_\Delta^{(N)})$$

i.e. if we look at the interaction as a function of the normalized fields we see that if $d < 4$ the "effective coupling constant" is $\lambda 2^{(d-4)N}$: this will allow the use of perturbation theory for N large: the fact that $\lambda 2^{(d-4)N} \xrightarrow[N\to\infty]{} 0$ is also called "asymptotic freedom".

2. Analysis of the Ultraviolet Problem in the Bernoulli Model to Order 1 in d = 2. The Lower Bound

The Bernoulli field is defined by setting $\alpha = 0$ (i.e. all the z's become independently distributed), see (1.1).

The theory of the ultraviolet problem for the Bernoulli field is quite easy but very instructive, already in the $d = 2$, $t = 1$ case. We first find a lower bound for

$$Z(I) = \int P_N(dz) \exp V_{1,I}^{(N)}(\phi) \tag{2.1}$$

by saying

$$Z(I) \geq \int P_N(dz) \Big(\prod_{i=o}^{N} \prod_{\substack{\Delta \in Q_i \\ \Delta \subset I}} \chi_\Delta^{B_i}(X^{(i)}) \Big) e^{V_{1,I}^{(N)}} \tag{2.2}$$

where χ_Δ^B are suitable characteristic functions: χ_Δ^B is the c.f. of the event, if $\Delta \in Q_k$:

$$\{X \mid |X_\Delta^{(k)}| < B(1+k)^4 \log(e+\lambda^{-1})\} \tag{2.3}$$

where $B > 0$ will be chosen later and is supposed at least such that: $H_4(b) \geq -\min(H_4) = 3/2$, $\forall b \geq B$. The $\log(e+\lambda^{-1})$ has been inserted to improve the estimates, automatically, at λ small, and is not the optimal choice.

To estimate the above integral we notice

$$Z(I) \geq \int P_{N-1}(dz) \Big(\prod_{i=o}^{N-1} \prod_{\substack{\Delta \in Q_{N-1} \\ \Delta \subset I}} \chi_\Delta^{B_i} \Big) \times$$

$$\times \prod_{\Delta' \in Q_{N-1}} \left(\int \chi_\Delta^{B_N}(X_\Delta^{(N)}) e^{-\lambda(1+\Gamma_N)^2 2^{-2N} H_4(X_\Delta)} e^{-z_\Delta^2} \frac{dz_\Delta}{\sqrt{\pi}}\right)^4 \quad (2.4)$$

Hence the estimate can be obtained applying the following easy lemma

Lemma1: Let $|X^1| \leq B^1$ and let $X = \frac{z + \sqrt{\Gamma}X^1}{\sqrt{1+\Gamma}}$, $B \geq B^1$; $B^1 \geq 1$, let $\bar{B} = B\sqrt{1+\Gamma} - \sqrt{\Gamma}B^1 \geq 1$, then:

$$\int \exp - \lambda(1+\Gamma)^2 H_4(X)[\chi^B(X) + \sigma(1-\chi^B(X))] \frac{e^{-z^2}}{\sqrt{\pi}} dz =$$

$$= \exp - \lambda\Gamma^2 H_4(X^1) + \varepsilon_{\Gamma,\sigma}(\lambda,B,\bar{B}) \qquad \sigma = 0,1 \quad (2.5)$$

and ε can be estimated by

$$|\varepsilon_{\Gamma,\sigma}(\lambda,B,\bar{B}| \leq (\lambda\Gamma^2 B^4 e^{C_1\lambda(1+\Gamma^2)B^4})^2 + C_2 e^{-C_3\bar{B}^2 + C_4(\lambda\Gamma^2 B^4)} \equiv \quad (2.6)$$

$$\equiv \eta_\Gamma(\lambda,B,\bar{B})$$

uniformly in $|X^1| \leq B^1$.

Proof: it is an easy exercise.

The application is immediate: the integrals in (2.4) are bounded by

$$\int \chi_\Delta^{B_N}(X_\Delta^{(N)}) e^{-\lambda(1+\Gamma_N)2^{-2N} H_4(X_\Delta)} \frac{e^{-z_\Delta^2}}{\sqrt{\pi}} dz_\Delta \geq$$

$$\geq \exp - \lambda 2^{-2N} \Gamma_N H_4(X_\Delta^{(N-1)}) + \eta_{\Gamma_N}(\lambda 2^{-2N}, B_N, \bar{B}_N) \quad (2.7)$$

where

$$\bar{B}_N = \sqrt{1+\Gamma_N}\, B_N - \sqrt{\Gamma_N}\, B_{N-1} = O(N^{3+1/2}B). \quad (2.8)$$

Since in (2.4) we are only interested in the case $|X_{\Delta'}^{(N-1)}| < B_{N-1}$.

Next notice that $\Gamma_N = (1+\Gamma_{N-1})$ (recall $d = 2$) and $4\cdot 2^{-2N} = 2^{-2(N-1)}$ so that (2.7), (2.4) imply, denoting

$$\chi_N^B = \prod_{i=o}^{N} \prod_{\substack{\Delta\in Q_N \\ \Delta\subset I}} \chi_\Delta^{B_i}$$

$$Z(I) \geq (\int \chi_N^B \, e^{V_{1,I}^{(N)}} P_N(d\phi)) \geq$$

$$\geq \int \chi_{N-1}^B \prod_{\Delta^1 \in Q_{N-1}} \exp - \lambda H_4(X_\Delta^{(N-1)}) \, 2^{-2N}(1+\Gamma_N) \, P_{N-1}(d\phi) \cdot$$

$$\cdot \exp - \eta_{\Gamma_N}(\lambda 2^{-2N}, B_N, \bar{B}_N) \, 2^{2N}|I| = \tag{2.9}$$

$$= \int \chi_{N-1}^B \, e^{V_{1,I}^{(N-1)}} \, P_{N-1}(d\phi) \exp - \eta_{\Gamma_N}(\lambda 2^{-2N}, B_N, \bar{B}_N).$$

Hence

$$Z(I) \geq \exp - |I| \sum_{j=o}^{\infty} \eta_{\Gamma_j}(\lambda 2^{-2j}, B_j, \bar{B}_j) \tag{2.10}$$

provided B is so chosen that $\forall j$, $\bar{B}_j \geq 1$ (e.g. $B = 1$).

It is clear that the above series converges and, from (2.6), we see that the sum can be bounded by

$$G \, \lambda^2(\log(e+\lambda^{-1}))^8 \tag{2.11}$$

for λ small enough.

Before continuing let us interpret the above calculation:

i) It is clear that

$$\int V_{1,O}^{(N)} \, (\phi^{(N)}) \, P_N(dz) \equiv V_{1,O}^{(N-1)}(\phi^{(N-1)}) \tag{2.12}$$

and it is natural to attempt to use this information to prove the bounds.

ii) The naive bound $Z(I) \leq \exp \max V_{O,\tau}^{(N)}$ gives $Z(I)$ exp + $\lambda(1+\Gamma_N)2^{-2N} \frac{3}{2} 2^{2N} |I| \to \infty$. However, we can take better advantage of the asymptotic freedom and of the fact that the $V_{1,I}^{(N)}$ is a sum of random variables little correlated.

We try to write

$$\int \exp V_1^{(N)}\, \hat{P}_N(dz) = \exp[\{\hat{E}_N(V_z^{(N)}) + \frac{1}{2!}\hat{E}_N^T(V_I^{(N)};1) +$$

$$+\ldots+\frac{1}{t!}\hat{E}_N^T(V_\tau^{(N)};t)\} + (\text{error})] \tag{2.13}$$

where $\hat{E}_N$ is the average with respect to $(\hat{P}_N dz) \equiv$ distribution of $(z_\Delta)_{\Delta\in Q_N}$, and $E_N^T(\cdot;s)$ truncated expectation of order s.

The fact that the z_Δ variables are not too correlated (in our case they are actually independent) will tell us that the error is of the order of the volume of I in units of 2^{-N}, i.e. const $|I|2^{dN}$ and not const $(|I|2^{dN})^{t+1}$ as it would be for strongly correlated fields $(z_\Delta)_{\Delta\in Q_N}$. The asymptotic freedom says that the constant is of the order of $(\lambda 2^{(d-4)N}(1+\Gamma_N))$.

Hence we see that the errors in (2.13) will be of the order

$$\lambda^{t+1}\, 2^{(t+1)(d-4)}\, 2^{dN}\, (1+\Gamma_N)^{2(t+1)} \tag{2.14}$$

and therefore summable over N as soon as $(t+1)(d-4) > d$, i.e. if $d = 2$ already for $t = 1$ (if $d = 3$ we should take $t = 3$).

iii) The calculation made before was just an attempt to estimate the error of (2.13) form below: the attempt was successful even though the result was not as good as the naive estimate (2.14) (e.g. we have not found a bound like λ^2 but only like $\lambda^2(\log(e+\lambda^{-1}))^4$, which is, however, of order larger than 1 as $\lambda \to 0$).

3. The Upper Bound in the Bernoulli Case

We proceed, again, recursively. The basic remarks are the following.

i) Let $\phi = (z_\Delta)_{\Delta\in U_i Q_i}$ be a random field.

Let

$$\bar{B}_N = \sqrt{1+\Gamma_N}\, B_N - \sqrt{\Gamma_N}\, B_{N-1} \qquad = O\,(B \log(e+\lambda^{-1})\,(1+N)^{3+1/2})$$

$$\tilde{B}_N = B_N \,/\, (1+N)^2 \qquad = O\,(B \log(e+\lambda)^{-1}(1+N)^2) \tag{3.1}$$

and define

$$D_N(\phi) = \{\Delta \,|\, \Delta \in Q_N \;,\; |X_\Delta| > B_N\} \tag{3.2}$$

$$R_N(\phi) = \{\Delta \,|\, \Delta \in Q_N \;,\; |z_\Delta| > \tilde{B}_N\} \tag{3.3}$$

and notice that $\forall\phi$:

$$R_N \supset D_N \setminus D_{N-1} \tag{3.4}$$

$$\Delta \in D_{N-1} \setminus (R_N \cup D_N) \Rightarrow |X_\Delta^{(N)}| \geq \frac{B_{N-1}\sqrt{\Gamma_N} - \hat{B}_N}{\sqrt{1+\Gamma_N}} \geq \frac{B_{N-1}}{2}. \tag{3.5}$$

ii) The positivity of $H_4(b)$, $b \geq B_N$, implies, c.f.i. (1.10):

$$V_{1,I}^{(N)}(\phi) \geq V_{1,I\setminus D_N}^{(N)}(\phi) \tag{3.6}$$

hence

$$Z(I) \geq \int e^{V_{1,I\setminus D_N}^{(N)}(\phi)}\, P_N(d\phi) \equiv Z_o^{(N)}(I). \tag{3.7}$$

Also, from (1.10)

$$V_{1,I\; D_N}^{(N)}(\phi) \leq V_{1,I\setminus(D_N \cup \hat{R}_N)}^{(N)} + 4 \,\#(\hat{R}_N)\, \frac{3}{2}\, \lambda\, 2^{-2N}(1+\Gamma_N) \tag{3.8}$$

where $\#(R_N)$ = number of tesserae in R_N, and $\hat{R}_N$ is the smallest set paved by Q_{N-1} containing R_N. Remark then that (3.8), (1.10), (3.5) imply

$$V_{1,I\; D_N}^{(N)}(\phi) \leq V_{1,I\setminus\hat{R}_N \cup D_{N-1}}^{(N)} + 4\,\#(\hat{R}_N)\, \frac{3}{2}\, \lambda\, 2^{-2N}(1+\Gamma_N) +$$

$$- \lambda\, 2^{-2N}(1+\Gamma_N)\, H_4(\bar{B}_N \sqrt{1+\Gamma_N})\, \#(D_{N-1}). \tag{3.9}$$

Hence

$$\int e^{V^{(N)}_{1,I\backslash D_N}(\phi)} dP_N \le \int e^{V^{(N)}_{1,I\backslash D_{N-1}\cup \hat{R}_N}} dP_N \, e^{4\#(R_N)\frac{3}{2}\lambda 2^{-2N}(1+\Gamma_N)}$$

(3.10)

$$\le \int \prod_{\Delta\in R_N} \bar{\chi}^{B}_{\Delta}(z) \prod_{\Delta\in R_N} (1-\bar{\chi}^{B}_{\Delta}(z)) e^{V^{(N)}_{1,I\backslash D_{N-1}\cup \hat{R}_N}} e^{4\#(R_N)\frac{3}{2}\lambda 2^{-2N}(1+\Gamma_N)} dP_N$$

where $\bar{\chi}^{B}_{\Delta}$ is the c.f. of the event

$$|z^{(N)}_{\Delta}| \ge \tilde{B}_N. \qquad (3.11)$$

Hence if R is a set paved by Q_N:

$$\le \sum_{R\subset I} \int \prod_{\Delta\in R} \bar{\chi}^{B}_{\Delta}(z) \prod_{\Delta\in R} (1-\bar{\chi}^{B}_{\Delta}(z)) \, e^{V^{(N)}_{1,I\backslash D_{N-1}\cup \hat{R}_N}} e^{4\#(R)\frac{3}{2}\lambda(1+\Gamma_N)2^{-2N}} dP_N$$

$$\le \sum_{R\subset I} \int \prod_{\Delta\in R} \bar{\chi}^{B}_{\Delta}(z) \, e^{V^{(N)}_{1,I\backslash D_{N-1}\cup \hat{R}}} e^{4\#(R)\cdot\frac{3}{2}\cdot\lambda(1+\Gamma_N)2^{-2N}} dP_N$$

(3.12)

$$= \sum_{R\subset I} \int \prod_{\Delta\in R}\Big(\int \bar{\chi}^{B}_{\Delta}(z) \, e^{4\#(R)\frac{3}{2}\lambda(1+\Gamma_N)2^{-2N}} \frac{e^{-z^2_\Delta}dz_\Delta}{\sqrt{\pi}}\Big) \cdot$$

$$\cdot \Big[\prod_{\Delta'\in I\backslash \hat{R}\cup D_N} \Big(\int \exp - \lambda(1+\Gamma_N)^2 H_4(X^{(N)}_{\Delta})2^{-2N} \frac{e^{-z^2}}{\sqrt{\pi}} dz\Big)^4\Big] dP_{N-1}$$

applying the lemma of the preceding section the last integral can be bounded by

$$\sum_{R\subset I} \int \Big(\int \prod_{\Delta\in R} \bar{\chi}^{B}(z) \, e \frac{-z^2_\Delta dz_\Delta}{\sqrt{\pi}}\Big) \cdot \exp - V^{(N-1)}_{1,I\backslash D_{N-1}\cup \hat{R}} \cdot \qquad (3.13)$$

$$\cdot \exp + \eta_{\Gamma_N}(\lambda 2^{-2N}, B_N, B_{N-1})|I| \exp + 4\#(R)\frac{3}{2}\lambda(1+\Gamma_N)^{2^{-2N}}$$

Applying

$$V^{(N-1)}_{1,I\setminus D_{N-1}\cup\hat{R}_N} \le V^{(N-1)}_{1,I\setminus D_{N-1}} + 4\#(R_N)(1+\Gamma_N)^2\lambda 2^{-2N} H_4(B_{N-1}) \tag{3.14}$$

we have found:

$$Z_o^{(N)}(I) \le Z_o^{(N-1)}(I)\cdot \sum_{R\subset I} (e^{8\#(R_N)(1+\Gamma_N)^2 2^{-2N}\lambda H_4(B_N)}$$

$$\exp - \tilde{c}\,\tilde{B}_N^2\,\#(R_N))\ \exp \eta_{\Gamma_N}(\lambda 2^{-2N}, B_N, B_{N-1})|I| \tag{3.15}$$

$$= Z_o^{(N-1)}(I)\ \exp(\eta_{\Gamma_N} + (\lambda 2^{-2N}, B_N, B_{N-1}) + \eta^1(\lambda,N))|I|$$

where

$$\eta^1(\lambda,N) = \log(1+\exp - \tilde{c}\tilde{B}_N^2 + 8(1+\Gamma_N)^2\lambda 2^{-2N} H_4(B_N)). \tag{3.16}$$

Hence

$$Z_o^{(N)}(I) \le \exp I \sum_{N=0}^{\infty} (\eta_{\Gamma_N}(\lambda 2^{-2N}, B_N, B_{N-1}) + \eta^1(\lambda,N) \tag{3.17}$$

and it is easily realized that the series in parenthesis behaves as $\lambda \to 0$ as (2.11) with a different G, say G'. Hence:

$$E(\lambda) \le (G+G')\ \lambda^2 \log(e+\lambda^{-1})^8. \tag{3.18}$$

4. Hierarchical Model to Order d = 2

We define the renormalized interaction to order t, for d = 2, as:

$$V^{(N)}_{t;I} = V^{(N)}_{1;I} - \sum_{k=2}^{t} \frac{\varepsilon^T(V^{(N)}_{1;I};k)}{k!}\ . \tag{4.1}$$

Consider for example t = 2.

It is natural to proceed as before: however, it will no longer be so easy because

$$V^{(N-1)}_{2,I} \neq \hat{E}^{T}_{N}(V^{(N)}_{2;I}) + \frac{1}{2} \hat{E}^{T}_{N}(V^{(N)}_{2;I};2). \tag{4.2}$$

Therefore we introduce, recursively

$$\tilde{V}^{(N-k)}_{2;I} = \left[\hat{E}^{T}_{N-k+1} (\tilde{V}^{(N-k+1)}_{2;I}) + \frac{1}{2} \hat{E}^{T}_{N-k+1} (\tilde{V}^{N-k+1)}_{2;I};2) \right]_{(2)} \tag{4.3}$$

$$\tilde{V}^{(N)}_{2;I} = V^{(N)}_{2;I} \tag{4.4}$$

where $[\ ]_{(2)}$ = truncation of a polynomial in λ to order λ^2.

To proceed as before one needs to first get an idea on how $\tilde{V}^{(N-k)}_{2;I}$ looks like: it is easy to see that (by induction)

$$\tilde{V}^{(N-k)}_{2;I} = \left[\hat{E}^{T}_{>N-k}(V^{(N)}_{2;N}) + \frac{1}{2} \hat{E}^{T}_{>N-k}(V^{(N)}_{2,N} ; 2) \right]_{(2)} \tag{4.5}$$

hence if $\tilde{\phi} = \phi^{[\leq N]}$, $\phi = \phi^{[\leq N-k]}$:

$$\tilde{V}^{(N-k)}_{2;I} = \int_I - \lambda :\phi^4_\xi: d\xi - \frac{\lambda^2}{2} 4! \int_I C^{(N)4}_{\xi\eta} d\xi\, d\eta \tag{4.6}$$

$$+ \frac{1}{2} \int_I \lambda^2 \int (:\tilde{\phi}^4_\xi: :\tilde{\phi}^4_\eta:) d\tilde{P}_{>N-k} d\xi\, d\eta - \frac{1}{2}\int \lambda^2 :\phi^4: :\phi^4: d\xi\, d\eta.$$

The integral

$$\int :\tilde{\phi}^4_\xi: :\tilde{\phi}^4_\eta: d\tilde{P}_{>N-k} \tag{4.7}$$

can be graphically represented as the integral of the graph

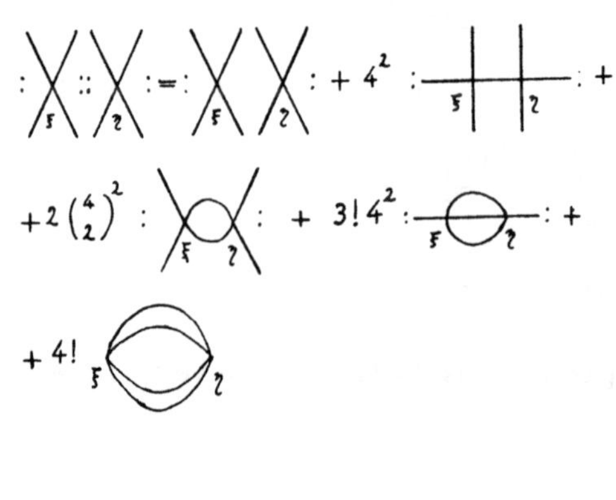

(4.8)

where each line represents $\tilde{\phi}$ and each internal line represents $C^{(N)}$. Integration over $\tilde{\phi}$ simply gives the same result described by (4.8) but with the lines interpreted as ϕ, now. The last term in (4.6) can be treated likewise and yields the same result as (4.8) with the external lines interpreted as $\tilde{\phi}$ and internal ones also interpreted as $C^{(N-k)}$.

Therefore we can write (4.5) as

$$V_{2;I}^{(N-k)} = V_{2;I}^{(N-k)} + 4^2 :$$ 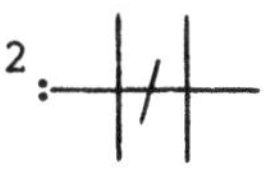 $$: 2\binom{4}{2}^2 :$$  $$:$$

$$+ 3!\, 4^2 :$$ 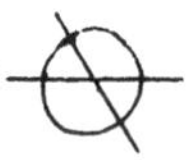 $$: \qquad (4.9)$$

when a bar on a graph means that it has to be computed as the difference of two graphs with the internal lines once considered as $C^{(N)}$ and the other time as $C^{(N-k)}$.

Since the model is herarchical it is clear that the structure of $\tilde{V}_{2,I}^{(N-k)}$ will be

$$\tilde{V}_{2;I}^{(N-k)} = \sum_{\Delta \in Q_{N-k}} \{ -\lambda\, 2^{-2(N-k)} (1+\Gamma_{N-k})\, H_4(X_\Delta^{(N-k)}) + \lambda^2\, C_{N,N-k}\, f_{N,N-k}\, (X_\Delta^{(N-k)}) \} \qquad (4.10)$$

and we are now concerned by finding some expressions for $f_{N,N-k}$.

We want to show that $f_{N,N-k}(X_\Delta^{(N-k)})$ can be written as

$$\lambda^2 (\bar{C}_{N,N-k}^{(6)}\, H_6(X_\Delta^{(N-k)}) + \bar{C}_{N,N-k}^{(4)}\, H_4(X_\Delta^{(N-k)}) + \bar{C}_{N,N-k}^{(2)}\, H_2(X_\Delta^{(N-k)}) + \bar{C}_{N,N-k}^{(o)})\, 2^{-4(N-k)} (1+\Gamma_{N-k})^6 \qquad (4.11)$$

and $\exists\, \bar{C}$

$$|\bar{C}_{N,N-k}^{(j)}| \le \bar{C} \qquad \forall\, N,k. \qquad (4.12)$$

If this were true we would have proven the lower bound: we could in fact proceed recursively using the following lemma:

Lemma 2: Let $|X^1| \leq B^1$ and let $X = \frac{z+\sqrt{\Gamma}X^1}{\sqrt{1+\Gamma}}$ $B^1 \geq 1$, $\bar{B} = B\sqrt{1+\Gamma} - B^1\sqrt{\Gamma} \geq 1$ then if V has the form

$$V(X) = \sum_{n=1}^{D} a^n X^n \tag{4.13}$$

and if $\|a\| = \sup_{n,\Delta} |a^n|$ then

$$\int e^{-z^2} \frac{dz}{\sqrt{\pi}} e^V \chi^B(X) = \exp\{ \sum_{j=o}^{t} \frac{\mathcal{E}^T(V;j)}{j!} + \varepsilon_\Gamma(B,B^1,\|a\|)\}$$

where, for suitable constants G_1, C_1, C_2, C_3, C_0

$$|\varepsilon_\Gamma(B, B^1, \|a\|)| \leq G_1\{(C_1 B^D \|a\| e^{2B^D \|a\| C_2})^{t+1} + \tag{4.14}$$
$$+ C_2 e^{-C_3 \bar{B}^2 + C_4 B^D \|a\|}\}$$

which allows to deduce

$$\int e^{\tilde{V}_{2,I}^{(N-k)}} \chi_{N-k}^B P_{N-k}(dz) = \int \exp\{ \sum_{j=1}^{2} \frac{\mathcal{E}_{N-k}^T(\tilde{V}_{2,I}^{(N-k)};j)}{j!} \tag{4.15}$$
$$+ \bar{\eta}(\lambda,N)|I|\} \chi_{N-k-1}^B P_{N-k-1}(dz)$$

where $\bar{\eta}(\lambda,N)$ is

$$G_1(C_1 B_N^G \lambda 2^{-2N} \bar{C} e^{2B_N^6 \lambda 2^{-2N} C_2 \bar{C}})^3 + \tag{4.16}$$
$$+ C_2 e^{-C_3 \bar{B}_N^2 + C_4 \bar{B}_N^6 \lambda 2^{-2N} \bar{C}}$$

which can be summed over N giving a result $\propto O(\lambda^3(\lg e+\lambda^{-1})^{18})$.

(4.15), (4.16) are not yet the result that we want because we have still to replace

$$\sum_{j=1}^{2} \frac{\hat{\mathcal{E}}_{N-k}(V_{2;I}^{(N-k)};j)}{j!}$$

by its second order truncation with respect to λ. This is done easily, since, clearly, the difference between these two quantities is

$$\leq (\lambda \bar{C} B^6_{N-k} 2^{-(N-k)})^3 \qquad (4.17)$$

and, therefore, it can be put into the error.

It remains to prove the bounds (4.11), (4.12). Consider for instance $\frac{3!4}{2}$: 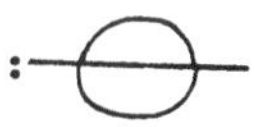 : which is the most difficult and dangerous:

$$\frac{3!4}{2} \lambda^2 \int_I d\xi \, d\eta \, (C^{(N)^3}_{\xi,\eta} - C^{(N-k)^3}_{T,\eta}) : \phi_\xi \phi_\eta : =$$

$$(4.18)$$

$$= \sum_{\substack{\Delta,\Delta^1 \subset I \\ \Delta,\Delta^1 \in Q_{N-k}}} \frac{4!\lambda}{2} \int_{\Delta\times\Delta^1} (C^{(N-k)}_{\Delta,\Delta^1})(C^{(N)^3}_{\xi,\eta} - C^{(N-k)^3}_{T,\eta}) : (X_\Delta X_{\Delta^1}) : d\xi d\eta$$

notice that $C^{(N)}_{\xi,\eta}$ and $C^{(N-k)}_{\xi,\eta}$ must be equal, unless ξ,η are in the same tessera of Q_{N-k}. Then the above term is

$$\sum_{\substack{\Delta \in Q_{N-k} \\ \Delta \subset I}} \frac{4!\lambda}{2} 2^{(d-2)(N-k)} (1+\Gamma_{N-k}) \, H_2(X_\Delta) \left(\int_{\Delta^2} (C^{(N)^3}_{\xi\eta} - C^{(N-k)^3}_{\xi\eta}) d\xi d\eta \right). \qquad (4.19)$$

So let us study

$$0 \leq \int_{\Delta^2} (C^{(N)^3}_{\xi\eta} - C^{(N-k)^3}_{\xi\eta}) \, d\xi d\eta = \sum_{i=N-k+1}^{N} \int_{\Delta^2} (C^{(i)^3}_{\xi\eta} - C^{(i-1)^3}_{\xi\eta}) d\xi d\eta$$

$$= \sum_{i=N-k+1}^{N} \sum_{\substack{\Delta^1 \in Q_{i-1} \\ \Delta^1 \subset \Delta}} \int_{(\Delta^1)^2} (C^{(i)^3}_{\xi\eta} - C^{(i-1)^3}_{\xi\eta}) \, d\xi d\eta \leq$$

$$\leq \sum_{i=N-k+1}^{N} (1+\Gamma_i)^3 \underbrace{2^{3(d-2)i}}_{\text{co-variance}^3} \underbrace{2^{-2d(i-1)}}_{\text{Volume element of } (\Delta^1)^2} \underbrace{2^{[(i-1)-(N-k)]d}}_{\#\text{elements in } \Delta} \underbrace{C}_{\text{constant}}$$

which can be bounded by

$$4 \tilde{C} \sum_{i=N-k+1}^{N} (1+\Gamma_i)^3 \, 2^{(2d-6)i} \, 2^{-d(N-k)} . \tag{4.20}$$

Hence the coefficient of $H_2(X_\Delta)$, called before $C_{N,N-k}$, can be bounded by

$$\frac{4!\lambda^2 4\tilde{C}}{2} (1+\Gamma_{N-k}) \sum_{i=N-k+1}^{N} 2^{(2d-6)i} (1+\Gamma_i) \, 2^{-2(N-k)} \tag{4.21}$$

and this is bounded by

$$\bar{C} \lambda^2 (1+\Gamma_{N-k})^4 \, 2^{+2(d-4)(N-k)} \qquad d = 2$$

$$\bar{C} \lambda^2 (1+\Gamma_{N-k})^4 \, 2^{+2(d-4)(N-k)} \Big(\sum_{i=N-k+1}^{N} 1\Big) \qquad d = 3 \tag{4.22}$$

hence, if d = 2, (4.11), (4.12) are proven.

In three dimensions this point would be the only different point in the whole proof. In order to avoid this problem the definition of renormalized hamiltonian will be changed so that the coefficient of $H_2(X_\Delta)$, just computed, turns out to be just zero.

This can be achieved by introducing in $V_{0,I}^{(N)}$ a "mass counterterm", i.e.

$$V_{0,I}^{(N)} = \lambda \int_I :\phi_\xi^{(\leq N)^4}: d\xi - \lambda^2 \frac{4\cdot 3!}{2} \int_{I\times I} C_{\xi\eta}^{[\leq N]^3} :\phi_\xi^{[\leq N]^2}: d\xi. \tag{4.23}$$

Then the interaction renormalized to order t becomes

$$V_{t,I}^{(N)} = \Big[V_{0,I}^{(N)} - \sum_{j=1}^{t} \frac{1}{j!} \mathcal{E}^T(V_{0;I}^{(N)};j \Big]_{(t)} \tag{4.24}$$

and the above scheme works for all $t \geq 3$. The only thing that has really to be ckecked is the bound on the coefficients $C_{N,N-k}$. This is long but straightforward.

5. The Upper Bound in the Case d = 2, t = 2

This is essentially a repetition of the argument in §3. However, some comments seem to be important.

In fact, the presence of very high order polynomials in $\tilde{V}^{(N)}$ puts some doubts about the possibility of removing the field cut-offs.

The reason why such terms do not cause problems is that at the beginning one can make use of the locality of the interaction to "eliminate" the interaction from the regions where the fields are large.

Let us recall the structure of $V_I^{(N)}$:

$$V_I^{(N)} = - \lambda(1+\Gamma_N)^2 \, 2^{-2N} \sum_{\Delta\subset I} H_4(X_\Delta) - \frac{\lambda^2}{2} \, \mathcal{E}^T(V_{O,I}^{(N)};2) \equiv$$

$$\equiv - \lambda(1+\Gamma_N)^2 \, 2^{-2N} \sum_{\Delta\subset I} H_4(X_\Delta) - \lambda^2 \, 2^{-2N} \sum_{\Delta\subset I} \gamma_\Delta \tag{5.1}$$

and γ_Δ has a bound like

$$\gamma_\Delta \equiv \frac{1}{2} \, 4! \int_{\Delta\times I} (C_{\xi\eta}^{[\leq N]})^4 \, \frac{d\xi d\eta}{2^{-2N}} \leq C < \infty. \tag{5.2}$$

Therefore, if we introduce the constants

$$\bar{B}_N = \sqrt{1+\Gamma_N} \, B_N - \sqrt{\Gamma_N} \, B_{N-1}$$

$$\tilde{B}_N = B_N \,/\, (1+N)^2 \tag{5.3}$$

and the regions

$$D_N(\phi) = \{\Delta | \; \Delta \in Q_N, \; |X_\Delta| > B_N\}$$
$$R_N(\phi) = \{\Delta | \; \Delta \in Q_N \quad |z_\Delta| > B_N\} \tag{5.4}$$

It is then, as already remarked, true that

$$R_N(\phi) \supset D_N(\phi) \backslash D_{N-1}(\phi) \tag{5.5}$$

$\forall\phi$, because in $D_N \backslash D_{N-1}$ one has $|z| > \bar{B}_N >> \tilde{B}_N$; also:

$$\Delta \in D_{N-1}(\phi) \backslash R_N(z) \Rightarrow |X_\Delta^{(N)}| \geq \frac{B_{N-1}\sqrt{\Gamma_N} - \tilde{B}_N}{\sqrt{1+\Gamma_N}} \geq \frac{1}{2} B_{N-1} >> 1.$$

Therefore

$$V_I^{(N)} \leq \hat{V}_I^{(N)} \equiv V_{I\backslash D_N}^{(N)} \tag{5.6}$$

and to obtain an upper bound it is enough to study

$$\int \exp \hat{V}_I^{(N)} P_N(dz). \tag{5.7}$$

Let us call χ_Δ^B the characteristic function of

$$\{z | \; |z_\Delta| < B \log(e+\lambda^{-1})\} \tag{5.8}$$

and put $\overset{\circ}{\chi}{}_\Delta^B = 1 - \chi_\Delta^B$. Then

$$\int \exp V_I^{(N)} \; P_N(dz) = \sum_{R \subset Q_N} \int \prod_{\Delta \in R} \overset{\circ}{\chi}{}_\Delta^{\tilde{B}_N} \prod_{\Delta \notin R} \chi_\Delta^{\hat{B}_N} e^{\hat{V}_I^{(N)}}. \tag{5.9}$$

We are still bothered by the fact that $\hat{V}_I^{(N)}$ contains "interaction" inside R where something still "goes wrong".

Therefore, we make a "generous" estimate: if

$$\hat{V}_I^{(N)} \leq \bar{C} \lambda e^\lambda (1+N)^2 \; H_4(B_N) \; 2^{-2N} \# (R) + V_{I\backslash D_N \cup R_N}^{(N)} \tag{5.10}$$

and if $\hat{R}$ denotes the smallest set paved by Q_{N-1} containing R

$$\hat{V}_I^{(N)} \leq \bar{C}^1 \lambda e^\lambda (1+N)^2 \; H_4(B_N) \; 2^{-2N} \# (R_N) + V_{I\backslash D_N \cup \hat{R}_N}^{(N)} \tag{5.11}$$

when $\prod_{\Delta\in R} \chi_{\Delta}^{\circ\tilde{B}_N} \prod_{\Delta\in R} \chi_{\Delta}^{\circ\tilde{B}_N} = 1$ and, under this condition, we see that

$$\hat{V}_I^{(N)} \leq \bar{C}^1 \lambda e^{\lambda} (1+N)^2 H_4(B_N) 2^{-2N} \# (R_N) + V^{(N)}_{I\setminus D_{N-1} \cup \hat{R}_N}. \quad (5.12)$$

It should be stressed, perhaps, that ν_{Δ} is a "non local" term (involving an integral over I) therefore modifications of the region where there is interaction produce errors due to the change of ν_{Δ}: such errors are very large, being proportional to the areas of the deleted regions. They have been absorbed in the errors as indicated above because of positivity or because $H_4(B_N)$ is large (this explains why we have not set $\bar{C}^1 = 4\bar{C}$, as it would be the case if this fact had been overlooked). Notice, however, $V^{(N)}_{I\setminus S} = \sum_{\Delta\subset I\setminus S} V^{(N)}_{\Delta}$, $\forall$ S paved by R_{N-1}.

We can now perform the integration over $z^{(N)}$; we proceed as usual

$$\int P_N(dz) e^{V^{(N)}_{I\setminus D_{N-1}} \hat{R}_N} \chi_R^{\circ\tilde{B}_N} \chi_{R^c}^{\tilde{B}_N} \leqq$$

$$\leqq \int P_{N-1}(dz)\, \chi_R^{\circ B_N}(z^{(N)}) \left(\prod_{\Delta\in R} \frac{e^{-(z_{\Delta}^{(N)})^2}}{\sqrt{\pi}} dz_{\Delta}^{(N)} \right) \cdot \prod_{\Delta' \not\subset \hat{R}UD_{B-1}}$$

$$\cdot \int e^{V_{\Delta}^{(N)} \frac{(X_{\Delta}^{(N-1)}\sqrt{\Gamma_N} + z)}{\sqrt{1+\Gamma_N}}} \frac{e^{-(z)^2}}{\sqrt{\pi}} dz)^4 \leqq \quad (5.13)$$

$$\leq \int P_{N-1}(dz)\, e^{(g_1 - g_2 \tilde{B}_N^2) \# (R)} \prod_{\Delta' \not\subset \hat{R}UD_{N-1}} e^{\tilde{V}^{(N-1)}_{I\setminus \hat{R}UD_{N-1}}} =$$

$$= e^{(g_1 - g_2\tilde{B}_N^2) \# (R)} \int P_{N-1}(dz) \exp \tilde{V}^{(N-1)}_{I\setminus D_{N-1} \cup \hat{R}} \exp \varepsilon_-(N)|I|$$

where $\varepsilon_-(N)$ is the estimate of the error in the lower bound (changed in sign). However, inside $I\setminus D_{N-1}$ the field $X^{(N-1)}$ is small ($<B_{N-1}$) hence the high order terms in $\tilde{V}^{(N-1)}$ which are not constant are really "neglegible" compared to the leading

term: so

$$\tilde{V}^{(N-1)}_{I\backslash D_{N-1}\cup\hat{R}} \leq \tilde{V}^{(N-1)}_{I\backslash D_{N-1}} + \#(R)2^{-2N}\bar{C}'''(1+N)^2 H_4(B_{N-1})\lambda e^{\lambda} \tag{5.14}$$

and since

$$\sum_R e^{(g_1-g_2\tilde{B}_N^2)\#(R_N)} e^{\#(R)2^{-2N}\bar{C}'''(1+N^2)H_4(B_{N-1})\lambda e^{\lambda}} =$$

$$= (1+e^{g_1-g_2\tilde{B}_N^2} e^{\lambda e^{\lambda}2^{-2N}\bar{C}'''(1+N)^2 H_4(B_{N-1})})^{|I|2^{2N}} \tag{5.15}$$

we have shown

$$\int e^{\hat{V}_I^{(N)}} P_N(dz) \leq \exp(\varepsilon_-(N)+\eta(N))|I| \cdot$$

$$\cdot\int \exp \tilde{V}^{(N-1)}_{I\backslash D_{N-1}} P_N(dz). \tag{5.16}$$

It is now natural to define

$$\hat{V}_I^{(h)} = \tilde{V}^{(h)}_{I\backslash D_h} \qquad h = 0,1,\dots,N \tag{5.17}$$

and, of course

$$\tilde{V}^{(h)}_{I\backslash D_h} \leq \tilde{V}^{(h)}_{I\backslash D_h\cup R_h} + \#(R_h)\,\hat{C}\,\lambda e^{\lambda}(1+h)^2 2^{-2h} H_4(B_h) \tag{5.18}$$

because the high orders do not matter since we are removing the fields only from regions where $|X^{(N)}|<B_N$!

Similarly

$$\tilde{V}^{(h)}_{I\backslash D_h} \leq \tilde{V}^{(h)}_{I\backslash D_h\cup R_h} + \#(R_h)\,\hat{C}^1\,\lambda e\lambda(1+h)^2\, 2^{-2h} H_4(B_h)$$

$$\leq \tilde{V}^{(h)}_{I\backslash D_{h-1}\cup\hat{R}_h} + \#(R_h)\,\hat{C}^1\,\lambda e\lambda(1+h)^2\, 2^{-2h} H_4(B_h) \tag{5.19}$$

where the last step is perhaps the most interesting. It is possible because in $D_{h-1}\backslash D_h\cup R_h$ the field $X^{(h)}$ though not too large ($X^{(h)}<B_h$) is still very large ($|X|^{(h)}>B_h/(1+h)^2$): hence the higher order terms are really small compared to the leading term which is positive because $B_h/(1+h)^2$ is very large. This is true $\forall\lambda$ if h is large enough: $h \geq h_\lambda$. Since the higher orders are of higher order in λ this will be true for all $h \geq 0$ if λ is small enough.

Finally integrating over $z^{(h)}$ the exp $\tilde{V}^{(h)}_{I\setminus Q_{n-1}\cup\hat{R}}$ when $\mathring{\chi}_R^{\tilde{B}_h}\chi_{R^c}^{\tilde{B}_h} = 1$ gives us a result

$$\leq \exp(g_1 - g_2\tilde{B}_h^2)\#(R)\int P_{h-1}(dz)\ \exp \tilde{V}^{(h-1)}_{I\setminus D_{h-1}\cup\hat{R}}\ \exp \varepsilon_-(h)|I| \tag{5.20}$$

by the very definition of $\tilde{V}^{(h-1)}$. Since outside D_{h-1} the higher order terms are really "higher order" we immediately realize that

$$\begin{aligned}\tilde{V}^{(h-1)}_{I\setminus D_{h-1}\cup\hat{R}} &\leq \tilde{V}^{(h-1)}_{I\setminus D_{h-1}} + (\#(R)\ \hat{C}''\ \lambda e^{\lambda}(1+h)^2\ 2^{-2h}\ H_4(B_{h-1}))\\ &= \hat{V}^{(h-1)}_I + (\text{error}).\end{aligned} \tag{5.21}$$

Therefore we are in a inductive position and we have achieved the estimate:

$$E_+(\lambda) = \sum_{h\geq h_\lambda}^{\infty} (\varepsilon_-(h) + \eta(h)) + \tilde{E}(\lambda) \tag{5.22}$$

where

$$\tilde{E}(\lambda) = \max_{\phi}\ \hat{V}^{(h_\lambda)}_I(\phi)|I|^{-1} \tag{5.23}$$

and $\tilde{E}(\lambda) = 0$ if λ is small enough.

Since $E_+(\lambda)$ is easily seen to have the wanted form (i.e. $\lambda^{-2} E_+(\lambda) \to 0$ as $\lambda \to 0$) the proof is finished.

We have now all the basic ingredients and mechanisms to attack more complex problems like the hierarchical Markov field or the Euclidean field.

6. The Euclidean Field Theory

I shall now directly consider the case of the three dimensional ϕ^4 theory without illustrating the case of the Markov field (see [3]).

We define a gaussian field Z distributed with covariance operator

$$A = C \frac{(1-D)(\gamma^2-D)}{(\gamma^2-1)} \tag{6.1}$$

where D is the Laplace operator and the constant C is chosen so that

$$E(Z_\xi^2) = \frac{1}{2} \tag{6.2}$$

i.e.

$$\frac{1}{(2\pi)^3} \frac{1}{C} \int_{R^3} \frac{(\gamma^2-1)\ d^3k}{(1+k^2)(\gamma^2+k^2)} = \frac{1}{2} . \tag{6.3}$$

Then the euclidean field ϕ is defined as

$$\phi_\xi = \sum_{N=0}^{\infty} \sqrt{\gamma^N}\ Z_\xi^{(N)} \tag{6.4}$$

where $Z^{(N)}$ are gaussian fields independently distributed so that $Z_\xi^{(N)}$ has the same distribution of $Z_{2^N\xi}$

$$Z_\xi^{(N)} \underset{\text{prob.}}{\sim} Z_{2^N\xi} . \tag{6.5}$$

The field (6.4) has covariance operator $A_o = C^{-1}(1-D)$ and (6.4), (6.5) is the probabilistic interpretation of the obvious relation

$$\frac{1}{1-D} = \sum_{N=0}^{\infty} \left(\frac{1}{\gamma^{2N}-D} - \frac{1}{\gamma^{2(N+1)}-D}\right) . \tag{6.6}$$

It is natural to introduce the cut-off and the normalized fields

$$\phi_\xi^{[\leq N]} = \sum_{k=0}^{N} \sqrt{\gamma^k}\ Z_\xi^{(k)} \qquad N = 0,1,\ldots \tag{6.7}$$

$$X_\xi^{(N)} = \frac{\phi_\xi^{[\leq N]}}{\sqrt{2E((\phi_\xi^{[\leq N]})^2)}} = \frac{Z_\xi^{(N)} + \sqrt{\Gamma_N}\ X_\xi^{(N-1)}}{\sqrt{1+\Gamma_N}} . \tag{6.8}$$

We then define the "third order renormalized interaction" $V_I^{(N)}$ as

$$V_{O,I}^{(N)} = -\lambda \int_I :(\phi_\xi^{[\le N]})^4 : d\xi \tag{6.9}$$

$$V_I^{(N)} = V_{O,I}^{(N)} - \frac{1}{2!}\langle (V_{O,I}^{(N)})^2\rangle_{(2)} - \frac{1}{2!}\langle (V_{O,I}^{(N)})^2\rangle_{(o)} - \frac{1}{3!}\langle (V_{O,I}^{(N)})^3\rangle_{(o)} \tag{6.10}$$

where

$$\frac{1}{2}\langle (V_{O,I}^{(N)})^2\rangle_{(2)} = \frac{\lambda^2\ 4\cdot 3!}{2} \int_{I^2} d\xi d\eta\ (C_{\xi\eta}^{[\le N]})^2\ :(\phi_\xi^{[\le N]})^2:$$

$$\frac{1}{2}\langle (V_{O,I}^{(N)})^2\rangle_{(o)} = \frac{\lambda^2\ 4!}{2} \int_{I^2} d\xi d\eta\ (C_{\xi\eta}^{[\le N]})^4 \tag{6.11}$$

$$\frac{1}{3!}\langle (V_{O,I}^{(N)})^3\rangle_{(o)} = \frac{\lambda^3}{3!}\ (\tfrac{4}{2})^3 (2!)^3 \int_{I^3} d\xi d\eta d\zeta\, (C_{\xi\eta}^{[\le N]})^2 (C_{\eta\xi}^{[\le N]})^2 (C_{\zeta\xi}^{[\le N]})^2$$

and

$$C_{\xi\eta}^{[\le N]} = \int \phi_\xi^{[\le N]}\ \phi_\eta^{[\le N]}\ P_N(dZ) \tag{6.12}$$

and $P_N = \prod_{i=o}^{N} \hat{P}_i$ denotes the probability distribution of $Z^{(o)}$, $Z^{(1)}, \ldots$.

Our problem is to show the existence of $E(\lambda)$ such that

$$\int \exp V_I^{(N)}\ P_N(dZ) = \exp \theta\, E(\lambda)\,|I| \tag{6.13}$$

where $\theta \in [-1,1]$ and

$$\lim_{\lambda\to 0} \lambda^{-3}\, E(\lambda) = 0. \tag{6.14}$$

This problem contains a new difficulty with respect to the already treated models: the field $Z^{(N)}$ is not built with simpler fields which are "independent on a scale γ^{-N}": however, the covariance of $Z^{(N)}$ decays exponentially over a scale γ^{-N}. This means, in colourful language, that the ultraviolet problem is now coupled with an infrared problem: the volume of I "felt by the high frequencies" will always be very large ($\sim \gamma^{3N}|I|$).

The idea of the proof of (6.13), (6.14) is the same as the one used in the former case. One integrates over the high frequencies showing that the result has the right form in order to apply a recursion argument.

Therefore it is natural to introduce, as in the hierarchical case, the quantity $\tilde{V}_I^{(h)}$ defined as

$$\tilde{V}_I^{(N)} = V_I^{(N)} \tag{6.15}$$

$$\tilde{V}_I^{(h)} = [\mathcal{E}_{h+1}^T(\tilde{V}_I^{(h+1)};1) + \frac{1}{2!}\mathcal{E}_{h+1}^T(\tilde{V}_I^{(h+1)};2) + \frac{1}{3!}\mathcal{E}_{h+1}^T(\tilde{V}_I^{(h+1)};3)]_{(3)}$$

with the notation naturally following the one previously used.

Before continuing it is good to get a feeling, and also a workable form, for $\tilde{V}_I^{(N)}$. This is the so called "small diagram theory".

The result is

$$\tilde{V}_I^{(h)} = -\lambda\int_I \gamma^{2h}(1+\Gamma_h)^2 :(X_\xi^{(h)})^4: d\xi -$$

$$-\lambda^2\int_{I\times I}\gamma^{4h}\bar{A}_{\xi\eta}\frac{:(X_\xi^{(h)}-X_\eta^{(h)})^2:}{(\gamma^h|\xi-\eta|)^{1/2}}e^{-\gamma^h d(\xi,\eta)} \tag{6.16}$$

$$+\sum_{\substack{p=0\\p+2q+3r=3}}^{3}\sum_{q=0}^{1}\sum_{r=0}^{1}(\lambda\gamma^{2h})^p(\lambda^2 h\gamma_h)^q(\lambda^3 h))^r \sum_{\substack{n_1\ldots n_p\\0\le\Sigma_i n_i+\Sigma_g m_i+m_r'\le 8}}\sum_{m_1\ldots m_q,\ m_r}$$

$$\int\bar{A}_{\xi_1\ldots\xi_p\eta_1\ldots\eta_q\eta_r'}^{n_1\ldots n_p\,m_1\ldots m_q,\,m_r'}e^{-\gamma^h d(\xi_1\ldots\xi_p\eta_1\ldots\eta_q)}(d\xi_1\ldots d\eta_q\,d\eta_1\ldots d\eta_2)\times$$

$$\times X_{\xi_1}^{n_1}\ldots X_{\xi_p}^{n_p}X_{\eta_1}^{m_1}\ldots X_{\eta_q}^{m_q}\quad X_{\eta_r'}^{m_r'}$$

where in the above formula the choices of the indices which give rise to meaningless expressions (e.g. if r = 0 or n = 0 η_1' or ξ_1 can be defined) should be interpreted by simply ignoring the indices which cannot make sense.

The coefficients $\bar{A}^{\cdots}_{\cdots}$ turn out to have the following properties: $\exists\ \bar{A} > 1$ such that

$$\bar{A}^{-1} \le \bar{A}_{\xi\eta}\ (\gamma^h|\xi-\eta|)^{3-1/2} \cdot \min(1,\gamma^h|\xi-\eta|) \le \bar{A}$$

$$\int_{\Delta_1\times\Delta_2\times\ldots} |A^{n_1\ldots}_{\xi_1\ldots}|\ \ d\xi_1 d\xi_2\ldots \le \bar{A}(\gamma^{-3h})^{p+q+r} \tag{6.17}$$

The calculation leading to (6.16) is a long but simple algebraic analysis. It can be most easily done using diagrams. Represent $V^{(N)}$ as

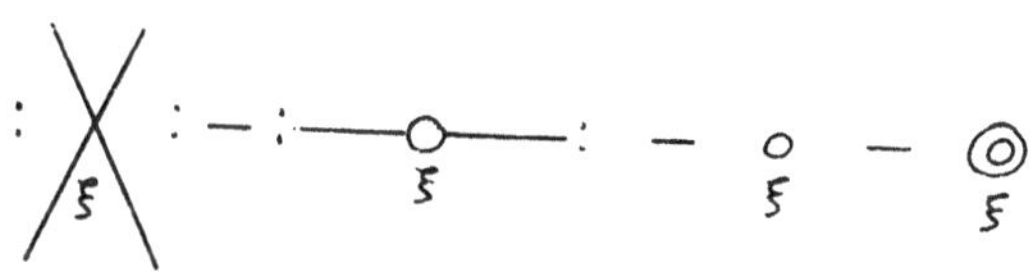

(6.18)

Then $\tilde{V}^{(h)}$ will have a representation like

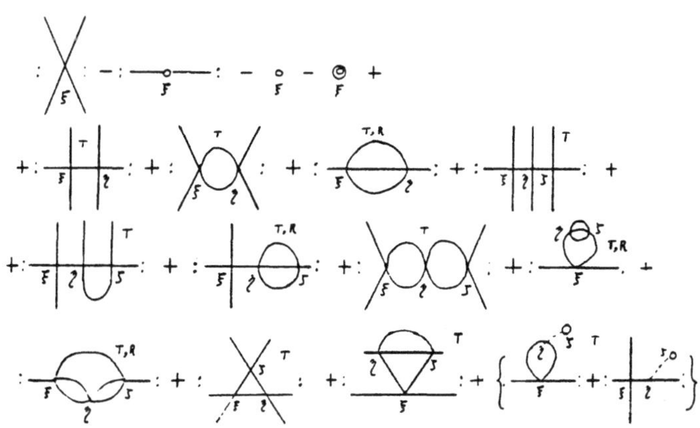

(6.19)

where the T means that "there is a hard line connecting all the points of the graph": the combinatorial coefficients and the actual combination of covariances entering into the above expressions have to be found by more careful analysis (this is reminded by the index R). See, for details, appendix B of [3], [5].

Look for instance at : 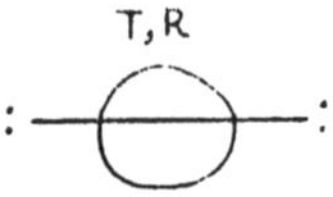 : it gives

$$+\frac{\lambda^2}{2}\,3!\,4^2\int_{I\times I}(C_{\xi\eta}^{[\leq N]^3}-C_{\xi\eta}^{[\leq h]^3})\;:(\phi_\xi\phi_\eta-\phi_\xi{}^2):\;d\xi d\eta = \tag{6.20}$$

$$=-\frac{\lambda^2 3!\,4^2}{4}\,\gamma^h(1+\Gamma_h)\int_{I\times I}(C_{\xi\eta}^{[\leq N]^3}-C_{\xi\eta}^{[\leq h]^3})\;:(X_\xi^{(h)}-X_\eta^{(h)})^2:\;d\xi d\eta$$

so

$$\bar{A}_{\xi\eta}=\frac{3!\,4^2}{4}\,(1+\Gamma_h)\,\gamma^{3h}\,\frac{(\gamma^h|\xi-\eta|)^{1/2}}{|\xi-\eta|^3}\,(e^{-|\xi-\eta|}e^{-\gamma^N|\xi-\eta|}-e^{-|\xi-\eta|}+e^{-\gamma^h|\xi-\eta|}) =$$

$$=\frac{3!\,4^2}{4}\,(1+\Gamma_h)\,(\gamma^h|\xi-\eta|)^{-3+1/2}\,e^{-\gamma^h|\xi-\eta|}\,(1-e^{-(\gamma^N-\gamma^h)|\xi-\eta|})\geq$$

$$\geq\frac{3!\,4^2}{4}\gamma(\gamma^h|\xi-\eta|)^{-3+1/2}\,e^{-\gamma^h|\xi-\eta|}\,(1-e^{-\gamma})\,\min(1,\gamma^N|\xi-\eta|) \tag{6.21}$$

hence

$$\int_{\Delta^2}\bar{A}_{\xi\eta}\leq[\frac{3!\,4^2}{4}\,\gamma(1-e^{-\gamma})]\gamma^{-6h}\int_{R^3}\frac{e^{-|x|}}{|x|^{3-1/2}}\,d^3x\;. \tag{6.22}$$

As another example we consider the graph: 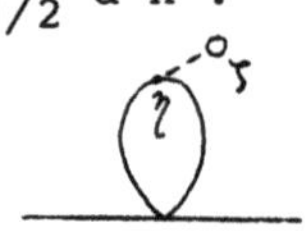 which gives rise to (if $C_i \equiv C^{[\leq i]}$):

$$\frac{\lambda^3}{2}\,4^2 3!\,\binom{2}{0}\binom{4}{0}\,2!\,\sum_{h+1}^{N}\int_{I^3}d\xi d\eta d\zeta\,(C_i^2(\xi,\eta)-C_{i-1}^2(\xi,\eta)\;C_i^3(\eta,\zeta)\;:\phi_\xi{}^2: \tag{6.23}$$

and this corresponds to $r = 1$, $m'_r = 2$ and the A coefficient is

$$\frac{\lambda^3}{2}\,4^2 3!\,\binom{2}{0}\binom{4}{0}\sum_{i=h+1}^{N}(1+\Gamma_h)\gamma^h\int_{I^2}d\eta d\zeta\,(C_i^2(\xi,\eta)-C_{i-1}^2(\xi,\eta))\,C_i^3(\eta,\zeta)$$

$$\leq \frac{\lambda^3}{2} \; 4^2 3! \; (1+\Gamma_h)\gamma^h \sum_{i=h+1}^{N} \int_{I^2} d\eta d\zeta^{5i} \frac{(1-e^{-\gamma^i|\xi-\eta|})}{(\gamma^i|\xi-\eta|\,)^2} \; e^{-\gamma^{i-1}|\xi-\eta|} \frac{(1-e^{-\gamma^i|\zeta-\eta|})}{(\gamma^i|\zeta-\eta|)^3}$$

$$\leq \frac{\lambda^3 4^2 3!}{2} \frac{\gamma}{\gamma-1} \sum_{i=h+1}^{N} \gamma^{-i} \int_{R^3} \frac{e^{-|x|}}{|x|^2} dx \int_{R^3} \frac{1-e^{-|x|}}{|x|^3} e^{-\gamma^{-i}|x|} dx$$

$$\leq C \lambda^3 \gamma^h \sum_{i=h+1}^{N} i \, \gamma^{-i} \leq \bar{C} \lambda^3 h. \tag{6.24}$$

So we shall be able to bound the interaction $\tilde{V}^{(N)}$ when

$$\begin{aligned} &|X_\xi^{(N)}| \leq B_N && \forall \; \xi \in I \\ &|X_\xi^{(N)} - X_\eta^{(N)}| \leq B_N(\gamma^N|\xi-\eta|)^{1/4} && \forall \; \xi,\eta \in I, \quad \gamma^N|\xi-\eta|<|. \end{aligned} \tag{6.25}$$

It is therefore natural, following the scheme of the Bernoulli case to introduce the following regions

$$D_N^g(\phi) = \{\xi| \; |X_\xi^{(N)}| > B_N(1+\gamma^N \; d(\xi,I)\}$$

$$\begin{aligned} D_N^r(\phi) = \{\xi| \; \exists \; \eta, \; |\gamma^N(\xi-\eta)|<1 \text{s.t.} \; |X_\xi^{(N)}-X_\eta^{(N)}|> \\ > (1+\gamma^N \; d(\xi,I)) \; |\xi-\eta|^{1/4}\} \end{aligned} \tag{6.26}$$

$R_N(\phi) = \{\Delta|\Delta \in Q_N$ such that $\exists \; \xi \in \Delta$ where

$|Z_\xi^{(N)}|>B_N^1 \; (1+\gamma^N \; d(\xi,I))$ or $\exists \; \xi,\eta \in \Delta$ where $|Z_\xi^{(N)}-Z_\eta^{(N)}|\geq$

$\geq B_N^1 \; (\gamma^N|\xi-\eta|)^{1/4} \; (1+\gamma^N(|(\xi,I))$ and $|\xi-\eta|\gamma^N \leq 1\}$.

Notice that D_N^g, D_N^r are "arbitrary" open sets while $R_N(\phi)$ is pavable by Q_N, by definition: i.e. R_N is defined in a substantially different way from D_N^g and D_N^r. Alternatively we could also define D_N^g and D_N^r to be the smallest Q_N-pavable sets containing the above defined sets: this would not be, however, a natural definition as it was,"by accident",in the hierarchical case.

We also define

$$\hat{V}_I^{(h)}(\phi) = \{\text{same as (6.16) with } I\backslash D_h^g(\phi) \text{ replacing every-} \tag{6.27}$$
$$\text{where } I, \text{ and } (I\backslash D_h^g(\phi))^2\backslash D_h^r(\phi) \text{ replacing } I\times I\}.$$

Calling $\tilde{I} = I\backslash D_h^g(\phi)$ this means that we replace I by $\tilde{I}$ everywhere and, then, $\tilde{I}\times\tilde{I}$ by $(\tilde{I}\times\tilde{I})/D_h^r(\phi)$ or $\tilde{I}\times\tilde{I}\times\tilde{I}$ by the set obtained by taking out of $\tilde{I}\times\tilde{I}\times\tilde{I}$ all triples containing a pair in D_N^r. The A-coefficients that are put into the integrals defining (6.27) are those relative to the region I.

The above definition of $\hat{V}$ is slightly simpler than the one of ref. [5]: it seems, however, as good. On the other hand it makes the proof of the basic inequalities (6.28) ÷ (6.30) below slightly harder, (see [5]). The functions $\hat{V}$ verify some inequalities, as in the Bernoulli case: suppose that B is large enough, (i.e. B larger than the largest zero of $H_4(b)$), then $\exists\ h_\lambda$, $\bar{C}$, and

$$\text{i)}\quad V_I^{(N)} \leq \hat{V}_I^{(N)} \tag{6.28}$$

$$\text{ii)}\quad \hat{V}_I^{(h)}(\phi)\leq(\bar{C}\ \lambda\ e^{\bar{C}\lambda}\ \gamma^{2h}\ B_h^{\bar{C}})\gamma^{-3h}\ \#(R_h(\phi)\cap I) + H^{(h)}_{S\backslash\hat{R}_h(\phi)} \tag{6.29}$$

where $\hat{R}_h(\phi) = \{\Delta|\ \Delta\in Q_h,\ d(\Delta,R_h(\phi^{(h)}))\leq(B_h^1)^3\gamma^h\}$ and $H_S^{(h)}$ is defined as (6.27), replacing I everywhere in the integration regions, by $S\backslash D_{h-1}^g(\phi)$ and, then, $(I\backslash D_h^g(\phi))^2$ by $(S\backslash D_{h-1}^g(\phi))^2\backslash D_{h-1}^r$ (in the same sense), and using A-coefficients associated with the region S.

$$\text{iii)}\quad \left\{\sum_{k=1}^{3} \frac{\mathcal{E}_h^T(H_{I\backslash\hat{R}_h(\phi)};K)}{k!}\right\} - \tilde{V}_I^{(h-1)} \leq (\bar{C}\lambda e^{\bar{C}\lambda}B_{h-1}^{\bar{C}}\gamma^{-h})^4\gamma^{3h}|I| \tag{6.30}$$

iv) $h_\lambda \equiv 0$ if is small enough.

All the above properties of $\tilde{V}_I^{(h)}$, $\hat{V}_I^{(h)}$ have obvious analogues in the Bernoulli model and are what we need to copy the argument already explained in the Bernoulli case to the Field theory of Euclidean fields: the only missing tool is an algorithm for estimating the error involved in the integration of the highest frequency field.

Such algorithm is provided by the following lemma of probability theory which allows to conclude that even the results of the integrations are as if the model was hierarchical. Let Z be the standard field with (6.1) as covariance operator and let:

$$U_J(Z) = \sum_{P=1}^{S} \sum_{\Sigma_i n_i \le D} \int_{J^P} A^{n_1 \dots n_p}_{\xi_1 \dots \xi_p} e^{-\kappa d(\xi_1, \dots, \xi_p)} Z^{n_1}_{\xi_1} \dots Z^{n_p}_{\xi_p} \, d\xi_1, \dots d\xi_p + \tag{6.31}$$

$$+ \sum_{q^1=1}^{S^1} \sum_{\Sigma_i m_i \le D^1} \int_{J^{2q}} A^{m_1 \dots m_q}_{\xi_1 \xi_1' \dots \xi_q \xi_q'} e^{-\kappa d(\xi_1 \dots \xi_q')} d\xi_1 \dots d\xi_q' \prod_{j=1}^{q} \left(\frac{Z_{\xi_j} - Z_{\xi_j'}}{|\xi_i - \xi_j'|} \right)^{m_j}$$

where the A's are such that

$$||\underline{A}|| = \sup \int_{\Delta_1 x \dots x \Delta_p} |A^{n_1 \dots n_p}_{\xi_1 \dots \xi_p}| \, d\xi_1 \dots d\xi_p + \tag{6.32}$$

$$+ \sup \int_{\Delta_1 x \Delta_1' x \dots x \Delta_q x \Delta_q'} |A^{m_1 \dots m_q}_{\xi_1 \xi_1' \dots \xi_q \xi_q'}| \, d\xi_1 \dots d\xi_q' < + \infty.$$

Let $\chi^B_\Delta(Z)$ be the c.h.f. of the event

$$\{Z| \ |Z_\xi| < B \quad \forall \ \xi \in \Delta, \ \ |Z_\xi - Z_\eta| < B \ |\xi - \eta|^{1/4}, \quad \forall \ \xi \ \eta \in \Delta, \ |\zeta - \eta| \le | \} \tag{6.33}$$

$\Delta \in Q_o$. Let $\overset{\circ}{\chi}{}^B_\Delta = 1 - \chi^B_\Delta$. Then

<u>Lemma:</u> Let γ be close enough to 1. Given $t \ge 0$ integer $\exists$ B^*, G, G', ρ, ρ', ρ^N, γ_1, γ_2 depending only on t, D, D', κ such that if B B* and if $R \subset Q_o$, $J \subset I$ and $H_{J \backslash \hat{R}}$ is defined from (6.31) as after (6.29)

$$\int \overset{\circ}{\chi}{}^B_R \, \chi^B_{Q_o \backslash R} \, e^{H_{J \backslash \hat{R}}} P(dZ) \le$$

$$\le \exp \left(|I| \ \delta(B, ||A||) + \delta'(B \, ||A||) |R| \right)$$

$$\exp\,(\gamma_1-\gamma_2 B^2)|R| \quad \exp\{\sum_{k=o}^{t} \frac{E^T(U_J;k)}{k!}\}(t) \tag{6.34}$$

$$\int \chi^B_{Q_o}\, e^{U_J}\, P(dZ) \geq \exp - |I|\delta(B,\|\underline{A}\|)\, \exp\{\sum_{k=1}^{t} \frac{E^T(U_J;k)}{k!}$$

where

$$\delta(B,\|\underline{A}\|) = G((\|\underline{A}\|B^r\, e^{\rho''(\|A\|B^\rho)})^{t+1} + e^{-\rho''B^2+\rho'B^\rho}\|\underline{A}\|) \tag{6.35}$$

$$\delta'(B,\|\underline{A}\|) = G'(\|\underline{A}\|B \quad e^{\rho''\|A\|B^\rho}). \tag{6.36}$$

In the next section we show how the ultraviolet estimate follows from the above lemma.

7. The Upper and Lower Bounds

To find a lower bound we simply write

$$\int e^{V_I^{(N)}} P_N(dZ) \geq \int e^{V_I^{(N)}}, \prod_{i=o}^{N} \chi(D_i(X^{(i)}) = \emptyset)\, P_N(dZ) = Z_-(N) \tag{7.1}$$

where $\chi(D_N(\phi) = \phi_N)$ is the characteristic function of the event $D_N(\phi) = \emptyset$. Then

$$Z_-(N) = \int \prod_{i=o}^{N-1} \chi(D^{(i)}(\phi) = \emptyset)(\int\chi(D^{(N)}(\phi) = \emptyset)\ \hat{P}_N(dZ)$$

$$\exp V_I^{(N)})\ P_{N-1}(dZ). \tag{7.2}$$

Since $V_I^{(N)}$ has the structure (6.31) after an appropriate change of scale and $\exists\ \bar{C} > o$ s.t.

$$\|A\| \leq \bar{C}\, B_{N-1}^{\bar{C}}\, \lambda\gamma^{-N}\, e^{\lambda\,\bar{C}} \tag{7.3}$$

we can apply the basic lemma to conclude

$$Z_-(N) \geq \exp - \quad \delta\,(\bar{C}\, B_{N-1}^{\bar{C}}\, \lambda\gamma^{-N}\, e^{\lambda\,\bar{C}};\ B_N)\ \gamma^{3N}|I|\ \cdot$$

$$\int \prod_{i=o}^{N-1} \chi(D^i(\phi) = \emptyset) \quad \exp\, \tilde{V}_I^{(N-1)} \tag{7.4}$$

and since $\tilde{V}_I^{(N-1)}$ has again the structure (6.31) we can indefinitely apply the lemma to conclude that

$$Z_-(N) \geq \exp-\left(\sum_{N=0}^{\infty} \delta(\bar{C}\, B_{N-1}^{\bar{C}} \lambda\gamma^{-N} e^{\bar{C}\lambda}\ ;\ B_N)\gamma^{3N}\right)|I|. \tag{7.5}$$

To find the upper bound we use first (6.28) to obtain

$$\int e^{V_I^{(N)}} P_N(dZ) \leq Z_+(N) = \int e^{\hat{V}_I^{(N)}} P_N(dZ). \tag{7.6}$$

Then we use the decomposition of 1

$$1 = \sum_{R\subset Q_N} \overset{\circ}{\chi}{}_R^{B_N}\ \chi_{Q_N\setminus R}^{B_N} \tag{7.7}$$

and write

$$Z_+(N) = \sum_{R\subset Q_N} \int \overset{\circ}{\chi}{}_R^{B_N}\ \chi_{Q_N\setminus R}^{B_N}\ e^{\hat{V}_I^{(N)}}\ P_N(d\mathbf{Z}) \tag{7.8}$$

and using (6.29)

$$Z_+(N) \leq \sum_{R\subset Q_N} \exp\{\bar{C}\lambda e^{\bar{C}\lambda}\gamma^{2N}\, B_N^{\bar{C}}\gamma^{-3N} \# (R\cap I)\}\cdot \exp H_{I\setminus R_N}^{(N)}(\phi)\cdot \tag{7.9}$$

Using now the basic lemma, after a change of scale (noticing that $H_S^{(N)}$ has, $\forall S$, the appropriate form (6.31)) one finds by (6.34), (6.30):

$$\begin{aligned} Z_+(N) \leq{} & \exp + \delta(\bar{C}\, B_{N-1}^{\bar{C}}\ \lambda\gamma^{-N}\ e^{+\lambda\ \bar{C}};\ B_N^1)|I|\ \cdot \\ & \cdot\left(\sum_{R\subset Q_N} \exp\{\delta'(\bar{C}\lambda e^{\bar{C}\lambda}\ \gamma^{-N}\ B_{N-1}^{\bar{C}};\ B_N^1)+\gamma_1-\gamma_2(B_N^{1^2})\}\#(R)\right). \\ & \cdot \exp\ (\bar{C}\ B_{N-1}^{\bar{C}}\ \lambda\gamma^{-N}\ e^{\lambda\ \bar{C}})|I|\quad Z_+(N-1) \end{aligned} \tag{7.10}$$

and, by our construction, $Z_+(N-1)$ has the same structure as $Z_+(N)$ and the argument can be iterated , as long as $h \geq h_\lambda$:

$$Z_+(N) \leq Z_+(h_\lambda)\quad \exp \sum_{h=h_\lambda}^{\infty} (\varepsilon(h) + \eta(h))|I| \tag{7.11}$$

where

$$\begin{aligned} \varepsilon(h) &= [\delta(\bar{C}\, B_{h-1}^{\bar{C}}\ \lambda\gamma^{-h}\ e^{\bar{C}\lambda}\ ;\ B_h^1)] + \bar{C}\, B_{h-1}^{\bar{C}}\ \lambda\gamma^{-h}e^{\lambda\bar{C}} \\ \eta(h) &= \log\left(1 + \exp\delta'(\bar{C}\lambda e^{\bar{C}\lambda}\gamma^{-N}\, B_N^{\bar{C}}) + \gamma_1 - \gamma_2\, B_N^1\right). \end{aligned} \tag{7.12}$$

Finally

$$Z_+(h_\lambda) \le \exp \max_\phi \hat{V}^{(h_\lambda - 1)}(\phi). \tag{7.13}$$

Hence

$$Z_+(N) \le \exp |I| \{ \sum_{h=h_\lambda}^{\infty} (\varepsilon(h) + \eta(h)) + \tilde{e}(\lambda)) \} \tag{7.14}$$

where $\tilde{e}(\lambda)|I|$ is an estimate for $\sup \hat{V}^{(h_\lambda - 1)}(\phi)$. Since $h_\lambda \equiv 0$, if λ is small enough, we see that one can take:

$$E_+(\lambda) = \sum_{h=h_\lambda}^{\infty} (\varepsilon(h) + \eta(h)) \tag{7.15}$$

and this concludes the proof since it is easy to see that

$$\lim_{\lambda \to 0} \lambda^{-3} E_\pm(\lambda) = 0 . \tag{7.16}$$

So it only remains to prove the basic lemma, last of the § 6.

8. Concluding Remarks

The proof of the lemma of § 6 is discussed in ref. [5] where it is reduced to a general support property of free fields roughly saying that the probability that a gaussian field, associated with an elliptic operator "with positive mass", is everywhere in a region I bounded by B times the square root of its covariance is bounded below by

$$\exp - C_1 e^{-C_2 B^2} |I|$$

$\forall I \subset R^d$, if C_1, C_2 are suitably chosen and B is larger than B* (I-independent). This support property, very easy to prove in the hierarchical models using techniques of statistical mechanics or gaussian processes, is much harder in the continuum cases. It is proven in ref. [4].

As it is well known the results presented here were all well known see the basic references quoted in [3], [5].

The method, however, seems to be somewhat different (although the basic idea is, of course, the phase space cell decomposition of Glimm-Jaffe, [6]).

It seems to me that this idea is closely related to that in the analysis of the structure of the functional integrals developped by Wilson [7]: the relation should appear clear in our proof.

The proof discussed here is in the works [3], [5].

The above presentation is mainly taken out of my notes from lectures given at the Ecole Polytechnique in October-November 1978, in Palaiseau.

References

[1] G. Gallavotti, Memorie dell' Accademia dei Lincei, XV, 23, 1978.

[2] G. Gallavotti, Annal. di Matematica pura e applicata, CXX, 1, 1979.

[3] G. Benfatto, M. Cassandro, G. Gallavotti, F. Nicolò, E. Presutti, E. Olivieri, E. Scacciatelli, Comm. Math. Phys. 59, 143, 1978.

[4] G. Benfatto, G. Gallavotti, F. Nicolò, "Elliptic equations and gaussian processes", to appear in J. Funct. Analysis, preprint IHES, January 1979.

[5] G. Benfatto, M. Cassandro, G. Gallavotti, F. Nicolò, E. Olivieri, E. Presutti, E. Scacciatelli, Comm. Math. Phys. 71, 95, 1980.

[6] J. Glimm: Comm. Math. Phys. 8, 12, 1968.
J. Glimm, A. Jaffe: Fortschr. d. Physik, 21, 327, 1973.

[7] K. Wilson, Phys. Rev. B4, 3174 and 3184, 1971.

Summary

Raphael Høegh-Krohn

We have come to the end of our program and the end of this week that was called "Bielefeld Encounters in Physics and Mathematics II", and it is time to make a summary and try to see what we have achieved.

Let me start by thanking the organizing committee which consisted of Sergio Albeverio, Philippe Blanchard, Erwin Brüning, Gustav Sommer, Peter Stichel, Othmar Steinmann and chairman Ludwig Streit for the fantastic and unique arrangement that they have made for us here in Bielefeld. Every detail has been cared for and looked after to ensure our wellbeing. When the restaurants at the University were closed on Saturday and we worried how everyone was going to get fed it turned out that we worried without any cause. The chairman had brought his whole family, wife and children, with grill and hot dogs to feed us all. But even though the material arrangements were perfect the arrangements I really had in mind were those of the scientific program for this conference. We are now in the middle of a period where the strong interaction between quantum field theory - statistical mechanics - probability theory and C*-algebra are paying off at a high rate and we are grateful to the whole committee for the harmonious way in which they have managed to blend these four fields of science here this week. It is, however, no accident that this successful blend has been made here at ZiF, in fact it was the forerunner of this encounter, the first Encounter in Physics and Mathematics here at ZiF a couple of years ago, that helped set off this fruitful interaction between physicists and mathematicians. I believe it is very important that scientists from these fields have the opportunity to meet here.

The philosophical problem underlying our activity is of course what could be called the main philosophical problem of our time, namely the consistency of the two basic philosophies on which we base our understanding of nature, i.e. the philosophy of quantum mechanics and the philosophy of special relativity. But we have come to understand that we cannot solve this problem overnight and that any real progress must necessarily be slow because it can only come through greater understanding and this understanding grows only gradually. There is, however, one way in which one can speed up this process and that is through exchange of understanding in such a conference as this one. For this reason I hope that we will be able to come back again and again here to ZiF. It has been a fantastic experience for me to see how statistical mechanics - probability theory - quantum field theory and C*-algebra blend and support each other in this Encounter. We see that quantum field theory which started its existence as a domain of study within the theory of C*-algebras has now become a sub-domain of statistical mechanics which again has become a sub-domain of probability theory. For instance, we learned from Jona-Lasinio that the question whether the quantum field has the strong Markov property with respect to random stopping contours, is not only meaningful but important, and furthermore that the stochastic formulation of quantum mechanics is not only a mathematical nicety but an essential tool for explicit computations in quantum mechanics. Another extremely interesting work in this direction is the study of renormalization of quantum fields presented by Gallavotti where one uses exact support properties of the free measure on $S^1(R^3)$ as a main tool. From Fröhlich's beautiful lecture we learned that our knowledge in quantum gauge fields is again expressed in probabilistic terms. Albeverio shows us, again by probabilistic methods, that the exponential interaction is trivial in four dimensions.

But this strongly interacting field of physics and mathematics we have been exposed to here does not consist only of physicists doing probability theory. Now it consists to an equal amount of probabilists doing physics. We hear from Higuchi that the interphase in the Ising model is a Brownian bridge and Hida analyzes everything with white noise. Preston studies Lattice systems and

Föllmer interacting Markov chains.

However, all integration is not probabilistic, or more properly, time is not always imaginary. There is also the oscillatory integral, and from Klauder's beautiful lecture we learned how to deal with it in connection with coherent states, and Blanchard shows us how to compute the zeta function for Schrödinger operators by using path integrals. I think we should not forget the merits of the oscillatory path integrals in all this progress of applying probability theory. The oscillatory integral has clear analytic advantages that remain to be explored. As an example of this, I will mention the strong result of Enss for the scattering of the Schrödinger operator. Jaffe on the other hand uses probabilistic methods in his beautiful lecture to get deep results about the scattering operator in quantum field theory.

In C*-algebras, on the other hand, we meet a new type of integration which is the non-commutative integration theory associated with the KMS-condition that Takesaki told us about. Again we should point out that the motivation for this beautiful theory comes from statistical mechanics. The automorphisms that arise are simply the time flows of the physical system, and the infinitesimal generators of these are the derivations studied by Bratteli in his lecture.

It looks to me as if this strong mixing of probability theory with everything will continue even more in the future. The connections between probability theory and quantum field theory will be even stronger. I would tell any student of physics to start with probability theory and any teacher of probability theory to study quantum fields. Because we have to realize that the boson fields are in reality diffusion fields and must be studied as such. However, the probabilist should know that the future lies in more dimensional Markov processes. Also for C*-algebras, I think the future lies in the direction of probability theory, or to be more precise, probability theory should be a branch of the theory of C*-algebras.

It has been a fantastic experience to sit here this week and see again and again months or maybe even years of hard work, condensed into an hour's talk or half an hour's seminar, explode on the blackboard. The amount of information which is thus condensed and transmitted through carefully prepared lectures is unbelievable and for this I want to thank everyone who contributed with a lecture or seminar.

Last but not least, I would like to thank the very professional staff here at ZiF for helping us with everything and thus making this Conference possible. I will conclude with the hope that we will be back here at ZiF with the same topics, expanded, however, to include speakers on turbulence and amorphous solids, for Encounters III to bring in the harvest of this conference.

MIX
Papier aus verantwortungsvollen Quellen
Paper from responsible sources
FSC® C105338

If you have any concerns about our products,
you can contact us on
ProductSafety@springernature.com

In case Publisher is established outside the EU,
the EU authorized representative is:
Springer Nature Customer Service Center GmbH
Europaplatz 3, 69115 Heidelberg, Germany

Printed by Libri Plureos GmbH
in Hamburg, Germany